Probability and Mathematical Statistics (Continued)

SHORACK and WELLNER · Empirical Processes with Applications to Statistics
STAUDTE and SHEATHER · Robust Estimation and Testing
STOYANOV · Counterexamples in Probability
STYAN · The Collected Papers of T. W. Anderson: 1943–1985
WHITTAKER · Graphical Models in Applied Multivariate Statistics
YANG · The Construction Theory of Denumerable Markov Processes

Applied Probability and Statistics

ABRAHAM and LEDOLTER · Statistical Methods for Forecasting
AGRESTI · Analysis of Ordinal Categorical Data
AGRESTI · Categorical Data Analysis
ANDERSON and LOYNES · The Teaching of Practical Statistics
ANDERSON, AUQUIER, HAUCK, OAKES, VANDAELE, and WEISBERG · Statistical Methods for Comparative Studies
ASMUSSEN · Applied Probability and Queues
*BAILEY · The Elements of Stochastic Processes with Applications to the Natural Sciences
BARNETT · Interpreting Multivariate Data
BARNETT and LEWIS · Outliers in Statistical Data, *Second Edition*
BARTHOLOMEW, FORBES, and McLEAN · Statistical Techniques for Manpower Planning, *Second Edition*
BATES and WATTS · Nonlinear Regression Analysis and Its Applications
BELSLEY · Conditioning Diagnostics: Collinearity and Weak Data in Regression
BELSLEY, KUH, and WELSCH · Regression Diagnostics: Identifying Influential Data and Sources of Collinearity
BHAT · Elements of Applied Stochastic Processes, *Second Edition*
BHATTACHARYA and WAYMIRE · Stochastic Processes with Applications
BIEMER, GROVES, LYBERG, MATHIOWETZ, and SUDMAN · Measurement Errors in Surveys
BLOOMFIELD · Fourier Analysis of Time Series: An Introduction
BOLLEN · Structural Equations with Latent Variables
BOX · R. A. Fisher, the Life of a Scientist
BOX and DRAPER · Empirical Model-Building and Response Surfaces
BOX and DRAPER · Evolutionary Operation: A Statistical Method for Process Improvement
BOX, HUNTER, and HUNTER · Statistics for Experimenters: An Introduction to Design, Data Analysis, and Model Building
BROWN and HOLLANDER · Statistics: A Biomedical Introduction
BUCKLEW · Large Deviation Techniques in Decision, Simulation, and Estimation
BUNKE and BUNKE · Nonlinear Regression, Functional Relations and Robust Methods: Statistical Methods of Model Building
CHATTERJEE and HADI · Sensitivity Analysis in Linear Regression
CHATTERJEE and PRICE · Regression Analysis by Example, *Second Edition*
CLARKE and DISNEY · Probability and Random Processes: A First Course with Applications, *Second Edition*
COCHRAN · Sampling Techniques, *Third Edition*
COCHRAN and COX · Experimental Designs, *Second Edition*
CONOVER · Practical Nonparametric Statistics, *Second Edition*
CONOVER and IMAN · Introduction to Modern Business Statistics
CORNELL · Experiments with Mixtures, Designs, Models, and the Analysis of Mixture Data, *Second Edition*
COX · A Handbook of Introductory Statistical Methods
COX · Planning of Experiments
CRESSIE · Statistics for Spatial Data
DANIEL · Applications of Statistics to Industrial Experimentation
DANIEL · Biostatistics: A Foundation for Analysis in the Health Sciences, *Fifth Edition*
DAVID · Order Statistics, *Second Edition*
DEGROOT, FIENBERG, and KADANE · Statistics and the Law
*DEMING · Sample Design in Business Research
DILLON and GOLDSTEIN · Multivariate Analysis: Methods and Applications
DODGE and ROMIG · Sampling Inspection Tables, *Second Edition*
DOWDY and WEARDEN · Statistics for Research, *Second Edition*
DRAPER and SMITH · Applied Regression Analysis, *Second Edition*
DUNN · Basic Statistics: A Primer for the Biomedical Sciences, *Second Edition*

*Now available in a lower priced paperback edition in the Wiley Classics Library.

Applied Probability and Statistics (Continued)

DUNN and CLARK · Applied Statistics: Analysis of Variance and Regression, *Second Edition*
ELANDT-JOHNSON and JOHNSON · Survival Models and Data Analysis
FLEISS · The Design and Analysis of Clinical Experiments
FLEISS · Statistical Methods for Rates and Proportions, *Second Edition*
FLEMING and HARRINGTON · Counting Processes and Survival Analysis
FLURY · Common Principal Components and Related Multivariate Models
GALLANT · Nonlinear Statistical Models
GROSS and HARRIS · Fundamentals of Queueing Theory, *Second Edition*
GROVES · Survey Errors and Survey Costs
GROVES, BIEMER, LYBERG, MASSEY, NICHOLLS, and WAKSBERG · Telephone Survey Methodology
HAHN and MEEKER · Statistical Intervals: A Guide for Practitioners
HAND · Discrimination and Classification
HEIBERGER · Computation for the Analysis of Designed Experiments
HELLER · MACSYMA for Statisticians
HOAGLIN, MOSTELLER, and TUKEY · Exploratory Approach to Analysis of Variance
HOAGLIN, MOSTELLER, and TUKEY · Exploring Data Tables, Trends and Shapes
HOAGLIN, MOSTELLER, and TUKEY · Understanding Robust and Exploratory Data Analysis
HOCHBERG and TAMHANE · Multiple Comparison Procedures
HOEL · Elementary Statistics, *Fifth Edition*
HOGG and KLUGMAN · Loss Distributions
HOLLANDER and WOLFE · Nonparametric Statistical Methods
HOSMER and LEMESHOW · Applied Logistic Regression
IMAN and CONOVER · Modern Business Statistics
JACKSON · A User's Guide to Principle Components
JOHN · Statistical Methods in Engineering and Quality Assurance
JOHNSON · Multivariate Statistical Simulation
JOHNSON and KOTZ · Distributions in Statistics
Discrete Distributions
Continuous Univariate Distributions—1
Continuous Univariate Distributions—2
Continuous Multivariate Distributions
JUDGE, GRIFFITHS, HILL, LÜTKEPOHL, and LEE · The Theory and Practice of Econometrics, *Second Edition*
JUDGE, HILL, GRIFFITHS, LÜTKEPOHL, and LEE · Introduction to the Theory and Practice of Econometrics, *Second Edition*
KALBFLEISCH and PRENTICE · The Statistical Analysis of Failure Time Data
KASPRZYK, DUNCAN, KALTON, and SINGH · Panel Surveys
KISH · Statistical Design for Research
KISH · Survey Sampling
LAWLESS · Statistical Models and Methods for Lifetime Data
LEBART, MORINEAU, and WARWICK · Multivariate Descriptive Statistical Analysis: Correspondence Analysis and Related Techniques for Large Matrices
LEVY and LEMESHOW · Sampling of Populations: Methods and Applications
LINHART and ZUCCHINI · Model Selection
LITTLE and RUBIN · Statistical Analysis with Missing Data
MAGNUS and NEUDECKER · Matrix Differential Calculus with Applications in Statistics and Econometrics
MAINDONALD · Statistical Computation
MALLOWS · Design, Data, and Analysis by Some Friends of Cuthbert Daniel
MANN, SCHAFER, and SINGPURWALLA · Methods for Statistical Analysis of Reliability and Life Data
MASON, GUNST, and HESS · Statistical Design and Analysis of Experiments with Applications to Engineering and Science
MILLER · Survival Analysis
MONTGOMERY and PECK · Introduction to Linear Regression Analysis, *Second Edition*
NELSON · Accelerated Testing, Statistical Models, Test Plans, and Data Analyses
NELSON · Applied Life Data Analysis
OCHI · Applied Probability and Stochastic Processes in Engineering and Physical Sciences
OSBORNE · Finite Algorithms in Optimization and Data Analysis
PANKRATZ · Forecasting with Dynamic Regression Models

Continued on back end papers

Exploring the Limits of Bootstrap

Exploring the Limits of Bootstrap

Edited by
RAOUL LEPAGE
Michigan State University, East Lansing, Michigan

LYNNE BILLARD
University of Georgia, Athens, Georgia

A Wiley-Interscience Publication
JOHN WILEY & SONS, INC.
New York • Chichester • Brisbane • Toronto • Singapore

***Library of Congress Cataloging in Publication Data*:**

Exploring the limits of bootstrap / edited by Raoul LePage, Lynne Billard.

p. cm.– (Wiley series in probability and mathematical statistics. Applied probability and statistics)
"A Wiley-interscience publication".
Includes bibliographical references and index.
ISBN 0-471-53631-8

1. Estimation theory. 2. Distribution (Probability theory) 3. Sampling (Statistics) 4. Nonparametric statistics. I. LePage. Raoul, 1938- . II. Billard, L. (Lynne), 1943- III. Series.
QA276.8.E97 1992
519.5'44--dc20 91-31546
CIP

Printed and bound in the United States of America by Braun-Brumfield, Inc.

10 9 8 7 6 5 4 3 2 1

CONTRIBUTORS

MIGUEL A. ARCONES, Department of Mathematics, University of Connecticut, Storrs, Connecticut

K. B. ATHREYA, Departments of Mathematics and Statistics, Iowa State University, Ames, Iowa

WILLIAM A. BAILEY, Kemper National Insurance Companies, Long Grove, Illinois

EDWARD J. BEDRICK, Department of Mathematics and Statistics, University of New Mexico, Alberquerque, New Mexico.

P. J. BICKEL, Department of Statistics, University of California, Berkeley, California

L. BILLARD, Department of Statistics, University of Georgia, Athens, Georgia

DAVID BROWNSTONE, Department of Economics, University of California, Irvine, California

D. S. BURDICK, Institute of Statistics and Decision Sciences, Duke University, Durham, North Carolina

SOMNATH DATTA, Department of Statistics, University of Georgia, Athens, Georgia

BRADLEY EFRON, Department of Statistics, Stanford University, Stanford, California

C. D. FUH, Institute of Statistical Science, Academia Sinica, Taipei, Taiwan

EVARIST GINÉ, Department of Mathematics, University of Connecticut, Storrs, Connecticut

PETER HALL, Department of Statistics, Faculty of Economics and Commerce, Australian National University, Canberra, Australia

R. HELMERS, Centre for Mathematics and Computer Science, Amsterdam, The Netherlands

JOE R. HILL, EDS Research, Alberquerque, New Mexico

JOHN J. HSIEH, Department of Preventative Medicine and Biostatistics, University of Toronto, Toronto, Canada

P. JANSSEN, Limburgs Universitair Centrum, Universitaire Campus, Diepenbeek, Belgium

JOHN G. KINATEDER, Department of Statistics, Michigan State University, East Lansing, Michigan

VICTOR KIPNIS, Department of Economics, University of Southern California, Los Angeles, California

S. N. LAHIRI, Department of Statistics, Iowa State University, Ames, Iowa

RAOUL LEPAGE, Department of Statistics, Michigan State University, East Lansing, Michigan

REGINA Y. LIU, Department of Statistics, Rutgers University, New Brunswick, New Jersey

J. S. MARRON, Department of Statistics, University of North Carolina, Chapel Hill, North Carolina

WILLIAM P. McCORMICK, Department of Statistics, University of Georgia, Athens, Georgia

B. C. MITCHELL, Institute of Statistics and Decision Sciences, Duke University, Durham, North Carolina

DIMITRIS N. POLITIS, Department of Statistics, Purdue University, West Lafayette, Indiana

JOSEPH P. ROMANO, Department of Statistics, Stanford University, Stanford, California

NICHOLAS SCHORK, Departments of Statistics and Medicine, University of Michigan, Ann Arbor, Michigan

KESAR SINGH, Department of Statistics, Rutgers University, New Brunswick, New Jersey

R. R. SITTER, Department of Mathematics and Statistics, Carlton University, Ottawa, Canada

MALCOLM S. TAYLOR, US Army Ballistic Research Laboratory, Aberdeen Proving Ground, Maryland

JAMES R. THOMPSON, Department of Statistics, Rice University, Houston, Texas

ROBERT TIBSHIRANI, Department of Preventative Medicine and Biostatistics and Department of Statistics, University of Toronto, Toronto, Canada

XIN M. TU, Department of Biostatistics, Harvard School of Public Health, Boston, Massachusetts

N. VERAVERBEKE, Limburgs Universitair Centrum, Universitaire Campus, Diepenbeek, Belgium

DONGSHENG, TU, Center for Mutivariate Analysis, Pennsylvania State University, University Park, Pennsylvania

CONTENTS

PREFACE

Remarkable developments in statistics have followed Efron's introduction of the bootstrap. In rapid succession, a great many problems have proven amenable to this new method which advances the notion that by sampling from our data we can approximate the sampling variations which produced that data in the first place.

The hard work of developing a theory of bootstrap for parametric, non-parametric, time series, and other areas of statistics, continues at a brisk pace in a stream of fundamental papers by top researchers, most of whom are represented in this volume. There have been some extraordinary revelations.

We have learned that virtually every sample of modest size taken from a normal population can tell us the shape of the Student's t-distributions. A computer can be programmed to do it without any reference to the mathematical form of a t-distribution, and can just as easily apply the same elementary method to a sample scatterplot in order to estimate the sampling errors of the regression coefficients. This is accomplished easily and naturally, even for the more complicated case of a sample drawn from a finite population. We might say that in such cases computer + bootstrap = basic statistics, without much need for specialized formulas or tables.

Consider our astonishment when we learned that simply bootstrapping the t-statistic will in many cases outperform the normal approximation. With some caution, we say that in such cases computer + bootstrap + basic statistics = second order efficient statistics of the kind few statisticians are able to do without digging into papers on Edgeworth expansions.

The main features of these developments can be appreciated by almost anyone having an understanding of basic statistics. Those with deeper preparations will find rich reward.

In this volume you will find original research papers addressing the theme *Exploring the Limits of Bootstrap*, most of which were presented at a special topics meeting of the Institute of Mathematical Statistics held in East Lansing, in May 1990, at the suggestion of Lynne Billard. Raoul LePage selected the theme and agreed to act as organizer and Program Chair. Nearly all of the top researchers on bootstrap are represented here, together with many newcomers, some of whom are working in areas of application. Their contributions to this volume address the full range of issues surrounding the bootstrap, among them: consistency of bootstrap in general circumstances including M-estimation, U-statistics and empirical processes; invariance principles; comparison of bootstraps; efficient bootstrap simulation; bootstrapping long-tailed errors; bias-reduction; applicability to dependent samples including Markov processes and time series (moving and circular block bootstrap); bootstrap bandwidth selection in density estimation; bootstrap applied to complex problems. Many of the topics receive attention in more than one paper.

As a compliment to these new developments, we have a series of application papers dealing with such topics as: exploratory regression; linear model selection; hazard processes; computing bootstrap distributions sans resampling; nonparametric density estimation; resampling with jackknife.

Taken together, these papers represent some of the finest thinking on the subject of bootstrap and offer a rich guide to the literature.

We wish to thank the Institute of Mathematical Statistics for its sponsorship of the meeting, and for its continuing support of research on bootstrap as represented by the preponderance of basic papers you will find referenced to the Annals of Statistics or the Annals of Probability. We also wish to thank the Interface Foundation of North America for cooperation extended during the joint meeting with INTERFACE '90, and Joseph C. Gardiner who assisted with local arrangements for the IMS. Special thanks are due Frank C. Hoppensteadt, Dean of the College of Natural Science, and Paul M. Hunt, Vice Provost and head of Academic Computing, who supported efforts to bring both IMS-BOOTSTRAP and INTERFACE '90 conferences to Michigan State University. William Noble kindly assisted with special local arrangements for young researchers and foreign visitors.

Special thanks are also due to the National Science Foundation and the Office of Naval Research who through their support of INTERFACE '90 indirectly encouraged participation in the bootstrap conference as well.

Raoul LePage, Editor and Program Chair
Lynne Billard, Editor and Program Secretary

Part 1
Introduction

Introduction to Bootstrap

Bradley Efron and Raoul LePage
Stanford University and Michigan State University

Introduction. The following problem arises in almost every data analysis: the statistician observes some data x, and from x constructs an estimate $\hat{\theta} = t(x)$ for a parameter of interest θ. In the most familiar case, x consists of real-valued observations $x_1, .., x_n$ independently sampled from an unknown probability distribution F; the parameter of interest θ is the true mean of F, say $\mu(F) = \int x \, dF(x)$; and the statistic t(x) is the sample mean $\bar{x}$.

The statistician's first job is to select the estimator $\hat{\theta} = t(x)$. Having selected $\hat{\theta}$, a second and in some ways more formidable task arises: to assess the accuracy of $\hat{\theta}$ as an estimator of the true value θ. The *standard error* of $\hat{\theta}$, the square root of its variance,

$$se\{\hat{\theta}\,;\,F\} = [var_F\{t(x)\}]^{1/2}\,, \tag{1}$$

is the most common measure of accuracy for estimators $\hat{\theta}$ that are unbiased, or nearly so.

Elementary statistics courses focus on formulas for the standard error of $\hat{\theta} = \bar{x}$, the mean. Let $\sigma^2(F)$ be the variance of F,

$$\sigma^2(F) = \int_{-\infty}^{\infty} \{x - \mu(F)\}^2 \, dF(x)\,.$$

A simple but powerful formula relates $se(\bar{x}\,;\,F)$ to $\sigma^2(F)$,

$$se\{\bar{x}\,;\,F\} = [\sigma^2(F)/n]^{1/2}\,. \tag{2}$$

This looks useless for practical purposes since $\sigma^2(F)$ is itself a function of the unknown distribution F. However a simple unbiased estimate exists for $\sigma^2(F)$,

$$\hat{\sigma}^2(F) = \frac{\Sigma_{i=1}^{n}(x_i - \bar{x})^2}{(n - 1)}\,. \tag{3}$$

Exploring the Limits of Bootstrap. Edited by Raoul LePage and Lynne Billard.
 ISBN: 0471-53631-8.

Substituting (3) into (2) gives an estimated standard error for $\overline{x}$,

$$\hat{se}\{\overline{x}\} = [\frac{\sum_{i=1}^{n}(x_i - \overline{x})^2}{n \cdot (n - 1)}]^{1/2} . \tag{4}$$

What about estimators $\hat{\theta}$ more complicated than $\overline{x}$, for example a median, a sample correlation coefficient, or the maximum eigenvalue of a sample covariance matrix? Formulas like (2) don't exist for most estimators other than the mean. Much of statistical history concerns finding approximations to formula (2) for more general estimators (often for "mean-like" statistics such as the ratio of two means), and avoiding estimators that do not have such formulas.

Jackknife. In 1958 John Tukey revolutionized error estimation with his "jackknife" method, built upon Quenouille's older technique for estimating biases. Tukey's method avoided formula (2) entirely, going directly to a generalization of (4). Suppose the data x consists of n independent and identically distributed observations x_i; from some unknown distribution F, which we can write diagrammatically as

$$F \overset{iid}{\rightarrow} (x_1, x_2, \dots , x_n) = x . \tag{5}$$

Let $x_{(i)}$ be the data set with the i-th datum removed,

$$x_{(i)} = (x_1, x_2, \dots , x_{i-1}, x_{i+1}, \dots , x_n) ,$$

and let $\hat{\theta}_{(i)}$ equal $t(x_{(i)})$, the statistic $\hat{\theta}$ reevaluated for the deleted-point data set $x_{(i)}$. The jackknife estimate of standard error is

$$se_{jack}\{\hat{\theta}\} = [\frac{n-1}{n} \sum_{i=1}^{n}(\hat{\theta}_{(i)} - \overline{\theta}_{(\cdot)})^2]^{1/2} \tag{6}$$

where $\overline{\theta}_{(\cdot)}$ equals $\sum_{i=1}^{n} \hat{\theta}_{(i)} /n$.

It is easy to verify that formula (6) reduces to formula (2) when $\hat{\theta}$ equals $\overline{x}$. The beauty of Tukey's jackknife is that it automatically produces a standard error estimate for even the most complicated estimator $\hat{\theta}$. All that is required is the ability to recompute $\hat{\theta}$, n times, once for each deleted-point vector x_{i}. The jackknife marked a decisive switch toward computation, and away from the sort of routine theorizing that statisticians traditionally did in extending formula (2) to the complications of real-life statistical practice.

Tukey's formula didn't eliminate statistical theory of course. Rather it focused attention on the theory justifying (6) as an accurate assessment of standard error. The jackknife turned out to work poorly on very un-smooth estimators like the sample median, but otherwise to give generally trustworthy results. See Miller (1964).

A major disappointment in the development of jackknife theory concerned "studentization." Standard errors are often used to set approximate (1-α)-level confidence intervals for θ, of the form

$$\hat{\theta} \pm z_{\alpha/2} \cdot \hat{se} \tag{7}$$

where z_p is the standard normal percentile $z_{0.025} = -1.960$, etc. In the case of estimating the true mean $\mu(F)$ with the sample mean $\bar{x}$, where $\hat{se}$ is given by (4), and where F is assumed to be a normal distribution, Student's famous result says that $z_{\alpha/2}$ in (7) should be replaced by $t_{\alpha/2,n-1}$, the student's t percentile with degrees of freedom ν=n-1 . Considerable efforts were made to find the correct degrees of freedom for $\hat{se}_{jack}$, in order to get approximate confidence intervals better than $\hat{\theta} + z_{\alpha/2}\ \hat{se}_{jack}$, but to no conclusive end.

Bootstrap. Efron's bootstrap (1979) began as an attempt to better understand the jackknife. This involved reexamining formula (2) which had been bypassed in going to (4). Suppose that in (2) the empirical distribution $\hat{F}$ of the observed data, that is the distribution

$$\hat{F}: \text{ probability } 1/n \text{ on } x_i \text{ for } i = 1, 2, \ldots, n, \tag{8}$$

is substituted for the unknown true distribution F. Since $\sigma^2(\hat{F}) = \overline{(x-\bar{x})^2}$, this gives the standard error estimate

$$\text{se}\{\bar{x}\,,\,\hat{F}\} = [\overline{(x-\bar{x})^2}/n]^{1/2}\,,$$

almost the same as the traditional estimate (4).

Efron's (1979) bootstrap paper makes these main points:

- That substituting $\hat{F}$ for F in formula (1) gives a reasonable estimate of standard error for any estimator , namely

$$\text{se}_{\text{boot}}\{\hat{\theta}\} \equiv \text{se}\{\hat{\theta}\,;\,\hat{F}\} = [\text{var}_{\hat{F}}\{t(x^*)\}]^{1/2}\,. \tag{9}$$

Here x^* indicates a hypothetical data vector generated by i.i.d. sampling from the distribution $\hat{F}$,

$$\hat{F} \overset{\text{iid}}{\rightarrow} (x_1^*, \ldots, x_n^*) = x^*, \tag{10}$$

as distinct from the data vector x actually observed.

- That there is a simple computer algorithm for estimating $\text{se}_{\text{boot}}\{\hat{\theta}\}$, i.e. for substituting $\hat{F}$ for F in (1), even though no formula like (2) exists for most $\hat{\theta}$.

- That se_{boot} agrees asymptotically with se_{jack}, and as a matter of fact the jackknife is a linear approximation to the more computer-intensive bootstrap process.

- That the bootstrap is a reasonable error estimator in its own right, easier to understand than the jackknife, and easier to extend to data structures beyond (5).

- That the bootstrap method could potentially be applied to problems of statistical error assessment beyond biases and standard errors, in particular the setting of approximate confidence intervals, but only if further progress were made in understanding the bootstrap's inferential basis.

Consistency of bootstrap. Progress came swiftly. Bickel and Freedman (1981) formulated the problem of the asymptotic consistency of bootstrap in a simple way. We may have some statistics of interest $\hat{\theta}(n;(x_1, \dots, x_n))$ which, when normalized, possess a limit distribution L_G, whenever the data x are iid from any distribution G belonging to some "neighborhood" of the true distribution F. We write this,

$$\{ \text{ the distribution of } \hat{\theta}(n;(x_1, \dots, x_n);G) \} \Rightarrow L_G . \tag{11}$$

Given a sample $x = (x_1, \dots, x_n)$ iid F, we would like bootstrap to be able to estimate the distribution L_F by resampling x and applying $\hat{\theta}$ with $\hat{F}$ in place of F, i.e. we want

$$\{ \text{ conditional distribution of } \hat{\theta}(n;(x_1^*, \dots, x_n^*);\hat{F}) \textbf{ given } x \} \Rightarrow L_F . \tag{12}$$

Bickel and Freedman (1981) isolate a set of three general conditions which together imply the bootstrap consistency result (12) in the circumstances (some of them quite general) studied thus far:

- The limit distribution L_G is continuously dependent upon G, the sampling distribution of the data.

- The convergence (11) is uniform for G belonging to the "neighborhood of F."

- With probability one, $\hat{F}$ will eventually remain within the "neighborhood of F."

Their paper should be consulted for the precise statement of these results. In the paper, a number of examples are developed which exhibit the failure of (12) due to violation of each of the three conditions individually.

Based on this work, bootstrap was rapidly shown to apply in a broad range of standard applications, including t-statistics, empirical, and quantile processes (Bickel and Freedman, 1981); multiple regression (Freedman, 1981); and stratified sampling (Bickel and Freedman, 1984).

Pivoting and Edgeworth expansions. At the same time, an unexpected major advance in the theory of bootstrap was obtained by Singh (1981), independently of Bickel and Freedman, who showed that one of Efron's bootstrap methods, the pivot or bootstrap-t method, in many cases accomplishes "Edgeworth correction." That is, for estimating a point of the sampling distribution of the studentized statistic, bootstrap automatically produces answers as good as one-term Edgeworth expansions, without requiring special theoretical calculations.

To put these results in some perspective, consider confidence intervals for the population mean μ obtained by using each of three distribution approximations:

- Central limit theorem approximation for the distribution of $(\bar{x}-\mu)/\hat{se}$, with $\hat{se}$ given by (4).

- Percentile (naive) bootstrap distribution of $\bar{x}^*-\bar{x}$, conditional on data x, as an approximation for the distribution of $\bar{x}-\mu$.

- Percentile-t bootstrap distribution of $(\bar{x}^*-\bar{x})/\hat{se}^*$, conditional on data x, as an approximation for the distribution of $(\bar{x}-\mu)/\hat{se}$.

Great interest is generated by the fact that one-sided α-intended-level confidence intervals based on the first two methods, i.e. the central limit theorem and naive bootstrap, have actual coverage probabilities which typically differ from α by $O(1/\sqrt{n})$. The coverage probability achieved by the percentile-t method, on the other hand, is typically correct to $O(1/n)$. This improvement is rooted in the Edgeworth correction effect.

The situation for symmetric two-sided confidence intervals is somewhat different since $1/\sqrt{n}$ terms cancel in the first two methods as well. In the case of the central limit method,

$$P(\ \mu \in \bar{x} \pm z_{a/2} \cdot \hat{se}\ ) = 1\text{-}\alpha + O(1/n), \tag{13}$$

while for the naive bootstrap interval $I(x) = \bar{x} \pm W(x)$ obtained by choosing $W(x)$ to satisfy,

$$P(\, \overline{x}^* \in I(x) \mid x\,) \approx 1\text{-}\alpha\,, \tag{14}$$

we likewise have,

$$P(\, \mu \in I(x)\,) = 1\text{-}\alpha + O(1/n). \tag{15}$$

Finally, for the percentile-t, choosing V(x) so that,

$$P(\, \mid \frac{\overline{x}^*\text{-}\overline{x}}{\hat{\sigma}^*} \mid < V(x) \mid x\,) \approx 1\text{-}\alpha\,, \tag{16}$$

where $\hat{\sigma}^*$ is the sample standard deviation of $(x_1^*, \ldots, x_n^*)$, we also have

$$P(\, \mu \in \overline{x} \pm V(x)\,\hat{\sigma}\,) = 1\text{-}\alpha + O(1/n)\,. \tag{17}$$

Results of this type set in motion a stream of research aimed at producing bootstrap confidence intervals taking advantage of Edgeworth-correction effects without introducing unpleasant side effects, such as excessive width. See Efron (1987). Various bias corrections and other methods have been introduced. Hall (1988) compares seven different bootstrap methods utilizing Edgeworth expansions and makes the intriguing finding that "the bootstrap approximation is so good that bootstrap versions of erroneous theoretical critical points are also erroneous."

New directions. Some effort has gone into studying the precise nature of bootstrap's failure for long tailed errors. Athreya (1987) studied bootstrapping $\overline{x}$ when L_G are stable laws of index $\alpha<2$, and established that the limit distribution obtained by bootstrap is random. Recently, Giné and Zinn (1989) have shown that normal limits L_G are in some sense necessary when $\hat{\theta}(n; x)$ are normalized sums. Do bootstraps exist that can successfully cope with long-tailed errors in normalized sums, for how is one to know when the normal limits apply?

A potentially important development in bootstrap is "double-dip" bootstrap, i.e choosing among several competing estimators the one whose bootstrap-estimated sampling error is least; then again using bootstrap to assess the sampling error of this adaptively chosen estimator.

Other promising new lines of current research are directed toward extending bootstrap methods to dependent observations, such as Markov chains and time series, where the idea of moving-block bootstrap is receiving a lot of attention; and developing bootstrap methods for estimating the sampling error in density or spectral-density estimation.

Many of the papers presented at this conference concern the higher-order accuracy of bootstrap methods in these various contexts, and how the accuracy can be harnessed to the task of producing accurate statistical inference in an automatic way. "Automatic" here refers to shifting the burden of computation in any particular applied problem away from theoretical calculations and toward a computer algorithm. In particular the goal of automatically producing highly accurate confidence intervals seems to be moving towards a practical solution.

References

Athreya, K. B. (1987), Bootstrap of the mean in the infinite variance case. *Annals of Statistics,* **15**:724-731.

Bickel, P. J. and Freedman, D. A. (1981), Some asymptotic theory for the bootstrap. *Annals of Statistics,* **9**:1196-1217.

Bickel, P. J. and Freedman, D. A. (1984), Asymptotic normality and the bootstrap in stratified sampling. *Annals of Statistics,* **12**:470-482.

Efron, B. (1979), Bootstrap methods: Another look at the jackknife. *Annals of Statistics,* **7**:1-26.

Efron, B. (1987), Better bootstrap confidence intervals (with discussion). *Journal of the American Statistical Association,* **82**:171-200.

Freedman, D. A. (1981), Bootstrapping regression models. *Annals of Statistics,* **9**:1218-1228.

Giné, E. and Zinn, J. (1989), Necessary conditions for the bootstrap of the mean. *Annals of Statistics,* **17**:684-691.

Hall, P. (1988), Theoretical comparison of bootstrap confidence intervals (with discussion). *Annals of Statistics,* **16**:927-985.

Miller, R. G. (1964), A trustworthy jackknife. *Annals of Mathematical Statistics,* **35**:1594-1605.

Singh, K. (1981), On the asymptotic accuracy of Efron's bootstrap. *Annals of Statistics,* **9**:1187-1195.

Part 2

General Principles of The Bootstrap

On the bootstrap of M-estimators and other statistical functionals

Miguel A. Arcones*†
CUNY, Graduate Center
The University of Connecticut

Evarist Giné* ‡
CUNY, The College of Staten Island
The University of Connecticut

Abstract

Results on the bootstrap of empirical processes (Giné and Zinn (1990)) are applied to obtain the a.s. bootstrap CLT for M-estimators under weak differentiability hypotheses. Also, work of Dudley (1990) and Sheehy and Wellner (1988) on the bootstrap of differentiable functionals of the empirical process is reviewed and complemented.

*Work partially supported by NSF Grant No. DMS 9000132 and by PSC-CUNY Grants No. 669336 and No. 661376

†Current address: Mathematical Sciences Research Institute, 1000 Centennial Drive, Berkeley, CA 94720

‡Current address: The University of Connecticut, Department of Mathematics, Storrs, CT 06269-3009

Exploring the Limits of Bootstrap. Edited by Raoul LePage and Lynne Billard.
 ISBN: 0471-53631-8.

1 Introduction

Huber (1967) proves asymptotic normality of M-estimators under non-standard conditions on the basis of an asymptotic equicontinuity result for the empirical process indexed by a class of functions $\{\varphi(x,\theta) : \theta \in \Theta\}$ defined on $S \times \Theta, (S, \mathcal{S})$ a general measurable space and $\Theta \subset \mathbf{R}^d$. Thus, it was to be expected that the recent theory of empirical processes over general classes of functions would have something to contribute to the theory of M-estimators. In fact, Pollard (1985) extends Huber's results using modern empirical process theory. In a somewhat related development, Dudley (1990) and Sheehy and Wellner (1988) (see also Gill (1989)) introduce different notions of differentiable (von Mises) functionals in the general empirical process context, thus extending the classical notion of Fréchet and compact differentiable statistical functionals.

Giné and Zinn (1990) prove a general bootstrap CLT for empirical processes: under measurability conditions, if the empirical process indexed by a class of functions $\mathcal{F}$ satisfies the central limit theorem (CLT) then it satisfies also the bootstrap CLT in probability (a.s. if moreover the envelope of $\mathcal{F}$ is square integrable). Less general results for more general resampling procedures (such as the parametric bootstrap) are also available (Giné and Zinn (1991)).

In this article we combine Huber's and Pollard's work on M-estimators with Giné and Zinn's bootstrap CLT to obtain a.s. bootstrap limit theorems for M-estimators. Dudley and Sheehy and Wellner also used the bootstrap CLT for empirical processes to obtain bootstrap limit theorems for different types of differentiable functionals of the empirical processes. We give results that complement Dudley's, describe, with some complements and precisions, the work of Sheehy and Wellner and then apply these results to M-estimators once more.

We look at M-estimators twice in order to compare Huber's and Pollard's approach (the key term here is Pollard's "stochastic differentiability " condition) with the more classical approach based on different notions of differentiability (Reeds (1976), Gill (1989) and Sheehy and Wellner (1988) among others). It appears that, for M-estimators, both methods give similar results although the former is somewhat better: a condition on stochastic equicontinuity at $\theta(P)$ of the empirical process indexed by a certain class of functions in the former approach is replaced in the latter by the requirement that the empirical process over the same class of functions satisfy the CLT, a condition equivalent to stochastic equicontinuity over the whole domain Θ. Of course differentiability is more widely applicable.

The main results are the a.s. bootstrap CLT's for M-estimators, in Section 3. The proofs use three main tools from empirical processes namely the Giné-Zinn bootstrap CLT, an exponential inequality of Alexander (1984) (see Section 2(c)) and the "square root trick" exponential bound (LeCam (1986, page 546), Giné and Zinn (1984, 1986)). Concrete examples include the bootstrap of the spatial median and the bootstrap of k-means (this, for reasons of expediency, only in $\mathbf{R}$). (Can these two examples be handled with more "classical" methods i.e. proving some kind of differentiability with respect to the multivariate cdf? Perhaps, but we were not diligent enough to check this since our approach, that follows Pollard's, is so well adapted to the problem).

The results on differentiable functionals are in Section 4. The emphasis is on bootstrapping limit theorems for functionals which are differentiable in a very weak sense (only along a family of subsequences and at a fixed rate). Both the framework and the ideas of the proof of Theorem 4.6 belong to Sheehy and Wellner (1988) and Gill (1989). Our contribution to this subject consists only of providing some accurate proofs (particularly for the existence of a representation for which simultaneously the empirical process converges uniformly a.s. to the Gaussian limit and the bootstrap CLT holds a.s., which is given in Section 2) and observing that if the bootstrap sample size is of the order $0(n/\log\log n)$ then the bootstrap CLT actually holds a.s. (a phenomenon already pointed out before in similar situations - Yang (1985), Arcones and Giné (1990)).

2 Empirical processes.

In the first part of this section we give some notation and definitions about empirical processes and describe the results on the bootstrap of empirical processes of Giné and Zinn (1990, 1991). In a second part we give some technical results that will be useful in the study of the bootstrap for differentiable functionals. The third contains an application of Alexander's (1984) exponential inequalities to the almost sure behavior of empirical processes.

(a) The bootstrap of empirical processes. Let $(S, \mathcal{S}, P)$ be a probability space and let $\mathcal{F} \subset \mathcal{L}_2(S, \mathcal{S}, P)$ be a class of functions such that

$$F(s) := \sup\{|f(s)| : f \in \mathcal{F}\} < \infty \text{ for all } s \in S,$$

$$\sup\{|Pf| : f \in \mathcal{F}\} < \infty, \text{ where } Pf := \int f dP.$$

Under these two conditons, the maps

$$\delta_s : \mathcal{F} \to \mathbf{R} \text{ given by } \delta_s(f) = f(s)$$

and

$$P : \mathcal{F} \to \mathbf{R} \text{ given by } Pf = \int f dP$$

are *bounded.* So, $\delta_s, P \in l^\infty(\mathcal{F})$, the Banach space of real bounded functions on $\mathcal{F}$, equipped with the supremum norm

$$||x||_{\mathcal{F}} := \sup\{|x(f)| : f \in \mathcal{F}\}.$$

$l^\infty(\mathcal{F})$ plays the role of $D(-\infty, \infty)$ in the classical formulation of Donsker's convergence theorem for the empirical cdf (corresponding to the case $\mathcal{F} = \{I_{(-\infty,t]} : t \in \mathbf{R}\}$). The non-separability of $(l^\infty(\mathcal{F}), ||\cdot||_{\mathcal{F}})$, just like the non-separability of $D(0,1)$ for the sup norm topology, introduces some measurability complications when dealing with empirical processes. All of them can be satisfactorily dealt with under some assumptions on $\mathcal{F}$, in particular for $\mathcal{F}$ countable or for $\mathcal{F}$ image admissible Suslin (Dudley (1984)). We will write that $\mathcal{F}$ *is measurable* to mean that $\mathcal{F}$ satisfies the image admissible Suslin property.

Let $X_i : (S^N, \mathcal{S}^N, P^N) \to (S, \mathcal{S}, P)$ be the coordinate functionals of S^N. The X_i are i. i. d. (P). It will be convenient to have X_i defined on a larger probability space $(\Omega, \Sigma, Pr) = (S^N, \mathcal{S}^N, P^N) \times ([0,1], \mathcal{B}, \lambda)$. The *empirical measure* P_n^ω is defined as

$$P_n^\omega = n^{-1} \sum_{i=1}^{n} \delta_{X_i(\omega)},\ n \in N \tag{2.1}$$

and the (centered, normalized) *empirical process* $\nu_n = \nu_n^P = \nu_n^P(\omega)$ by

$$\nu_n = n^{1/2}(P_n - P),\ n \in N. \tag{2.2}$$

Both ν_n and P_n are random elements with values in $l^\infty(\mathcal{F})$. Let $\{G_P(f) : f \in \mathcal{F}\}$, the *P-Brownian bridge* indexed by $\mathcal{F}$, be the centered Gaussian process with covariance

$$EG_P(f)G_P(g) = Pfg - (Pf)(Pg),\ f, g \in \mathcal{F} \tag{2.3}$$

For further reference, we let

$$\rho_P^2(f,g) = P(f-g)^2 - (P(f-g))^2 = E(G_P(f) - G_P(g))^2, f, g \in \mathcal{F}. \tag{2.4}$$

Obviously, for every finite part $J \subset \mathcal{F}$,

$$(\nu_n(f) : f \in J) \to_{\mathcal{L}} (G_P(f) : f \in J)$$

by the finite-dimensional central limit theorem. An important question is whether this convergence can be made "uniform" over all of $\mathcal{F}$. To make this precise, we define (following Hoffmann-Jørgensen): if $\{X_n\}_{n=0}^\infty$ are $l^\infty(\mathcal{F})$-valued random elements and X_0 is measurable and has a separable support then

$$X_n \to_{\mathcal{L}} X_0 \tag{2.5}$$

in $l^\infty(\mathcal{F})$ iff

$$E^* H(X_n) \to EH(X_0) \tag{2.6}$$

for all $H : l^\infty(\mathcal{F}) \to \mathbf{R}$ bounded and continuous. E^* stands for outer integral. Note that if the process G_P has a version with bounded ρ_P-uniformly continuous trajectories, then it is measurable and has support $C_u(\mathcal{F}, \rho_P)$ (C_u = bounded uniformly continuous functions), which is separable in $(l^\infty(\mathcal{F}), ||\cdot||_\mathcal{F})$. If G_P has this property we say, for short, that G_P is *sample continuous*. Dudley's definition for "CLT for the empirical process uniform over $\mathcal{F}$" is: $\mathcal{F}$ is a *P-Donsker class* iff:
(i) G_P is sample continuous and
(ii) $\nu_n^P \to_\mathcal{L} G_P$ in $l^\infty(\mathcal{F})$.
To relate this definition to classical theory, just note that *Donsker's theorem* on weak convergence in $D(-\infty, \infty)$ of the empirical cdf can just be restated as: The class $\mathcal{F} = \{I_{(-\infty,t]} : t \in \mathbf{R}\}$ is P-Donsker for all $P \in \mathcal{P}(\mathbf{R})$. The use in statistics of this theorem rests in part on the continuous mapping theorem: if $\mathcal{F}$ is P-Donsker and $T : l^\infty(\mathcal{F}) \to \mathbf{R}$ is continuous then

$$T(\nu_n) \to_\mathcal{L} T(G_P) \tag{2.7}$$

(interpreted as regular convergence in distribution if $T(\nu_n)$ is a random variable, and in the sense of (2.5) and (2.6), with $l^\infty(\mathcal{F})$ replaced by $\mathbf{R}^d$, otherwise).

There are many interesting classes $\mathcal{F}$ besides $\{(-\infty, t] : t \in \mathbf{R}\}$, both for $S = \mathbf{R}$ or $\mathbf{R}^d$ and for more general S. We will encounter some below.

If one views ν_n^P as a measure-valued statistic, then its *simple bootstrap* is obviously $\nu_n^{P_n^\omega}$. In other words, if $X_{n1}^\omega, \ldots, X_{nn}^\omega$ are i.i.d. (P_n^ω), and if

$$\hat{P}_n^\omega = n^{-1} \sum_{i=1}^n \delta_{X_{ni}^\omega} \tag{2.8}$$

is the empirical measure constructed from the bootstrap sample $\{X_{ni}^\omega\}_{i=1}^n$ then the *bootstrap empirical process* is

$$\nu_n^{P_n^\omega} = n^{\frac{1}{2}}(\hat{P}_n^\omega - P_n^\omega) \tag{2.9}$$

We will write $\hat{\nu}_n$ or $\hat{\nu}_n^\omega$ for $\nu_n^{P_n^\omega}$.

In order to give a precise meaning to the bootstrap CLT (in probability and a.s.) for empirical processes, we should mention that convergence in law in $l^\infty(\mathcal{F})$ is metrizable in the following sense (e.g. Giné and Zinn (1991)): $X_n \to_\mathcal{L} X_0$ if and only if

$$d_{BL^*}(X_n, X_0) \to 0 \tag{2.10}$$

where

$$d_{BL^*}(X_n, X_0) = \sup\{|E^* H(X_n) - EH(X_0)| : ||H||_\infty \le 1, ||H||_{Lip} \le 1\}.$$

We then say that the *bootstrap CLT for the empirical process holds in probability* or that $\mathcal{F}$ is *pr-bootstrap P-Donsker* iff for all $\varepsilon > 0$

$$Pr^*\{\hat{d}_{BL^*}(\hat{\nu}_n, G_P) > \varepsilon\} \to 0. \tag{2.11}$$

Here $\hat{d}$ is d with E^* replaced by conditional expectation given the sample. The *bootstrap CLT for the empirical process holds a.s.* or $\mathcal{F}$ is *a.s.-bootstrap P-Donsker* iff

$$\hat{d}_{BL^*}(\hat{\nu}_n, G_P) \to 0 \text{ almost uniformly.} \tag{2.12}$$

Note that the bootstrap CLT for the empirical process gives at once, via the continuous mapping theorem, the bootstrap of a large variety of limit theorems.

The following result states that *the bootstrap always works* in this general context, even *without local uniformity* of the CLT in P.

Theorem 2.1 *(Giné and Zinn (1990)). Let $\mathcal{F}$ be a measurable class of functions on S. Then*
(i) $\mathcal{F}$ is P-Donsker $\Leftrightarrow$ $\mathcal{F}$ is pr-bootstrap P-Donsker.
(ii) $\mathcal{F}$ is P-Donsker and $PF^2 < \infty \Leftrightarrow \mathcal{F}$ is a.s.-bootstrap P-Donsker.

We refer to Giné and Zinn (1990) for the proof. There has recently been some interest in considering different bootstrap sample sizes. In the present situation this amounts to taking $\nu_m^{P_n^\omega}$ with m not necessarily equal to n. The proofs in Giné and Zinn (1990) can be modified to yield the following:

Proposition 2.2 *(Giné and Zinn, unpublished). Under the same measurability as in 2.1,*

(i) $\mathcal{F}$ is P-Donsker and $m_n \to \infty \Rightarrow d_{BL^*}(\nu_{m_n}^{P_n^\omega}, G_P) \to 0$ in Pr^*.
(ii) $\mathcal{F}$ is P-Donsker, $PF^2 < \infty$, $m_n \nearrow \infty$, $m_n/m_{2n} \geq c > 0$
$\Rightarrow \nu_{m_n}^{P_n^\omega} \to_{\mathcal{L}} G_P$ in $l^\infty(\mathcal{F})$ a.s.

We will not prove this proposition, but only mention the additional facts that must be combined with the proof of Theorem 2.1 in Giné and Zinn (1990) to yield Proposition 2.2. Their proofs are not difficult to implement (given the published background). These are:

1) If $\tilde{N}(m/n)$ is the difference of two independent Poisson variables with mean m/n, then $E\max_{k\leq n}|\tilde{N}_k(m/n)|/m^{\frac{1}{2}} \to 0$ where $\tilde{N}_k$ are i.i.d. with the law of $\tilde{N}$, and

$$\sup_{n,m>0}\int_0^\infty P\{|\tilde{N}(m/n)| > t\sqrt{m/n}\}^{\frac{1}{2}}dt < \infty$$

(4 is a bound for this integral).

2) The basic lemma of Ledoux, Talagrand and Zinn (Ledoux and Talagrand (1989)) admits the following modification: If $m_n \nearrow \infty$, $m_n/m_{2n} \geq c > 0$, $\{X_i\}$ are i.i.d. Banach space valued random variables and $\tilde{N}_i(m/n)$ are i.i.d. as in 1) and independent of $\{X_i\}$, then there is $K < \infty$ independent of $\{X_i\}$ such that

$$\limsup_{n\to\infty} n^{-\frac{1}{2}}\hat{E}||\sum_{i=1}^{n}(n/m_n)^{\frac{1}{2}}\tilde{N}_i(m_n/n)X_i||$$

$$\leq K\limsup_{n\to\infty} n^{-\frac{1}{2}}E||\sum_{i=1}^{n}(n/m_n)^{\frac{1}{2}}\tilde{N}_i(m_n/n)X_i|| \ a.s.$$

where $\hat{E}$ denotes conditional expectation given $\{X_i\}$.

The regularity of m_n in 2) seems to be required because the proof uses blocking of the sums (as in the proof of the LIL). However, we would not be surprised if Proposition 2.2 (ii) were true for arbitrary $m_n \to \infty$.

If it is known that $P \in \{P_\theta : \theta \in \Theta\}$ we may want to resample from $P_{\theta_n(\omega)}$ instead of P_n^ω, where $\theta_n(\omega)$ is an estimate of θ based on the sample. More generally, we may want to resample from $\tau_n(P_n)$ (like e.g. smoothing or symmetrization of P_n). For this situation there are no general results such as Theorem 2.1 (and there are counter examples: Beran (1984)). However, if there is uniformity in P in the CLT for ν_n^P then a very general resampling theorem holds.

A uniformly bounded class of functions $\mathcal{F}$ is *uniformly pregaussian* if the P-Brownian motions Z_P (covariance of $Z_P : EZ_P(f)Z_P(g) = Pfg$; its relationship to G_P is not unlike that of Brownian motion to Brownian bridge) are "nice" uniformly in $P \in \mathcal{P}_f(S)$, the p.m.'s on S with finite support. Note that if $P = \sum \alpha_i\delta_{s_i}$ then $Z_P = \sum \alpha_i^{\frac{1}{2}} g_i\delta_{s_i}$, g_i *i.i.d.*$(N(0,1))$. Concretely $\mathcal{F}$ uniformly bounded is UPG iff

i) $\sup_{P\in\mathcal{P}_f(S)} E||Z_P||_{\mathcal{F}} < \infty$

ii) $\lim_{\delta\to 0}\sup_{P\in\mathcal{P}_f(S)} E\sup_{P(f-g)^2\leq\delta^2, f,g\in\mathcal{F}} |Z_P(f) - Z_P(g)| = 0.$

Let $\mathcal{P}(S)$ denote the set of probability measures on $(S, \mathcal{S})$. It is proved in Giné and Zinn (1991) that $\mathcal{F}$ measurable and uniformly bounded is UPG if and only if:

1) conditions i) and ii) above hold with $\mathcal{P}_f(S)$ replaced by $\mathcal{P}(S)$ and

2) $\lim_{n\to\infty}\sup_{P\in\mathcal{P}(S)} d_{BL^*}(\nu_n^P, G_P) = 0.$

Note that 2) means precisely that the CLT for empirical processes indexed by $\mathcal{F}$ holds *uniformly* in $P \in \mathcal{P}(S)$. It is also proved there that if $\mathcal{F}$ is UPG then, letting $\mathcal{G} = \mathcal{F}\bigcup\mathcal{F}\cdot\mathcal{F} = \mathcal{F}\bigcup\{f\cdot h : f, h \in \mathcal{F}\}$,

$$||R_n - R_0||_{\mathcal{G}} \to 0 \Rightarrow d_{BL^*}(G_{R_n}, G_{R_0}) \to 0$$

for $R_n \in \mathcal{P}(S)$, $n = 1, \ldots$. Then a simple triangle inequality gives:

Theorem 2.3 *(Giné and Zinn (1991)). If $\mathcal{F}$ measurable and uniformly bounded is UPG then*

$$||R_n - R_0||_{\mathcal{G}} \to 0 \Rightarrow \nu_n^{R_n} \to_{\mathcal{L}} G_{R_0} \ \textit{in } l^\infty(\mathcal{F}). \tag{2.13}$$

Corollary 2.4 *Under the conditions of Theorem 2.3, if* $\|\tau_n(P_n) - P\|_{\mathcal{G}} \to 0$ *a.s. (in probability) then* $\nu_n^{\tau_n(P_n(\omega))} \to_{\hat{\mathcal{L}}} G_P$ *a.s. (in probability) in* $l^\infty(\mathcal{F})$.

In Corollary 2.4 the bootstrap sample size can be taken arbitrary as long as it tends to infinity: the same proof applies.

(b) Almost sure representations and the bootstrap. When dealing with differentiable functionals of the empirical process it is convenient to have versions for which convergence in law to G_P becomes a.s. convergence in $l^\infty(\mathcal{F})$ with the bootstrap CLT still holding. This was recognized by Gill (1989) in $\mathbb{R}$ and by Sheehy and Wellner (1988) in general. We prove here a version of Sheehy and Wellner's statement which follows by combining an important representation theorem of Dudley (1985) and Theorem 2.1 above. We recall the definition of perfect functions: a measurable function $g : (U, \mathcal{U}, P) \to (V, \mathcal{V})$ is perfect if $(P \circ g^{-1})^* = P^* \circ g^{-1}$ (see Dudley (1985) for different equivalent definitions). We also need the definition of almost uniform convergence (of non-measurable functions): $X_n \to X$ almost uniformly if $\|X_n - X\|_{\mathcal{F}}^* \to 0$ a.s. Almost uniform convergence (contrary to a.s. convergence) implies convergence in pr^*.

Theorem 2.5 *Let* $\mathcal{F}$ *be a P-Donsker measurable class of functions. Then there is a probability space* $(\tilde{\Omega}, \tilde{\Sigma}, \tilde{Pr})$ *and perfect functions* $g_n : (\tilde{\Omega}, \tilde{\Sigma}, \tilde{Pr}) \to (\Omega, \Sigma, Pr)$ *such that*

$$\tilde{Pr} \circ g_n^{-1} = Pr,\ n = 0, 1, \ldots,\ \ \nu_n \circ g_n \to G_P \circ g_0 \text{ almost uniformly} \tag{2.14}$$

and

$$\tilde{d}_{BL^*}(\nu_n^{P_n \circ g_n}, G_P) \to 0 \text{ almost uniformly.} \tag{2.15}$$

Proof By Theorem 2.1,

$$Y_n := \hat{d}_{BL^*}(\nu_n^{P_n}, G_P)^* \to 0 \text{ in probability .}$$

Therefore, $\mathcal{F}$ being P-Donsker, the $l^\infty(\mathcal{F}) \times \mathbb{R}$-valued random elements (ν_n, Y_n) converge in law to $(G_P, 0)$ (this is immediate from the definition of convergence in law). Then Dudley (1985), Theorem 4.1, implies the existence of a probability space $(\tilde{\Omega}, \tilde{\Sigma}, \tilde{Pr})$ and perfect functions g_n with $\tilde{Pr} \circ g_n = Pr$ such that

$$(\nu_n, Y_n) \circ g_n \to (G_P, 0) \circ g_0 \text{ almost uniformly.}$$

The result follows since $Y_n \circ g_n$ is the variable in (2.15). ∎

As another useful application of Dudley's representation theorem, we prove now a general continuous mapping theorem. Wellner (1989) has a similar statement (which is proved in van der Vaart and Wellner (1989)) and describes how this result is relevant for convergence of differentiable functionals. Our proof is different, closer to the proof of the continuous mapping theorem for a single functional in Pollard (1989).

Theorem 2.6 *Let $\{Z_n\}_{n=0}^{\infty}$ be random elements with values in a metric space (V, d). Suppose Z_0 has separable range and is measurable, and that $Z_n \to_{\mathcal{L}} Z_0$ (in the sense of (2.5), (2.6) with $l^\infty(\mathcal{F})$ replaced by V). Let V_0 be a separable Borel subset of V such that $Pr\{Z_0 \in V_0\} = 1$ and let $\bar{V}_0$ be its completion. For each $n \in N$ let $V_n \subset V$ be a set containing the range of Z_n. Let $H_0 : \bar{V}_0 \to \mathbf{R}$ be measurable and let $H_n : V_n \to \mathbf{R}$ be functions satisfying*

$$x_{n'} \in V_{n'}, n' \nearrow \infty, x_0 \in \bar{V}_0,\ d(x_{n'}, x_0) \to 0 \Rightarrow H_{n'}(x_{n'}) \to H_0(x_0) \qquad (2.16)$$

where we also denote by d the distance in the completion $\bar{V}$ of V. Then

$$H_n(Z_n) \to_{\mathcal{L}} H_0(Z_0).$$

Proof By Lusin's theorem (e.g. Dudley (1989, page 190)) and tightness of $\mathcal{L}(Z_0)$ in $\bar{V}_0$, for every $\tau > 0$ there exists $K \subset \bar{V}_0, K$ compact in $\bar{V}_0$, such that $Pr\{Z_0 \in K^c\} \leq \tau$ and H_0 is (uniformly) continuous on K. This and the convergence hypothesis on $\{H_n\}$ implies that for all $\varepsilon > 0$ there exist $\delta > 0$ and $n_0 < \infty$ such that

$$\sup_{x_n \in V_n, x_0 \in K, d(x_n, x_0) < \delta} |H_n(x_n) - H_0(x_0)| < \varepsilon \text{ for } n \geq n_0.$$

Let g_n, g_0 be the perfect maps prescribed by Dudley's (1985) theorem for $Z_n \to_{\mathcal{L}} Z_0$. By the properties of these maps and the definition of convergence in law it suffices to prove

$$H_n(Z_n \circ g_n) \to H_0(Z_0 \circ g_0) \text{ almost uniformly}$$

(since then, by Dudley (1985, Theorem 3.5), $H_n(Z_n) \to_{\mathcal{L}} H_0(Z_0)$). Since for $m \geq n_0$

$$\{\sup_{n \geq m} |H_n(Z_n \circ g_n) - H_0(Z_0 \circ g_0)| > \varepsilon\}$$
$$\subseteq \{\sup_{n \geq m} d(Z_n \circ g_n, Z_0 \circ g_0) \geq \delta\} \bigcup \{Z_0 \circ g_0 \in K^c\}$$

we have

$$\tilde{Pr}\{\sup_{n \geq m} |H_n(Z_n \circ g_n) - H_0(Z_0 \circ g_0)|^* > \varepsilon\}$$
$$= \tilde{Pr}^*\{\sup_{n \geq m} |H_n(Z_n \circ g_n) - H_0(Z_0 \circ g_0)| > \varepsilon\}$$
$$\leq \tilde{Pr}^*\{\sup_{n \geq m} d(Z_n \circ g_n, Z_0 \circ g_0) \geq \delta\} + \tau$$

whose lim sup is dominated by τ because $Z_n \circ g_n \to Z_0 \circ g_0$ almost uniformly. Since τ is arbitrary, $H_n(Z_n \circ g_n) \to H_0(Z_0 \circ g_0)$ almost uniformly and the result follows. ■

(c) Almost sure behavior of empirical processes. In Section 3 we require a.s. limit theorems for the empirical process, such as

$$(\log \log n)^{\frac{1}{2}} \|\nu_n\|_{\mathcal{F}_n} \to_{a.s.} 0$$

usually with $\mathcal{F}_n = \{f - g : \rho_P(f,g) \leq \delta_n, f, g \in \mathcal{F}\}$ for suitable δ_n. This will be a consequence of an inequality of Alexander (1984) for classes of functions related to the Vapnik-Čërvonenkis (VC) property. A class of sets $\mathcal{C}$ is VC if there exists $n < \infty$ such that

$$\sup_{x \in S^n} \#\{\{x_1, \ldots, x_n\} \bigcap C : C \in \mathcal{C}\} < 2^n.$$

These classes have very interesting properties regarding the CLT because their metric entropies with respect to the $\mathcal{L}_2(P)$ distances are small uniformly in P (Dudley (1984)). These properties are inherited by *VC-subgraph* classes of functions: a class $\mathcal{F}$ of functions if VC-subgraph if the class of sets $\{\{(x,t) : 0 \leq t \leq f(x) \text{ or } f(x) \leq t \leq 0\} : f \in \mathcal{F}\}$ are VC. The same is true for VC-subgraph difference classes $\mathcal{G} = \{f - g : f, g \in \mathcal{F}\}$ with $\mathcal{F}$ VC-sub graph. A typical example of a VC-subgraph class of functions is $\{h(x,\theta) = h(x - \theta) : \theta \in \mathbf{R}\}$ if $h : \mathbf{R} \to \mathbf{R}$ is monotone. Uniformly bounded VC-subgraph classes are UPG and, more importantly, they satisfy the following exponential inequality due to Alexander (1984, Theorem 2.8 and 1985, Theorem 2.2):

Theorem 2.7 *(Alexander). Let $P \in \mathcal{P}(S)$ and let $\mathcal{F}$ be a uniformly bounded VC-subgraph difference class of functions on S. Let $\alpha = \sup_{f \in \mathcal{F}} Var_P f$, let $n \geq 1$ and $M > 0$. There exists a constant K (depending on certain characteristics of $\mathcal{F}$) such that if*

$$M^2 \geq K\alpha L(\alpha^{-1}) \text{ and } M \geq KL(n)/n^{\frac{1}{2}} \tag{2.17}$$

then

$$\begin{aligned} Pr\{\|\nu_n\|_{\mathcal{F}} > M\} &\leq 16 \exp(-M^2/8\alpha r^2) \\ &+ 16 \exp(-(4r)^{-1} M n^{\frac{1}{2}} L(M/n^{\frac{1}{2}} \alpha r)) \end{aligned} \tag{2.18}$$

where $Lx = \log(x \vee e)$, and $r = \sup\{|f(x)| : f \in \mathcal{F}, x \in S\}$.

Theorem 2.7 has the following consequence:

Theorem 2.8 *Let $P \in \mathcal{P}(S)$ and let $\mathcal{F}$ be a measuranble uniformly bounded VC-subgraph class of functions. Let $\mathcal{F}'_n = \{f - g : f, g \in \mathcal{F}, \rho_P^2(f,g) \leq c/(\log \log n)^{2+\delta}\}$ and $\tilde{\mathcal{F}}_n = \{f - g : f, g \in \mathcal{F}, \rho_P^2(f,g) \leq c/(\log \log n)^{1+\delta}\}$ for some $c < 0$ and $\delta > 0$. Then*

$$(\log \log n)^{\frac{1}{2}} \|\nu_n\|_{\mathcal{F}'_n} \rightarrow_{a.s.} 0 \text{ and } \|\nu_n\|_{\tilde{\mathcal{F}}_n} \rightarrow_{a.s.} 0. \tag{2.19}$$

Proof It suffices to show, for the first limit, that for all $\varepsilon > 0$,

$$\sum_{k=3}^{\infty} Pr\{\max_{2^k \leq n < 2^{k+1}} (\log \log n)^{\frac{1}{2}} ||\nu_n||_{\mathcal{F}'_n} > \varepsilon\} < \infty.$$

We can replace $n^{-1} \log \log n$ by $2^{-k} \log k$ and $\mathcal{F}'_n$ by $\mathcal{F}'_{2^k}$ so that the above series is dominated by

$$\sum_{k=3}^{\infty} Pr\{\max_{2^k \leq n < 2^{k+1}} ||\sum_{i=1}^{n} (f(X_i) - Pf)||_{\mathcal{F}'_{2^k}} > \varepsilon 2^{\frac{k}{2}}/(\log k)^{\frac{1}{2}}\}.$$

Inequality (2.18) gives

$$\max_{2^k \leq l < 2^{k+1}} Pr\{|| \sum_{i=l+1}^{2^{k+1}} (f(X_i) - Pf)||_{\mathcal{F}'_{2^k}} > \varepsilon 2^{(k-2)/2}/(\log k)^{\frac{1}{2}}\} \rightarrow 0$$

since

$$M^2/\alpha \geq \varepsilon^2 (\log k)^{1+\delta}/4c \text{ and } Mn^{\frac{1}{2}} \geq \varepsilon 2^{(k-2)/2}/(\log k)^{\frac{1}{2}}. \tag{2.20}$$

We can apply Ottaviani's inequality (e.g. Araujo and Giné (1980), page 111) and get the series above dominated by

$$2 \sum_{k=3}^{\infty} Pr\{|| \sum_{i=1}^{2^{k+1}} (f(X_i) - P(f)||_{\mathcal{F}'_{2^k}} > \varepsilon 2^{(k-2)/2}/(\log k)^{\frac{1}{2}}\}.$$

This series is convergent by (2.18) applied with $M = \varepsilon/2(\log k)^{\frac{1}{2}}$ and $\alpha < c/(\log k)^{2+\delta}$, essentially as in (2.20). A similar proof gives the second limit. ■

3 The a.s. Bootstrap of M-estimators

In this section we prove an a.s. bootstrap CLT for M-estimators under conditions close to the "non-standard" conditions of Huber (1967) and those of Pollard (1985). The proofs are based on methods from these two papers,

Theorem 2.1, Alexander's (1984) exponential bounds (through Theorem 2.8) and the "square root trick" (LeCam (1986); Giné and Zinn (1984, 1986)).

M-estimators can be formally defined in two ways, either by maximization (minimization) of $P_n g(\cdot,\theta)$ for some criterion function g, or as solutions of $P_n h(\cdot,\theta) = 0$ (or almost 0). As in Huber (1967), we will treat both types separately.

Let P be a probability measure on $(S, \mathcal{S})$, let $\Theta \subset \mathbf{R}^d$ be a G_δ set such that $0 \in \Theta^\circ$, and let

$$g : S \times \Theta \to \mathbf{R}$$

be a jointly measurable function. We let

$$G(\theta) = Pg(\cdot,\theta),\ G_n(\theta) = P_n g(\cdot,\theta) \text{ for } \theta \in \Theta$$

where P_n is the empirical measure based on $\{X_i\}_{i=1}^n$ $i.i.d.(P)$. We make the following assumptions:

(A.1) $G(0) = \sup_{\theta\in\Theta} G(\theta)$.

(A.2) There is a symmetric positive definite bilinear form A_G such that, for θ in a neighborhood of zero,

$$G(\theta) = G(0) - \frac{1}{2}A_G(\theta,\theta) + o(|\theta|^2/\log|\log|\theta||). \tag{3.1}$$

Without loss of generality we will take $A_G = I$, i.e. $A_G(\theta,\theta') = \theta\cdot\theta'$.

(A.3) There exists $\Delta : S \to \mathbf{R}^d$ such that $P|\Delta|^2 < \infty$ and such that, if we let $r(x,\theta) = |\theta|^{-1}[g(x,\theta) - g(x,0) - \theta\cdot\Delta(x)], x \in S,\ \theta \neq 0$ $r(x,0) \equiv 0$ then there is $m < \infty$ and functions $r_i(x,\theta), i = 1,\dots m$, such that $r = \sum_{i=1}^m r_i$ and for some $K > 0$ the classes of functions $\mathcal{F}_i = \{r_i(\cdot,\theta) : |\theta| < K\}$ are measurable uniformly bounded VC-subgraph classes.

(A.4) $Pr_i^2(\cdot,\theta) \leq 1/(\log|\log|\theta||)^{2+\delta}$ for some $\delta > 0$, all $i = 1,\dots,m$ and all θ in a neighborhood of 0.

Remark 3.1 Conditions (A.3), (A.4) do not imply that g is differentiable, but they do give a sort of uniform stochastic differentiability of $\nu_n(g)$ at 0 (Pollard (1985)) since

$$\lim_{\delta\to 0}\limsup_{n\to\infty} E \sup_{|\theta|<\delta} |\nu_n r(\cdot,\theta)| = 0$$

by the asymptotic equicontinuity condition associated to the CLT for empirical processes (Dudley (1984); see also Giné and Zinn (1986)). Note that (A.3) implies that $\mathcal{F} := \{r(\cdot,\theta) : |\theta| \leq K\}$ is P-Donsker for all P (e.g. Alexander (1987)).

The estimator θ_n of $\theta(=0)$ is defined as any random variable θ_n satisfying

$$P_n g(\cdot,\theta_n) + o(n^{-1}) > \sup_{\theta\in\Theta} P_n g(\cdot,\theta). \tag{3.2}$$

where $o(n^{-1})$ is independent of ω. Such a random variable always exists if $(S,\mathcal{S},P)$ is complete by e.g. the section theorem in Cohn (1980, Corollary 8.5.4, page 286). Similarly, the *bootstrap* $\hat\theta_n^\omega$ of θ_n is defined, for each $n \in N$ and $\omega \in \Omega$, as a random variable satisfying

$$\hat P_n^\omega g(\cdot,\hat\theta_n^\omega) + o(n^{-1}) > \sup_{\theta\in\Theta} \hat P_n^\omega g(\cdot,\theta). \tag{3.3}$$

We also assume the following consistency properties:
(A.5) $\theta_n \to_{a.s.} 0$,
(A.6) $\hat\theta_n^\omega - \theta_n(\omega) \to_{\hat P_r} 0$ a.s.
Sufficient conditions for consistency are given in Theorem 3.5 below.

Theorem 3.2 *Let $g, P, \theta_n, \hat\theta_n$ satisfy (A.1) - (A.6), (3.2) and (3.3). Then*

$$\lim_{n\to\infty} \hat{\mathcal{L}}(n^{1/2}(\hat\theta_n^\omega - \theta_n(\omega))) =_{a.s.} \lim_{n\to\infty} \mathcal{L}(n^{1/2}\theta_n) \tag{3.4}$$

which is $N(0, A_G^{-1}(Cov\Delta)A_G^{-1})$.

Proof We proceed in three steps.

Claim 1 There exists $c < \infty$ such that, letting $a_n = (n^{-1}\log\log n)^{1/2}$,

$$\limsup_{n\to\infty} |\theta_n|/a_n < c \ a.s. \tag{3.5}$$

$$\hat Pr\{|\hat\theta_n|/a_n \le c\} \to_{a.s.} 1. \tag{3.6}$$

Proof Since $\theta_n \to_{a.s.} 0$ ((A.5)), (3.1) implies

$$G(\theta_n) - G(0) < -|\theta_n|^2/4$$

for all $n > n_0$, for some $n_0 < \infty$. Since $P|\Delta|^2 < \infty$, the law of the iterated logarithm (LIL) on $\mathbf{R}^d$ implies $\limsup_{n\to\infty} |(P_n - P)(\Delta)|/a_n \le d$ a.s. for some $d < \infty$. By (A.3), the class $\mathcal{F}$ verifies the LIL (Dudley and Philipp (1983); see also Alexander and Talagrand (1989)) so that $\limsup_{n\to\infty} ||P_n - P||_{\mathcal{F}}/a_n \le d$ a.s. for some $d < \infty$. These observations, together with (3.2) and the definition of r give that, for n large,

$$\begin{aligned}
0 &\le P_n(g(\cdot,\theta_n) - g(\cdot,0)) + o(n^{-1}) \\
&= (P_n - P)(g(\cdot,\theta_n) - g(\cdot,0)) + P(g(\cdot,\theta_n) - g(\cdot,0)) + o(n^{-1}) \\
&\le (P_n - P)(\theta_n \cdot \Delta(\cdot) + |\theta_n| r(\cdot,\theta_n)) - |\theta_n|^2/4 + o(n^{-1}) \\
&\le K|\theta_n| a_n - |\theta_n|^2/4 + o(n^{-1}) \ a.s.
\end{aligned}$$

for some $K < \infty$. This inequality implies (3.5) for $c = 4K$.

Let $\Delta_n = (P_n - P)(\Delta)$ and $\hat{\Delta}_n = (\hat{P}_n - P_n)(\Delta)$. As mentioned above, for some $c < \infty$,

$$\limsup_{n\to\infty} |\Delta_n|/a_n = c \ a.s. \tag{3.7}$$

Now, the bootstrap CLT for $\Delta(X_1)$ (Theorem 2.1, or Bickel and Freedman (1981)) gives that for any $c > 0$,

$$\hat{Pr}\{|\hat{\Delta}_n|/a_n \le c\} \to_{a.s.} 1. \tag{3.8}$$

Similarly, since $\mathcal{F}$ is a.s.-bootstrap P-Donsker,

$$\hat{Pr}\{||\hat{P}_n - P_n||_{\mathcal{F}}/a_n \le c\} \to_{a.s.} 1. \tag{3.9}$$

Now, proceeding as above, we have, for large n (by (A.5), (A.6))

$$0 \le |\hat{\theta}_n|[|\hat{\Delta}_n| + |\Delta_n| + ||\hat{P}_n - P_n||_{\mathcal{F}} + ||P_n - P||_{\mathcal{F}}] - |\hat{\theta}_n|^2/4 + o(n^{-1}),$$

and this, by (3.7)-(3.9) and the LIL for $\mathcal{F}$, gives (3.6).

Claim 2 The following limits hold for any $c < \infty$:

$$(\log\log n)^{1/2} \sup_{|\theta|\le ca_n} |\nu_n(r(\cdot,\theta))| \to_{a.s.} 0 \tag{3.10}$$

and

$$(\log\log n)^{1/2} \sup_{|\theta|\le ca_n} |\hat{\nu}_n(r(\cdot,\theta))| \to 0 \ \hat{Pr} - a.s., \ Pr - a.s. \tag{3.11}$$

Proof (3.10) is imediate from (A.3), (A.4) and Theorem 2.8 applied to $\mathcal{F}_i, i \le m$. In order to prove (3.11) using Alexander's bound, we must estimate the size of

$$\alpha_n = \sup\{P_n(r_1(\cdot,\theta))^2 : |\theta| \le ca_n\}.$$

For this we use the "square root trick" inequality in Giné and Zinn (1984, Lemma 5.2), which gives

$$Pr\{\alpha_n > 4(\log\log n)^{-(2+\delta)}\} \le (\log\log n)^{\tau} \exp(-n/(\log\log n)^{1+\delta'})$$

for some $\tau > 0$, $\delta' > 0$ and all n large enough. Therefore, eventually

$$\alpha_n \le 4(\log\log n)^{-(1+\delta)} \ a.s.$$

This, just as (A.4) in the non-random case, allows us to apply Theorem 2.8 conditionally on the sample, and obtain (3.11) the same way as (3.10).

Now we proceed to the proof of Theorem 3.2 using the above claims and an argument of Pollard (1985). By (3.1) and the definitions of Δ_n, $\hat{\Delta}_n$, θ_n and r we have:

$$
\begin{aligned}
0 &\leq \hat{P}_n(g(\cdot,\hat{\theta}_n) - g(\cdot,\Delta_n + \hat{\Delta}_n)) + o(n^{-1}) \\
&= (\hat{P}_n - P_n + P_n - P)(g(\cdot,\hat{\theta}_n) - g(\cdot,\Delta_n + \hat{\Delta}_n)) \\
&+ P(g(\cdot,\hat{\theta}_n) - g(\cdot,\Delta_n + \hat{\Delta}_n)) + o(n^{-1}) \\
&= (\hat{P}_n - P_n + P_n - P)[(\hat{\theta}_n - \Delta_n - \hat{\Delta}_n)\cdot\Delta(\cdot) \\
&+ |\hat{\theta}_n| r(\cdot,\hat{\theta}_n) - |\Delta_n + \hat{\Delta}_n| r(\cdot,\Delta_n + \hat{\Delta}_n)] \\
&- |\hat{\theta}_n|^2/2 + |\Delta_n + \hat{\Delta}_n|^2/2 \\
&+ o(|\hat{\theta}_n|^2/\log|\log|\hat{\theta}_n||) + o(|\Delta_n + \hat{\Delta}_n|^2/\log|\log|\Delta_n + \hat{\Delta}_n||) + o(n^{-1}).
\end{aligned}
$$

So, collecting terms and multiplying by n

$$
\begin{aligned}
& n|\hat{\theta}_n - \Delta_n - \hat{\Delta}_n|^2/2 \leq n^{1/2}|\hat{\theta}_n||n^{1/2}(\hat{P}_n - P_n + P_n - P)(r(\cdot,\hat{\theta}_n))| \\
&+ n^{1/2}|\Delta_n + \hat{\Delta}_n||n^{1/2}(\hat{P}_n - P_n + P_n - P)(r(\cdot,\Delta_n + \hat{\Delta}_n))| \\
&+ n\, o(|\hat{\theta}_n|^2/\log|\log|\hat{\theta}_n||) \\
&+ n\, o(|\Delta_n + \hat{\Delta}_n|^2/\log|\log|\Delta_n + \hat{\Delta}_n||) + o(1)
\end{aligned}
\tag{3.12}
$$

Now (3.6), (3.10) and (3.11) immediately give

$$n^{1/2}|\hat{\theta}_n||n^{1/2}(\hat{P}_n - P_n + P_n - P)(r(\cdot,\hat{\theta}_n))| \to_{\hat{P}_r} 0 \ a.s.$$

(3.7), (3.8), (3.10) and (3.11) show

$$n^{1/2}|\Delta_n + \hat{\Delta}_n||n^{1/2}(\hat{P}_n - P_n + P_n - P)(r(\cdot,\Delta_n + \hat{\Delta}_n))| \to_{\hat{P}_r} 0 \ a.s.$$

And the "o" terms also tend to zero in $\hat{P}r, Pr - a.s.$ by (3.6), (3.7) and (3.8). Thus, replacing these limits in (3.12) gives

$$n^{1/2}(\hat{\theta}_n - \Delta_n - \hat{\Delta}_n) \to_{\hat{P}_r} 0 \ a.s. \tag{3.13}$$

Nex we see that Δ_n can be replaced by θ_n in (3.13) again as a consequence of Theorem 2.8. The argument leading to (3.12), applied to the inequality

$$o \leq P_n(g(\cdot,\theta_n) - g(\cdot,\Delta_n)) + o(n^{-1})$$

(which holds by the definition of θ_n), yields

$$
\begin{aligned}
n|\theta_n - \Delta_n|^2/2 &\leq n^{1/2}|\Delta_n||\nu_n(r(\cdot,\Delta_n)) \\
&+ n^{1/2}|\theta_n||\nu_n(r(\cdot,\theta_n))| \\
&+ n\, o(|\theta_n - \theta|^2/\log|\log|\theta_n||) \\
&+ n\, o(|\Delta_n|^2/\log\log|\Delta_n||) + o(1).
\end{aligned}
$$

All the terms at the right side of this inequality tend to zero a.s. by (3.5), (3.7) and (3.10). Hence,

$$n^{1/2}(\theta_n - \Delta_n) \to_{a.s.} 0$$

and therefore, the limit (3.13) becomes

$$n^{1/2}(\hat{\theta}_n - \theta_n) - n^{1/2}\hat{\Delta}_n \to_{\hat{P}_r} 0 \ a.s.$$

By the bootstrap CLT in $\mathbf{R}^d$,

$$n^{1/2}\hat{\Delta}_n = \hat{\nu}_n(\Delta) \to_{\hat{\mathcal{L}}} N(0, Cov\Delta) \ a.s.$$

and therefore, $\{n^{1/2}(\hat{\theta}_n - \theta_n)\}$ converges in conditional law a.s. to the same limit. ■

Remark 3.3 If $o(|\theta|^2/\log|\log|\theta||)$ is replaced by $o(|\theta|^2)$ in (3.1), if (A.4) is replaced by $Pr^2(\cdot,\theta) \to 0$, if the condition on $\{r(\cdot,\theta) : |\theta| \leq K\}$ in (A.3) is replaced just by stochastic equicontinuity at 0, i.e. by

$$\lim_{\delta\to 0} \limsup_{n\to\infty} E \sup_{|\theta|\leq\delta} |\nu_n(r(\cdot,\theta))| = 0,$$

and "a.s." is replaced by "in probability" in (A.5) and (A.6), then the boostrap CLT holds in probability, that is

$$\hat{d}_{BL^*}(n^{1/2}(\hat{\theta}_n - \theta_n), N(0, A_G^{-1}(Cov\Delta)A_G^{-1}) \to 0 \ \textit{in probability}$$

(Romo (1990)). These modified conditions are quite close to Huber's (1967) and Pollard's (1985).

Next we give sufficient conditions for the consistency hypothesis (A.5) and (A.6) to hold. They are slightly stronger than those in Huber (1967). Let, as above, $P \in \mathcal{P}(S)$, $\Theta \subset \mathbf{R}^d$ be a G_δ set such that $0 \in \Theta^\circ, g : S \times \Theta \to \mathbf{R}$ jointly measurable, and assume, without loss of generality, that $G(0) = 0$ (where $G(\theta) = Pg(\cdot,\theta)$). The conditions that will imply consistency are as follows:

(B.1) $P|g(\cdot,\theta)| < \infty$ for each $\theta \in \Theta$.
(B.2) There exists $M > 0$ and a non-negative function $b(\theta), \theta \in \Theta$, such that if $C = \{\theta \in \Theta : |\theta| \leq M\}$ then
 (i) $\inf_{\theta\notin C} b(\theta) = b > 0$.
 (ii) $P\sup_{\theta\notin C} g(\cdot,\theta)/b(\theta) = a < 0$, and
 (iii) $\sup_{\theta\in C} |g(\cdot,\theta)| \in L_1(P)$.
(B.3) $\sup_{\theta\in C, |\theta|>\varepsilon} G(\theta) < 0$ for all $\varepsilon > 0$.
(B.4) The class of functions $\mathcal{G} = \{g(\cdot,\theta) : \theta \in C\}$ is a measurable P-Glivenko-Cantelli class.

Remark 3.4 A class of functions $\mathcal{G}$ is P-Glivenko-Cantelli (P-GC) if

$$||P_n - P||_{\mathcal{G}} \to_{a.s.} 0.$$

There are several necessary and/or sufficient conditions for $\mathcal{G}$ measurable to be P-GC (e.g. Giné and Zinn (1984), Dudley (1984) and references therein).

Theorem 3.5 *If $P, \mathcal{G}$ and Θ verify conditions (B.1)-(B.4) with $G(0) = 0$, and if θ_n and $\hat{\theta}_n$ verify the inequalities (3.2) and (3.3) then*

$$\theta_n \to_{a.s.} \tag{3.14}$$

and

$$\hat{\theta}_n - \theta_n \to_{\hat{P}r} 0 \ a.s. \tag{3.15}$$

Proof (3.14) is proved in Huber (1967). To prove (3.15) we first observe that by the bootstrap weak law of large numbers in $\mathbf{R}$ and by (B.2),

$$\hat{P}_n \sup_{\theta \notin C} g(\cdot, \theta)/b(\theta) \to_{\hat{P}r} P \sup_{\theta \notin C} g(\cdot, \theta)/b(\theta) \ a.s. \tag{3.16}$$

and

$$\hat{P}_n g(\cdot, 0) \to_{\hat{P}r} Pg(\cdot, 0) \ a.s. \tag{3.17}$$

So, by (3.16), $Pr - a.s.$, with $\hat{P}r$-probability tending to 1 as $n \to \infty$ we have that for all $\theta \notin C$,

$$\hat{P}_n g(\cdot, \theta) < (a + \varepsilon) b(\theta) \leq (a + \varepsilon) b < 0.$$

On the other hand, by (3.17) and the definition of $\hat{\theta}_n$,

$$\hat{P}r\{\hat{P}_n g(\cdot, \hat{\theta}_n) > -\delta\} \to_{a.s.} 1 \tag{3.18}$$

for all $\delta > 0$. Hence

$$\hat{P}r\{\hat{\theta}_n \notin C\} \to_{a.s.} 0. \tag{3.19}$$

Now, since G is P-GC ((B.4)) and (B.3)(iii) holds, the bootstrap Glivenko-Cantelli theorem in Giné and Zinn (1990) shows that

$$\left| \sup_{\theta \in C, |\theta| > \varepsilon} \hat{P}_n g(\cdot, \theta) - \sup_{\theta \in C, |\theta| > \varepsilon} G(\theta) \right|$$

$$\leq \sup_{\theta \in C, |\theta| > \varepsilon} |(\hat{P}_n - P) g(\cdot, \theta))| \to_{\hat{P}r} 0 \ a.s.$$

for all $\varepsilon > 0$. But by (3.18), (3.19) and (B.3) this is impossible unless $\hat{P}r\{|\theta| > \varepsilon\} \to_{a.s.} 0$. ■

We now consider the second type of M-estimators. Let $\Theta \subset \mathbf{R}^d$ with $0 \in \Theta^\circ, P \in \mathcal{P}(S)$, and $h : S \times \Theta \to \mathbf{R}^d$ jointly measurable be such that

(C.1) $Ph(\cdot,0)=0$ and $P(h(\cdot,0))^2<\infty$.
(C.2) $H(\theta):=Ph(\cdot,\theta)$ is "strongly" differentiable at zero with non-degerenate first derivative. Assuming (without loss of generality) $H'(0)=I$ the differentiability condition is as follows:

$$H(\theta)-H(\theta')=\theta-\theta'+o(|\theta-\theta'|) \text{ as } \theta\to 0 \text{ and } \theta'\to 0. \tag{3.20}$$

(C.3) The classes of functions $\mathcal{F}_i=\{h_i(\cdot,\theta)-h_i(\cdot,0):|\theta|\le M\}$ for some $M>0, i=1,\dots,d$, where h_i denotes the $i-th$ coordinate of h, satisfy the following property: there is $m<\infty$ and uniformly bounded measurable VC-subgraph classes $\mathcal{G}_{ij}=\{g_{ij}(\cdot,\theta):|\theta|\le M\}$ such that $h_i(\cdot,\theta)-h_i(\cdot,0)=\sum_{j=1}^m g_{ij}(\cdot,\theta)$.
(C.4) For $i\le d, j\le m, Var_P\, g_{ij}(\cdot,\theta)\le 1/(\log|\log|\theta||)^{1+\delta}$ for some $\delta>0$ and θ in a neighborhood of 0.
(C.5) There exist symmetric (P^n-completion) measurable functions $\theta(x_1,\dots,x_n)$ defined on the support of $P^n, n\in N$, such that if $\theta_n:=\theta(X_1,\dots,X_n)$, then

$$n^{1/2}P_nh(\cdot,\theta_n)\to_{a.s.}0 \quad\text{and}\quad \theta_n\to_{a.s.}0. \tag{3.21}$$

(C.6) For almost every $\omega\in\Omega$, there exist symmetric random variables $\theta_n^\omega=\theta_n^\omega(X_{n1}^\omega,\dots,X_{nn}^\omega)$ such that

$$n^{1/2}\hat{P}_n^\omega h(\cdot,\hat{\theta}_n^\omega)\to_{\hat{P}_r}0\ a.s. \text{ and } \hat{\theta}_n^\omega-\theta_n(\omega)\to_{\hat{P}_r}0\ a.s. \tag{3.22}$$

As before, consistency ((C.5)), ((C.6)) will be handled separately.

Theorem 3.6 *Under (C.1)-(C.6),*

$$\lim_{n\to\infty}\hat{\mathcal{L}}(n^{1/2}(\hat{\theta}_n-\theta))=_{a.s.}\lim_{n\to\infty}\mathcal{L}(n^{1/2}\theta_n) \tag{3.23}$$

which is $N(0,Cov_Ph(\cdot,0))$.

Proof First we show that

$$n^{1/2}Ph(\cdot,\theta_n)+n^{1/2}P_nh(\cdot,0)\to_{a.s.}0. \tag{3.24}$$

Since $H'(0)$ is not degenerate and $\theta_n\to_{a.s.}0$ ((3.20) and (3.21)) there exist $c<0$ and $n_0(\omega)<\infty$ $a.s.$ such that for $n>n_0$, $c|\theta_n|\le|H(\theta_n)|$. Hence, using (3.21) and the LIL for random vectors in $\mathbf{R}^d$ and for empirical processes (by (C.3), the P-empirical process indexed by $\mathcal{F}$ satisfies the LIL - see e.g. Dudley and Philipp (1983)), we have

$$n^{1/2}c|\theta_n|\le n^{1/2}|H(\theta_n)|$$

$$\le n^{1/2}|(P_n-P)h(\cdot,\theta_n)|+n^{1/2}|P_nh(\cdot,\theta_n)|$$

$$\leq n^{1/2}|(P_n - P)h(\cdot,0)| + n^{1/2}|(P_n - P)(h(\cdot,\theta_n) - h((\cdot,0))| + n^{1/2}|P_n h(\cdot,\theta_n)|$$
$$= 0((\log\log n)^{1/2})\ a.s. \tag{3.25}$$

Hence, for some $c < \infty$,

$$\limsup_{n\to\infty} n^{1/2}|\theta_n|/(\log\log n)^{1/2} \leq c\ a.s. \tag{3.26}$$

Theorem 2.8 gives that for all $c > 0$,

$$\sup_{|\theta|\leq ca_n} |n^{1/2}(P_n - P)(h(\cdot,\theta))| \to_{a.s.} 0 \tag{3.27}$$

which, by (3.26), implies

$$n^{1/2}(P_n - P)(h(\cdot,\theta_n) - h(\cdot,0)) \to_{a.s.} 0.$$

Hence, by (C.1) and (3.21),

$$n^{1/2}(Ph(\cdot,\theta_n) + P_n h(\cdot,0)) \to_{a.s.} 0$$

which is claim (3.24).

By the bootstrap CLT for empirical processes (Theorem 2.1) and (C.3), for every sequence $b_n \to \infty$.

$$\hat{Pr}\{n^{1/2}|(\hat{P}_n - P_n)(h(\cdot,\hat{\theta}_n) - h(\cdot,0))| \leq b_n\} \to_{a.s.} 1.$$

We can use this and (3.22) in a decomposition similar to (3.25), and obtain that there is $c < \infty$ such that

$$\hat{Pr}\{n^{1/2}|\hat{\theta}_n|/(\log\log n)^{1/2} \leq c\} \to_{a.s.} 1. \tag{3.28}$$

We consider now, in analogy with (3.24),

$$n^{1/2}(Ph(\cdot,\hat{\theta}_n) + \hat{P}_n h(\cdot,0))$$
$$= n^{1/2}\hat{P}_n h(\cdot,\hat{\theta}_n) - n^{1/2}((\hat{P}_n - P_n) + (P_n - P))(h(\cdot,\hat{\theta}_n) - h(\cdot,0))$$

The first term at the right hand side tends to zero in $\hat{Pr}, Pr - a.s.$ by (3.22). The next term, $\hat{\nu}_n(h(\cdot,\hat{\theta}_n) - h(\cdot,0))$, also converges to zero in $\hat{Pr}, Pr - a.s.$ by Theorem 2.1 (the bootstrap CLT implies, by (C.4), $\lim_{\delta\to 0}\limsup_n \hat{Pr}\{\sup_{|\theta|<\delta}|\hat{\nu}_n(h(\cdot,\theta) - h(\cdot,0))| > \varepsilon\} = 0\,a.s.$ for all $\varepsilon > 0$, and $\hat{\theta}_n \to_{\hat{Pr}} 0\ a.s.$ by (3.21) and (3.22)). And the same is true for the last term $\nu_n(h(\cdot,\hat{\theta}_n) - h(\cdot,0))$ by (3.27) and (3.28). Therefore,

$$n^{1/2}Ph(\cdot,\hat{\theta}_n) + n^{1/2}\hat{P}_n h(\cdot,0) \to_{\hat{Pr}} 0\ a.s.$$

This limit and (3.24) give, by subtraction,

$$n^{1/2}(H(\hat{\theta}_n) - H(\theta_n)) + n^{1/2}(\hat{P}_n - P_n)h(\cdot,0) \to_{\hat{Pr}} 0\ a.s. \tag{3.29}$$

This shows that for almost every ω the sequence $\{n^{1/2}(H(\hat{\theta}_n^{\omega}) - H(\theta_n(\omega)))\}$ is $\hat{P}r$-stochastically bounded: it converges weakly by the bootstrap CLT in $\mathbf{R}^d$. Since $\hat{\theta}_n - \theta_n \to_{\hat{P}_r} 0$ $a.s.$ ((3.22)), hypothesis (C.2) implies $\hat{P}r\{|\hat{\theta}_n - \theta_n| > 2|H(\hat{\theta}_n) - H(\theta_n)|\} \to_{a.s.} 0$. Hence the sequence $\{n^{1/2}|\hat{\theta}_n^{\omega} - \theta_n(\omega)|\}$ is $\hat{P}r$-stochastically bounded, $\omega - a.s.$ Then, $n^{1/2}o(|\hat{\theta}_n - \theta_n|) \to_{\hat{P}_r} 0$ $a.s.$ and (3.29) and (3.20) give

$$n^{1/2}(\hat{\theta}_n - \theta_n) + n^{1/2}(\hat{P}_n - P_n)h(\cdot, 0) \to_{\hat{P}_r} 0 \ a.s.$$

Now (3.23) follows by the bootstrap CLT in $\mathbf{R}^d$. ∎

The consistency conditions of Huber (1967, Case B) not only give consistency of θ_n, but also consistency of $\hat{\theta}_n$, i.e. (C.5) and (C.6). The proof is similar to that of Theorem 3.2.

Theorem 3.7 *If h is jointly measurable and if*

$$P_n h(\cdot, \theta_n) \to_{a.s.} 0, \text{ and } \hat{P}_n h(\cdot, \hat{\theta}_n) \to_{\hat{P}_r} 0 \ a.s.$$

for random variables $\theta_n, \hat{\theta}_n$, then

$$\theta_n \to_{a.s.} 0 \text{ and } \hat{\theta}_n - \theta_n \to_{\hat{P}_r} 0 \ a.s.$$

assuming the following conditions hold:

(1) h is P-a.s. continuous in θ,
(2) $H(\theta)$ exists for all $\theta \in \Theta$ and has a unique zero at $\theta = 0$,
(3) there exist a continuous function b on Θ, bounded away from zero and $M < \infty$ such that, letting $C := \{\theta : |\theta| \le M\} \bigcap \Theta$, we have $P \sup_{\theta \in \Theta} |h(\cdot, \theta)|/b(\theta) < \infty, \inf_{\theta \notin C} |H(\theta)|/b(\theta) \ge 1$ and $P \sup_{\theta \notin C} |h(\cdot, \theta) - H(\theta)|/b(\theta) < 1$, and
(4) $P \sup_{\theta \in C} |h(\cdot, \theta)| < \infty$ and $\{h(\cdot, \theta) : \theta \in C\}$ is a P-Glivenko-Cantelli class of functions.

Example 3.8 *(Bootstrapping spatial medians).*The spacial median of $P \in \mathcal{P}(\mathbf{R}^d)$ is the value $\theta(P)$ of θ that maximizes $Pg(\cdot, \theta)$ where

$$g(x, \theta) = |x| - |x - \theta|.$$

Since the case $d = 1$ is already studied in Bickel and Freedman (1981) (their proof contains some inaccuracies that can be fixed using e.g. Theorem 2.8) we will assume $d \ge 2$. The set of medians of P, which is convex, consists of a single point unless P is concentrated on a line (and has more than one median there). We assume *P is not concentrated on a line and has a bounded density in a neighborhood of* $\theta(P)$. From the fact that for $x \neq 0$

$$\left| |x| - |x-\theta| - \frac{\theta \cdot x}{|x|} + \frac{|\theta|^2}{2|x|} - \frac{|\theta \cdot x|^2}{2|x|^3} \right| \leq 2\frac{|\theta|^3}{|x|^2} \wedge 3\frac{|\theta|^2}{|x|}$$

it follows that

$$G(\theta) = G(0) - P\left(\frac{|\theta|^2}{2|x|} - \frac{|\theta \cdot x|^2}{2|x|^3}\right) + 0(|\theta|^3)$$

if $d \geq 3$ and the same expression with $0(|\theta|^3)$ replaced by $0(|\theta|^3 |\log|\theta||)$ if $d = 2$ (note that $\theta(P) = 0$ implies $P\theta \cdot x/|x| = 0$ for all $\theta \in \mathbf{R}^d$). Hence (A.2) holds. Pollard (1984, page 153) shows that

$$\mathcal{F} = \{r(x,\theta) = [|x| - |x-\theta| - \theta \cdot x/|x|]/|\theta| : |\theta| \neq 0\} \bigcup \{r(x,0) \equiv 0\}$$

is the sum of two VC-subgraph classes, so that (A.3) is satisfied. Moreover, it is easy to see that

$$|r(x,\theta)| \leq 2 \wedge 2|\theta|/|x|.$$

Hence, if P has a density f on $\{|x| < \delta\}$ bounded by M, we have

$$Er^2(x,\theta) \leq K(d,M)(|\theta| \vee |\theta|^2)$$

where $K(d,M)$ is a constant depending on d and M. So, (A.4) holds too. Finally, taking $b(\theta) = |\theta|$ conditions (B.1)-(B.3) are satisfied, and it is easy to see (as in Pollard (1984, page 153)) that $\{|x| - |x-\theta| : |\theta| \leq M\}$ is a bounded VC-subgraph class, hence a Glivenko-Cantelli class. The conclusion from Theorem 3.2 is then that *the central limit theorem for the spatial median can be bootstrapped a.s.* assuming P is not concentrated on a line and P has a bounded density in a neighborhood of $\theta(P)$ (since in applications $\theta(P)$ is unknown, this last condition should be strengthened, in practice, to P having bounded density).

Example 3.9 ($k - means\, in\, \mathbf{R}$). Given P in $\mathbf{R}$ a k-mean of $P, 1 \leq k < \infty$, is an ordered set of k points in $\mathbf{R}, \theta_1 \leq \ldots \leq \theta_k$, that minimizes the function $P[\min_{i \leq k}(x-\theta_i)^2 - x^2]$ (Hartigan (1978), Pollard (1981, 1982), Cuesta and Matrán (1988)). This notion has also a meaning in $\mathbf{R}^d$ and even in infinite dimensions. Since our purpose is only to illustrate the use of the previous theorems, we restrict our discussion to the simpler one-dimensional case. We make the following assumptions on P:

(1) $Px^2 < \infty$
(2) P has a unique k-median μ and it consists of k distinct points $(\mu_1, \ldots, \mu_k)$ with $\mu_1 < \ldots < \mu_k$.
(3) P has a differentiable density at the points $(\mu_i + \mu_{i+1})/2,\ i = 1, \ldots, k-1$.

We let $\Theta = \{(\theta_1, \ldots, \theta_k) \in \mathbf{R}^k : \theta_1 \le \ldots \le \theta_k\}$, $g(x,\theta) = \min_{i\le k}(x-\theta_i)^2$ (by hypothesis (1) the $-x^2$ term is unnecessary) and $G(\theta) = Pg(\cdot,\theta)$, $\theta \in \Theta$. By a compactness argument there always exists, for each $(x_1, \ldots, x_n) \in \mathbf{R}^n$, a point θ in Θ that minimizes $(n^{-1}\sum_{i=1}^n \delta_{x_i})g(\cdot,\theta)$. Then, by the section theorem in Cohn (1980, Cor. 8.5.4) there exists a universally measurable selection $\theta(x_1, \ldots, x_n)$. We let $\theta_n^\omega := \theta(X_1(\omega), \ldots, X_n(\omega))$ be our estimator of μ, and call it θ_n. For each $\omega \in \Omega$ and $n \in N, \hat{\theta}_n = \hat{\theta}_n^\omega := \theta(X_{n1}^\omega, \ldots, X_{nn}^\omega)$ is the bootstrap of θ_n.

Pollard (1982) proved a CLT for $\theta_n - \mu$. We will show here that Theorem 3.2 implies that Pollard's CLT can be bootstrapped a.s. under conditions (1)-(3). (See Romo (1990) for the bootstrap in probability in $\mathbf{R}^d$ and under somewhat weaker conditions.)

For consistency we follow Cuesta and Matrán (1988). They show that if Z_n, Z_0 are B-valued random variables, B a uniformly convex Banach space, such that $Z_n \to_{a.s.} Z_0, Z_0$ has a unique k-mean $\mu := \theta(Z)$ and $E\min_{i\le k}\|Z_n - \mu_i\|^2 \to E\min_{i\le k}\|Z_0 - \mu_i\|^2$, then $\theta(Z_n) \to \theta(Z_0)$. They apply this result and a Skorohod representation to show consistency of θ_n under hypotheses (1) and (2) that is, condition (A.5). This argument bootstraps as follows: Let $\mathcal{C} = \{(-\infty, t] : t \in \mathbf{R}\}$. Then by the bootstrap law of large numbers,

$$\|\hat{P}_n^\omega - P\|_{\mathcal{C}} \to_{\hat{P}_r} 0 \ a.s. and$$

$$\hat{P}_n^\omega \min_{i\le k}(x-\mu_i)^2 \to_{\hat{P}_r} P\min_{i\le k}(x-\mu_i)^2 \ a.s.$$

Therefore for each ω fixed and for every subsequence there is a further subsequence $\{n'\}$ for which these two limits hold a.s. Let ω' be in the set where convergence occurs (a set depending on ω). By Skorohod's representation there are random variables $Y_n^{\omega,\omega'}(\omega'')$, $n = 0, 1, \ldots$, on $(\Omega'', \Sigma'', P'')$ such that $P'' \circ (Y_n^{\omega,\omega'})^{-1} = \hat{P}_n^\omega(\omega')$ and $P'' \circ (Y_0^{\omega,\omega''})^{-1} = P$, and $Y_n^{\omega,\omega'} \to Y_0^{\omega,\omega'}\ P'' - a.s.$ Also $P''\min_{i\le k}(Y_{n'}^{\omega,\omega'} - \mu_i)^2 = \hat{P}_{n'}^\omega(\omega')\min_{i\le k}(x-\mu_i)^2 \to_{a.s.} P\min_{i\le k}(x-\mu_i)^2$. Therefore their result gives that $\theta(P''\circ(Y_{n'}^{\omega,\omega'})^{-1}) \to \mu$ that is, $\theta(\hat{P}_{n'}^\omega(\omega')) \to \mu$. Hence condition (A.6) is satisfied.

It is easily checked that if P has a differentiable density f at $(\mu_i + \mu_{i+1})/2$, $i = 1, \ldots, k-1$, then $G(\theta)$ is three times differentiable at $\theta = \mu$, hence condition (A.2) holds. If we let $M_1 = (-\infty, (\mu_1 + \mu_2)/2]$, $M_j = ((\mu_j + \mu_{j-1})/2, (\mu_j + \mu_{j+1})/2]$, $j = 2, \ldots, k-1$, and $M_k = ((\mu_{k-1} + \mu_k)/2, \infty)$, and if we define $A_j, j = 1, \ldots, k$, in the same way but replacing μ by θ then we have:

$$\Delta(x) = (-2(x-\mu_1)I_{M_1}(x), \ldots, -2(x-\mu_k)I_{M_k}(x))$$

and

$$r(x,\theta) = \sum_{i,j}(I_{M_iA_j}(x))(2(\theta_i - \theta_j)x + \mu_i^2 - 2\mu_i\theta_i + \theta_j^2)/|\theta - \mu|$$

(Pollard (1982)). Using the obvious facts that

(a) if $|\mu - \theta| \leq k^{-1/2} \min(\mu_{i+1} - \mu_i)/2$ then $M_i A_j = \phi$ unless $|i - j| \leq 1$,
(b) $I_{M_i A_{i+1}} \subset [(\mu_i + \mu_{i+1})/2 - |\theta - \mu|, (\mu_i + \mu_{i+1})/2]$ and
$I_{M_i A_{i-1}} \subset [(\mu_i + \mu_{i+1})/2, (\mu_i + \mu_{i+1})/2 + |\theta - \mu|]$,

we obtain

$$|r(x,\theta)| \leq |\mu - \theta| + 4 \sum_{i=1}^{k-1} (I_{M_i A_{i+1}} + I_{M_{i+1} A_i}) |\theta_i - \theta_{i+1}|.$$

This inequality implies that $\sup_{|\theta| \leq K} \|r(x,\theta)\|_\infty < \infty$ for all $K < \infty$ and that $Pr^2(\cdot,\theta) \leq C|\theta - \mu|$ for all θ, in a neighborhood of μ and for some $C < \infty$. Hence (A.4) holds. For each θ, $r(x,\theta)$ is the sum of k^2 or less functions which are linear on an interval and zero outside it. Hence condition (A.3) is also satisfied.

Theorem 3.2 applies and it follows that, under conditions (1)-(3) on P, the CLT for k-means of Pollard (1982) bootstraps a.s. (for $d = 1$).

Theorem 3.6 applies to monotone functions h with a significant simplification of the hypotheses:

Theorem 3.10 *Let $h : \mathbf{R} \to \mathbf{R}$ be a bounded monotone function and let P be a probability measure on $\mathbf{R}$. We let $h(x,\theta) := h(x - \theta)$, $x, \theta \in \mathbf{R}$. Assume:*

(D.1) Letting $H(\theta) = Ph(\cdot,\theta)$, we have $H(0) = 0, H'(0) = 1$ and

$$\lim_{r,s \to 0} (H(r) - H(s))/(r - s) = H'(0). \tag{3.30}$$

(D.2) There is a neighborhood U of 0 such that for all $\theta \in U$,
$Var_P(h(x,\theta) - h(x,0)) \leq 1/(\log|\log|\theta||)^{1+\delta}$ for some $\delta > 0$.
(D.3) If C is the set of discontinuity points of $h(x,\theta(P)) = h(x,0)$ and $C_\delta = \{x : \exists\, y \in C \text{ s.t. } |x - y| < \delta\}$, then P is continuous on C_δ for some $\delta > 0$.

Then

$$\lim_{n \to \infty} \hat{\mathcal{L}}(n^{1/2}(\hat{\theta}_n - \theta_n)) =_{a.s.} \lim_{n \to \infty} \mathcal{L}(n^{1/2}\theta_n) = N(0, Var_P h). \tag{3.31}$$

where

$$\theta_n = \inf\{\theta : P_n h(\cdot,\theta) > 0\} \quad \text{and} \quad \hat{\theta}_n = \inf\{\theta : \hat{P}_n h(\cdot,\theta) > 0\}.$$

Proof By (D.1), 0 is the only zero of the function $H(\theta)$. Since $P_n h(\cdot, \varepsilon) \to H(\varepsilon) > 0$ *a.s.* for all $\varepsilon > 0$ it follows that eventually $\theta_n \le \varepsilon$ *a.s.* Likewise $\theta \ge -\varepsilon$ *a.s.*, i.e.

$$\theta_n \to_{a.s.} 0. \tag{3.32}$$

The same argument using the bootstrap law of large numbers gives

$$\hat{\theta}_n - \theta_n \to_{\hat{P}_r} 0 \ a.s. \tag{3.33}$$

Let $|\theta| < \delta/2$. The sample points X_i satisfying $X_i - \theta \in C_{\delta/2}$ are all a.s. different by continuity of P on C_δ. Moreover, $h(x)$ is continuous at $x = X_i - \theta$ if $X_i - \theta \notin C_{\delta/2}$. Hence the function $P_n h(\cdot, \theta)$ has a jump at θ of size at most $2\|h\|_\infty/n$ *a.s.* This proves, by (3.32) and (3.33), that

$$n^{1/2} P_n h(\cdot, \theta_n) \to_{a.s.} 0 \text{ and } n^{1/2} \hat{P}_n h(\cdot, \hat{\theta}_n) \to_{\hat{P}_r} 0 \ a.s.$$

So, conditions (C.5) and (C.6) are satisfied.

(D.1) is just (C.1) and (C.2). (D.2) is (C.4). Finally (C.3) is satisfied because of the monotinicity of h: any class of sets ordered by inclusion is obviously VC. ■

Theorem 3.6 and its corollary Theorem 3.10 contain the bootstrap CLT for the most usual M-estimators in particular for the median, Huber's estimators, etc. For instance, Theorem 3.10 applies to

$$h(x) = -kI_{(-\infty,-k)}(x) + xI_{[-k,k]}(x) + kI_{(k,\infty)}(x)$$

under minimal conditions on P, namely that $P\{k\} = P\{-k\} = 0$ and $P(-k, k) \neq 0$ (assuming $Ph(\cdot, 0) = 0$).

4 The bootstrap of differentiable functionals

Let T be a real valued functional defined on a set $\mathcal{P} \subset \mathcal{P}(S)$ and let $\mathcal{F}$ be a class of measurable functions on $(S, \mathcal{S})$ with everywhere finite envelope F ($F(s) = \sup\{|f(s)| : f \in \mathcal{F}\}$) contained in $\mathcal{L}_2(S, \mathcal{S}, P)$ for each $P \in \mathcal{P}$. Dudley's (1990) definition of differentiability is as follows: T is (Fréchet) differentiable at P for $\mathcal{F}$ with derivative $f = f_{P,T}$ iff $Pf^2 < \infty$ and

$$T(Q) = T(P) + \int f d(Q - P) + o(\|Q - P\|_{\mathcal{F}}) \tag{4.1}$$

as $\|Q - P\|_{\mathcal{F}} \to 0$, $Q \in \mathcal{P}$. If $S = \mathrm{R}$ and $\mathcal{F} = \{I_{(-\infty,t]} : t \in \mathrm{R}\}$ this is the usual definition of Fréchet differentiability of statistical functionals. Dudley (1990) further defines: T is C^1 at $P \in \mathcal{P}$ for $\mathcal{F}$ if there is a $\|\cdot\|_{\mathcal{F}}$ neighborhood U

of P in $\mathcal{P}$ such that for all $Q \in U$, T is differentiable at Q with derivative f_Q uniformly continuous in Q in the sense that

$$\|f_R - f_Q\|'_{\mathcal{F},U}$$

$$:= \sup\{|\int (f_R - f_Q)d(Q_1 - Q_2)|/\|Q_1 - Q_2\|_{\mathcal{F}} : Q_1, Q_2 \in U, \|Q_1 - Q_2\|_{\mathcal{F}} > 0\}$$

tends to zero as $\|R - Q\|_{\mathcal{F}} \to 0$ for R and Q in U. Here is Dudley's bootstrap limit theorem for a single functional T (he also considers families of functionals).

Theorem 4.1 *(Dudley (1990)). Let $\mathcal{P}$ be a convex class of probability measures containing all the p.m.'s with finite support. Let $\mathcal{F}$ be a measurable P-Donsker class of functions on S such that $PF^2 < \infty$. Let T, defined on $\mathcal{P}$, be a C^1 functional for $\mathcal{F}$ at P such that the derivatives $f_{Q,T}$ are in $\mathcal{F}$ for all Q in a neighborhood of P (for $\|\cdot\|_{\mathcal{F}}$). Then*

$$\lim_{n\to\infty} \hat{\mathcal{L}}(n^{1/2}(T(\hat{P}_n) - T(P_n))) =_{a.s.} \lim_{n\to\infty} \mathcal{L}(n^{1/2}(T(P_n) - T(P)))$$
$$= N(0, Var_P f_{P,T}).$$

(In the second limit, since $T(P_n)$ may not be measurable, weak convergence is in the sense of Hoffmann-Jørgensen -(2.6) with $l^\infty(\mathcal{F})$ replaced by $\mathbf{R}$.)

The conditions of Theorem 4.1 can be relaxed if we allow the bootstrap CLT to hold only in (outer) probability. Moreover, the parametric or semiparametric bootstrap also holds if $\mathcal{F}$ is UPG.

Theorem 4.2 *Let P be a class of probability measures on $(S, \mathcal{S})$ containing all the p.m.'s with finite support. Let T be a functional on $\mathcal{P}$ differentiable at $P \in \mathcal{P}$. Let $\mathcal{F}$ be a measurable P-Donsker class of functions on S. Then*

$$\hat{d}_{BL^*}(n^{1/2}(T(\hat{P}_n) - T(P_n)), G_P(f_{P,T})) \to 0 \text{ in } Pr^*.$$

Proof Since $n^{1/2}\int f d(\hat{P}_n - P_n) \to_{\hat{\mathcal{L}}} G_P(f)$ *a.s.* ($f = f_{P,T}$ as above) by the bootstrap CLT in $\mathbf{R}$, it suffices to show

$$Pr^*\{\hat{E}(|n^{1/2}(T(\hat{P}_n) - T(P_n)) - n^{1/2}\int fd(\hat{P}_n - P_n)| \wedge 1) > \delta\} \to 0 \quad (4.2)$$

for all $\delta > 0$. Consider the set

$$A(n, M, \delta) = \{n^{1/2}\|P_n - P\|_{\mathcal{F}} \le M\} \bigcap \{\hat{d}_{BL^*}(\|\hat{\nu}_n\|_{\mathcal{F}}, \|G_P\|_{\mathcal{F}}) \le \delta/3\}.$$

On this set

$$\hat{Pr}\{\|\hat{\nu}_n\|_{\mathcal{F}} > M\} \le \delta/3 + Pr\{\|G_P\|_{\mathcal{F}} > M - 1\}.$$

Given $\eta > 0$, let M be such that $Pr\{\|G_P\|_{\mathcal{F}} > M-1\} < (\delta/6) \wedge (\eta/2)$ and let $n_0 < \infty$ be so that for $n > n_0, Pr^* A(n, M, \delta)^c < \eta$, which exists because $\mathcal{F}$ is P-Donsker and pr-bootstrap P-Donsker (Theorem 2.1). For $n > n_0$, $\omega \in A(n, M, \delta)$ and ω' such that $\|\hat{\nu}_n^{\omega}(\omega')\|_{\mathcal{F}} \leq M$, we can apply (4.1) to $T(\hat{P}_n) - T(P)$ and $T(P_n) - T(P)$, to get (assuming, without loss of generality, that $|o(t)|$ is monotone in t)

$$\begin{aligned} &|n^{1/2}(T(\hat{P}_n) - T(P_n)) - \int f d(\hat{P}_n - P_n))| \\ &= |n^{1/2} o(\|\hat{P}_n - P\|_{\mathcal{F}}) - n^{1/2} o(\|P_n - P\|_{\mathcal{F}})| \\ &\leq n^{1/2} |o(\|\hat{P}_n - P_n\|_{\mathcal{F}} + \|P_n - P\|_{\mathcal{F}})| + n^{1/2}|o(\|P_n - P\|_{\mathcal{F}})| \\ &\leq 2n^{1/2}|o(2M/n^{1/2})| \to 0 \text{ as } n \to \infty. \end{aligned} \tag{4.3}$$

Hence, from some n on, the probability in (4.2) is dominated by

$$Pr^* A(n, M, \delta)^c + Pr^*\{A(n, M, \delta) \bigcap [\hat{P}r\{\|\hat{\eta}_n\|_{\mathcal{F}} > M\} > \delta/2]\},$$

a sequence whose lim sup as $n \to \infty$ is not larger that ν. ∎

Remark 4.3 The above proof works also for T taking values in a separable Banach space B if we further assume that $f_{P,T}(X)$, with $\mathcal{L}(X) = P$, satisfies the CLT in B.

The same proof, using Theorem 2.3 instead of 2.2, gives:

Theorem 4.4 *Let $\mathcal{F}$ be a measurable UPG class of functions on S. Let $\{P_\theta : \theta \in \Theta\}$ be a family of probability measures on $(S, \mathcal{S})$ indexed by an open set $\Theta \subset \mathbf{R}$. Let T be a map defined on a set $\mathcal{P} \subset \mathcal{P}(S)$ containing $\{P_\theta\}$ and the p.m.'s with finite support, such that: (a) $T(P_\theta) = \theta$, $\theta \in \Theta$, and (b) T is differentiable at P_θ. Let $\theta_n = T(P_n)$. Then, letting $\mathcal{G} = \mathcal{F} \bigcup \mathcal{F} \cdot \mathcal{F}$, we have the following implication:*

$$\|P_{\theta_n} - P_\theta\|_{\mathcal{G}}^* \to 0 \textit{ in probability and}$$

$$\begin{aligned} &\{n^{1/2}\|P_{\theta_n} - P_\theta\|_{\mathcal{F}}^*\} \textit{ stochastically bounded} \\ &\Rightarrow \hat{d}_{BL^*}[n^{1/2}(T(\hat{P}_n^\theta) - \theta_n), G_{P_\theta}(f)] \to 0 \textit{ in } Pr^*, \end{aligned}$$

where $\hat{P}_n^\theta$ is the empirical measure constructed from X_{ni}^θ i.i.d.(P_{θ_n}).

Note that: (a) Θ could be a subset of a separable Banach space containing θ in its interior, and the theorem would still hold, as in Remark 4.2. (b) the consistency conditions on P_{θ_n} are quite natural: for any $\mathcal{F}$ UPG they are certainly satisfied by P_n hence, if $\{P_{\theta_n}\}$ dos not satisfy these two conditions, resampling from P_{θ_n} may not make much sense.

Gill (1989) and Sheehy and Wellner (1988) approach (Hadamard) differentiability via Theorem 2.5. Using versions $\tilde{P}_n$ of P_n, $\tilde{\hat{P}}_n$ of $\hat{P}_n$ and $\tilde{G}_P$ of G_P so that simultaneously $\|\tilde{\nu}_n - \tilde{G}_P\|_{\mathcal{F}} \to 0$ and $\hat{d}_{BL^*}(\tilde{\hat{\nu}}_n, \tilde{G}_P) \to 0$, *a.s.* we have in (4.3) $n^{1/2}o(\|\tilde{P}_n - P\|_{\mathcal{F}}) \to 0$ $\tilde{\omega} - a.s.$ and, further using Dudley's theorem to get that for each $\tilde{\omega}$ fixed (in a set of $\tilde{P}r$-measure one) $\|\tilde{\hat{\nu}}_n - \tilde{\tilde{G}}_P\|_{\mathcal{F}} \to 0$ *a.s.*, we also have $n^{1/2}o(\|\tilde{\tilde{\hat{P}}}_n - \tilde{P}_n\|_{\mathcal{F}} + \|\tilde{P}_n - P\|_{\mathcal{F}}) \to_{a.s.} 0$. The bootstrap CLT in pr for $n^{1/2}(T(\hat{P}_n) - T(P_n))$ is then obtained by passing back to the original variables via Theorem 3.5 of Dudley (1985). Making this argument precise, particularly if $T(x_1, \ldots, x_n) = T(n^{-1}\sum_{i=1}^n \delta_{x_i})$ is not measurable, requires some extra care since one must prove almost uniform convergence of the functional applied to the versions, something not always handled with rigor. A question of some interest is *how weak a differentiability requirement can we impose on T and still obtain a pr-bootstrap limit theorem.* The following definition (which is a slight modification of one in Gill (1989) and Sheehy and Wellner (1988)) will help to provide an answer.

Definition 4.5 *Let $\mathcal{F}$ be a set of measurable functions on S, $P \in \mathcal{P}$, $\mathcal{C}$ a family of convergent sequences $x = \{x_n\} \subset l^\infty(\mathcal{P})$ and $\mathcal{H}$ = linear span of $\bigcup_{x\in\mathcal{C}}(\{x_n\} \bigcup \{\lim x_n\})$. Let $T = \{T_n\}_{n=0}^\infty$ be a family of $\mathbf{R}^d$-valued functionals defined on subsets of $l^\infty(\mathcal{F})$ such that the domain of T_0 contains P and the domain of T_n contains the set $\{P + n^{1/2}x_n : x_n$ is the n-th term of $x \in \mathcal{C}\}$. Then T is $n^{-1/2}$-*differentiable along $\mathcal{C}$ at P *if there exists a linear continuous map* $T'_P : \mathcal{H} \to \mathbf{R}^d$, the derivative of T at P, *such that*

$$n^{1/2}[T_n(P + n^{-1/2}x_n) - T_0(P) - n^{-1/2}T'_P(x_n)] \to 0 \tag{4.4}$$

for all $\{x_n\} \in \mathcal{C}$. By continuity, $T'_P(x_n)$ can be replaced by $T'_P(x_0)$ in (4.4) if $x_0 = \lim x_n$.

Denote by $\mathcal{M}_{\mathcal{F}}$ the set of measures of finite total variation on $(S, \mathcal{S})$ which are in $l^\infty(\mathcal{F})$, and by $\mathcal{P}_{\mathcal{F}}$ the set of probability measures in $\mathcal{M}_{\mathcal{F}}$ (i.e. $Q \in \mathcal{P}_{\mathcal{F}}$ iff $\|Qf\|_{\mathcal{F}} < \infty$). For $x_n \in \mathcal{M}_{\mathcal{F}}$, $x_n \to x$ will mean $\|x_n - x\|_{\mathcal{F}} := \sup_{f\in\mathcal{F}} |x_n(f) - x(f)| \to 0$. We let $\mathcal{C}_u(\mathcal{F}, e_P)$ be the space of uniformly continuous functions on $(\mathcal{F}, e_P)$, where $e_P^2(f,g) = P(f-g)^2$.

The first and second parts of the following theorem are taken from Sheehy and Wellner (1988). It strictly contains Theorem 4.1 up to some measurability (which can possibly be removed), and also the main result in Yang (1985).

Theorem 4.6 *Let $\mathcal{F}$ be P-Donsker, let*

$$\mathcal{C} = \{\{x_n\} : x_n \in M_{\mathcal{F}}, P + n^{-1/2}x_n \in \mathcal{P}_{\mathcal{F}}, \lim_{n\to\infty} x_n \in \mathcal{C}_u(\mathcal{F}, e_P)\} \tag{4.5}$$

and let $T = \{T_n\}_{n=0}^\infty$ be a sequence of maps from subsets of $l^\infty(\mathcal{F})$ into $\mathbf{R}^d$ such that the domain of T_0 contains P and the domain of T_n contains $\mathcal{P}_{\mathcal{F}}$ for $n = 1, 2, \ldots$, and such that T is $n^{-1/2}$-differentiable along $\mathcal{C}$ at P. Then *(i)* $n^{1/2}(T_n(P_n) - T_0(P)) \to_{\mathcal{L}} T'_P(G_P)$.

(ii) If $\mathcal{F}$ is measurable and if the maps $(s_1, \ldots, s_n) \to T_n(n^{-1}\sum_{i=1}^n \delta_{s_i})$ *from* S^n *into* $\mathbf{R}^d$ *are measurable,* $n = 1, \ldots,$ *then* $\hat{d}_{BL^*}(n^{1/2}(T_n(\hat{P}_n) - T_n(P_n)), T'_P(G_P))$ *is measurable for all n and*

$$\hat{d}_{BL^*}(n^{1/2}(T_n(\hat{P}_n) - T_n(P_n)), G_P) \to 0 \text{ in probability.} \tag{4.6}$$

(iii) If $\mathcal{F}$ is measurable and $PF^2 < \infty$, *if T is* $n_k^{-1/2}$*-differentiable at P along sequences* $\{x_{n_k}\}$ *satisfying the definition of* $\mathcal{C}$ *with n replaced by* n_k, *and if the sequence* $N = N_n$ *is regular* ($N_n \nearrow \infty, N_n/N_{2n} \geq c > 0, n \in N$), *and satisfies* $N_n = 0(n/\log\log n)$, *then*

$$N^{1/2}(T_N(\hat{P}_{N,n}) - T_n(P_n)) \to_{\hat{\mathcal{L}}} T'_P(G_P) \text{ a.s.} \tag{4.7}$$

Remark 4.7 If $\mathcal{F}$ is UPG no regularity for the sequence N in (iii) is required: see Proposition 2.2 and the next proof. For $\mathcal{F}$ UPG it is also possible to prove an analogue of Theorem 4.3 under the weaker differentiability of Theorem 4.6

Proof of Theorem 4.6 To prove (i) we set in Theorem 2.6 $(V, d) = (l^\infty(\mathcal{F}), \|\cdot\|_{\mathcal{F}})$, $Z_n = \nu_n$, $Z_0 = G_P$, $V_n = \{x \in \mathcal{M}_{\mathcal{F}} : P + n^{-1/2}x \in \mathcal{P}_{\mathcal{F}}\}$, $V_0 = \mathcal{C}(\mathcal{F}, e_P), H_n(x) = n^{1/2}(T_n(P + n^{1/2}x) - T_0(P))$ and $H_0(y) = T'_P(y)$, $x \in V_n, y \in V_0$. Then (i) follows directly from that theorem. (This nice proof is Wellner's (1989).)

The proof of (ii) uses the representation Theorem 2.5 twice in the way outlined after Theorem 4.4. Since (ii) is just Theorem 3.3A in Sheehy and Wellner (1988) and their proof is accurate under the present measurability conditions, we will only prove (iii). (Note that d_{BL^*} is in fact a sup over a countable number of functions since $\mathbf{R}^d$ is σ-compact and the Lipschitz functions on a comptact set are a separable set for the sup norm.)

Under the hypothesis of (iii) the empirical process indexed by $\mathcal{F}$ satisfies the compact law of the iterated logarithm (Dudley and Philipp (1983), Theorem 1.2) with limit set $K = \{u_h(f) = \int fh dP : \int h^2 dP \leq 1, h$ in the $\mathcal{L}_2(P)$ - closed linear span of $\{f - Pf\}\}$ (Kuelbs (1976)). Note that $K \subset \mathcal{C}_u(\mathcal{F}, e_P)$. So we have that Pr-a.s.

$$\begin{aligned}&\text{every subsequence of } \{\nu_n(\omega)/(2\log\log n)^{1/2}\} \text{ has}\\&\text{a further subsequence that converges in } l^\infty(\mathcal{F}) \text{ to}\\&\text{some } x_0 \in K \ (x_0 \text{ depending on the subsequence}).\end{aligned} \tag{4.8}$$

Note also that Proposition 2.2, implies

$$d_{BL^*}(\hat{\nu}^\omega_{N,n}, G_P) \to 0 \text{ a.s.} \tag{4.9}$$

Let ω be such that (4.8) and (4.9) hold, and let $g_n, n = 0, 1, \ldots$, be the perfect functions of Dudley's theorem for $\{\hat{\nu}^{\omega}_{N,n}\}$ and G_P. Then

$$\|\hat{\nu}^{\omega}_{N,n}(g_n(\tilde{\omega}')) - G_P(g_0(\tilde{\omega}'))\|_{\mathcal{F}} \to 0 \ \tilde{P}r' - a.s. \tag{4.10}$$

Take now $\tilde{\omega}'$ such that (4.10) holds. Every subsequence of the natural numbers has a further subsequence $\{n'\}$ such that

$$N_{n'}/(n'/\log\log n') \to c \in [0, \infty) \text{ and } \nu_{n'}(\omega)/(2\log\log n')^{1/2} \to x_0$$

for some $x_0 \in K$. Hence $\{\hat{\nu}^{\omega}_{N',n'}(g_n(\tilde{\omega}')) + (N'/n')^{1/2}\nu_{n'}(\omega)\}$ and $\{(N'/n')^{1/2}\nu_{n'}(\omega)\}$ converge to points in $\mathcal{C}_u(\mathcal{F}, e_P)$ (where we let $N' := N_{n'}$), and we obtain,

$$N'^{1/2}(T_{N'}(\hat{P}^{\omega}_{N',n'}(g_{n'}(\tilde{\omega}'))) - T_{N'}(P_{n'}(\omega))$$

$$= N'^{1/2}[T_{N'}(P + N'^{-1/2}(\hat{\nu}^{\omega}_{N',n'}(g_{n'}(\tilde{\omega}')) + (N'/n')^{1/2}\nu_{n'}(\omega)))$$

$$-T_{N'}(P + N'^{-1/2}(N'/n')^{1/2}\nu_{n'}(\omega))] \to T'_P(G_P(g_0(\tilde{\omega}')))$$

by (4.4), (4.10) and the linearity of T'_P. Hence, this limit holds along the whole sequence $\{n\}$. Moreover, since $N^{1/2}(T_N(\hat{P}^{\omega}_{N,n}) - T_N(P_n(\omega)))$ is measurable in ω' for each ω (in fact, finite-valued measurable) the convergence above takes place $\tilde{P}r'$-almost uniformly and therefore, by Dudley (1985, Theorem 3.5),

$$N^{1/2}(T_N(\tilde{P}^{\omega}_{N,n}) - T_N(P_n)) \to_{\mathcal{L}} T'_P(G_P).$$

Since this holds for almost every ω, (iii) is proved. ■

Remark 4.8 In the application to M-estimators that follows, we require Theorem 4.6 under $n^{-1/2}$-differentiability of T along a class of sequences $\bar{\mathcal{C}}$ strictly contained in $\mathcal{C}$. Suppose $\mathcal{F}$ is P-Donsker and $\mathcal{G}$ is P-Glivenko-Cantelli (see Remark 3.4 for the definition of P-GC class), and let

$$\bar{\mathcal{C}} = \{\{x_n\} : x_n \in \mathcal{M}_{\mathcal{F}} \bigcap \mathcal{M}_{\mathcal{G}},\ P + n^{1/2}x_n \in \mathcal{P}_{\mathcal{F}},$$

$$\mathcal{F} - \lim_{n\to\infty} x_n \in \mathcal{C}_u(\mathcal{F}, e_P),\ n^{-1/2}\|x_n\|_{\mathcal{G}} \to 0\} \tag{4.11}$$

Then if T is $n^{-1/2}$-differentiable along $\bar{\mathcal{C}}$ the conclusions (i) and (ii) in Theorem 4.6 are also valid and so is (iii) assuming T is $n_k^{-1/2}$-differentiable along sequences $\{x_{n_k}\}$ satisfying the definition of $\bar{\mathcal{C}}$ with n replaced by n_k. The proof is as that of Theorem 4.6, with some natural changes that se sketch. For $H_n(x_n) = N^{1/2}(T_n(P + n^{1/2}x_n) - T_0(P))$ and $H_0(x_0) = T'_P(x_0)$, we apply Theorem 2.6 with $(V, d) = (l^{\infty}(\mathcal{F}) \times l^{\infty}(G))$, $\|(x, y)\| = \|x\|_{\mathcal{F}} + \|x\|_{\mathcal{G}}$), $V_n = \{(x_n, n^{-1/2}x_n)\}, V_0 = \{(x, 0) : x \in \mathcal{C}_u(\mathcal{F}, e_P)\}$. This gives (i). The proof of (ii) is as in Sheehy and Wellner (1988) but now one uses Theorem 2.5 for $(\nu_n, n^{-1/2}\nu_n) \to_{\mathcal{L}} (G_P, 0)$ (in $l^\infty(\mathcal{F}) \times l^\infty(\mathcal{G})$) which yields $(\nu_n, n^{-1/2}\nu_n) \circ g_n \to (G_P, 0) \circ g_0$ almost uniformly and $\hat{d}_{BL^*,V}[(\hat{\nu}_n, n^{-1/2}\hat{\nu}_n)^{g_n(\tilde{\omega})}, (\tilde{G}_P, 0)] \to 0$ almost uniformly by the bootstrap CLT and LLN, and then Dudley's representation theorem for each $\tilde{\omega}$ fixed, on the sequence $\{(\hat{\nu}_n, n^{1/2}\hat{\nu}_n)^{g_n(\tilde{\omega})}\}$. Similarly in (iii), using both the bootstrap CLT and the bootstrap LLN, we have not only (4.9) but

$$\hat{d}_{BL^*,V}[(\hat{\nu}^{\omega}_{N,n}, N^{-1/2}\hat{\nu}^{\omega}_{N,n}), (G_P, 0)] \to 0 \ a.s. \tag{4.12}$$

and the proof follows in an analogous way.

Remark 4.9 By Proposition 2.2 (i), statement (ii) in 4.6 also holds for any bootstrap sample size $N = N_n \to \infty$.

Example 4.10 *(On the bootstrap in probability of M-estimators).* Let $\Theta \subset \mathbf{R}^d$ be a G_δ set such that $\theta \in \Theta^\circ$, let $g : S \times \Theta \to \mathbf{R}$ and let P satisfy the consistency conditions (B.1) to (B.4) from Section 3 and, instead of (A.1)-(A.4), the following:

(A.1)' $G(0) = \sup_{\theta\in\Theta} G(\theta)$.

(A.2)' There is a symmetric positive definite bilinear form A_G such that $G(\theta) = G(0) - \frac{1}{2}A_G(\theta,\theta) + o(|\theta|^2)$ as $\theta \to 0$.
Without loss of generality, we assume $A_G(\theta, \theta') = \theta \cdot \theta'$.

(A.3)' There exists $\Delta : S \to \mathbf{R}^d$ such that $P|\Delta|^2 < \infty$ and such that if $r(x,\theta) = |\theta|^{-1}[g(x,\theta) - g(x,0) - \theta \cdot \Delta(x)], x \in S, \theta \neq 0$,
$r(x,0) \equiv 0$
then, for some $K > 0$, the class of functions
$\mathcal{F} = \{r(\cdot,\theta) : |\theta| \le K\} \bigcup \{\Delta_1, \dots, \Delta_d\}$
is a measurable P-Donsker class

(A.4)' $Pr^2(\cdot,\theta) \to$ as $\theta \to 0$.

In order to apply Theorem 4.6 and Remark 4.8 to obtain bootstrap results for the M-estimators corresponding to g we let: $\mathcal{F}$ as in (A.3)', $\mathcal{G} = \{\sup_{|\theta|\ge M} g(\cdot,\theta)/b(\theta), g(\cdot,0)\} \bigcup \{g(\cdot,\theta) : |\theta| \le M\}$, and $\bar{\mathcal{C}}$ as in (4.11) for these $\mathcal{F}$ and $\mathcal{G}$. Let $0 < f(n) = o(1/n)$ as $n \to \infty$. Then, as mentioned in Section 3, there is a universally measurable map $\theta_n : S^n \to \Theta$ such that $n^{-1}\sum_{i=1}^n g(s_i, \theta_n) \ge \sup_{\theta\in\Theta} n^{-1}\sum_{i=1}^n g(s_i,\theta) - f(n)$. We can then define for each Q in $\mathcal{P}_{\mathcal{G}}, T_n(Q)$ as: $T_n(n^{-1}\sum_{i=1}^n \delta_{s_i}) = \theta_n(s_1, \dots, s_n)$ and otherwise as any θ_Q satisfying $Qg(\cdot,\theta_Q) \ge \sup_{\theta\in\Theta} Q(\cdot,\theta) - f(n)$. Then the functionals T_n satisfy the measurability hypothesis of Theorem 4.6 and $T_n(P_n) = \theta_n, T_n(\hat{P}_n) = \hat{\theta}_n$ are respectively an M-estimator and its bootstrap. Finally T_0 is defined only at P as $T_0(P) = 0$. So the conclusions of Theorem 4.6 will follow if we prove that $T = \{T_n\}_{n=0}^\infty$ is $n^{-1/2}$-differentiable along $\bar{\mathcal{C}}$ at P. To do this we just apply methods of Huber (for consistancy) and Pollard (for the CLT):

Step 1. $\{x_n\} \in \bar{\mathcal{C}} \Rightarrow \gamma_n := T_n(P + x_n/n^{1/2}) \to 0$:
This follows immediately from

$$(P + n^{1/2}x_n)(\cdot \sup_{|\theta| \geq M} g(\cdot, \theta)/b(\theta)) \to a$$

$$\sup_{\varepsilon < |\theta| < M} (P + n^{-1/2}x_n)(g(\cdot, \theta)) \to 0$$

and

$$n^{-1/2}x_n(g(\cdot, 0)) \to 0$$

as in the proof of Theorem 3.5.

Step 2. $\{x_n\} \in \bar{\mathcal{C}} \Rightarrow \gamma_n = 0(n^{-1/2})$:
By Step 1, (A.2)' and the definition of T_n we have

$$0 \leq f(n) + (P + n^{-1/2}x_n)(g(\cdot, \gamma_n) - g(\cdot, 0))$$

$$\leq f(n) - |\gamma_n|^2/4 + \gamma_n n^{-1/2}x_n(\Delta) + |\gamma_n| n^{-1/2}x_n(r(\cdot, \gamma_n)).$$

Hence

$$n^{1/2}|\gamma_n|/4 \leq x_n(\Delta) + x_n(r(\cdot, \gamma_n)) + (f(n))^{1/2}n^{1/2}$$

which is uniformly bounded.

Step 3. $\{x_n\} \in \bar{\mathcal{C}} \Rightarrow n^{1/2}\gamma_n - x_n(\Delta) \to 0$:
By the definition of T_n

$$0 \leq f(n) + (P + n^{-1/2}x_n)(g(\cdot, \gamma_n) - g(\cdot, n^{-1/2}x_n(\Delta)))$$

so that, (A.2)' and (A.3)' give, as in (3.12),

$$|n^{1/2}\gamma_n - x_n(\Delta)|^2/2 \leq nf(n) + no(|\gamma_n|^2) + no(|n^{-1/2}x_n(\Delta)|^2)$$

$$+|x_n(\Delta)||x_n(r(\cdot, n^{-1/2}x_n(\Delta))| + n^{-1/2}|\gamma_n||x_n(r(\cdot, \gamma_n))|.$$

Now, $nf(n) \to 0$ by the definition of $f, no(|\gamma_n|^2) \to 0$ by Step 2 and $no(|n^{-1/2}x_n(\Delta)|^2) \to 0$ since $\{x_n(\Delta)\}$ converges. For the next to last term we observe $|x_n(\Delta)| \to |x_0(\Delta)|$ and $x_n(r(\cdot, n^{-1/2}x_n(\Delta))) \approx x_0(r(\cdot, n^{-1/2}x_n(\Delta))) \approx x_0(r(\cdot, 0)) = 0$ by (A.4)' and because $x_0 \in \mathcal{C}_u(\mathcal{F}, e_P)$. The last term tends to zero by the same argument. Hence

$$n^{1/2}\gamma_n - x_n(\Delta) \to 0,$$

i.e. T is differentiable and $T'_P(x) = x(\Delta)$ for $x \in l^\infty(\mathcal{F})$.

The conclusion is that (for $A_G = I$), under conditions (B.1)-(B.4) and (A.1)'-(A.4)'

$$n^{1/2}(\hat{\theta}_n - \theta_n) \to_{\mathcal{L}} N(0, Cov\Delta) \text{ in probability}$$

and

$$n^{1/2}(\hat{\theta}_{N,n} - \theta_n) \to_{\hat{\mathcal{L}}} N(0, Cov, \Delta) \text{ a.s.}$$

if $N = N_n$ is a regular sequence such that $N_n = 0(n/\log\log n)$.

Note that the conditions (B.1)-(B.4) and (A.1)'-(A.4)' are weaker than the conditions for the validity of Theorem 3.2 (the a.s. bootstrap) and not much stronger than Pollard's (1985). Pollard's (1985) "stochastic differentiability condition" is thus close to being equivalent to the differentiability condition (4.4) for $\bar{\mathcal{C}}$, which does not involve any randomness.

Theorem 4.6 also applies to M-estimators defined as zeros of $P_n h(\cdot, \theta)$ (actually with a proof simpler that in the previous example, that we omit). The result is the following: Assuming that the consistency conditions (1)-(4) of Theorem 3.7 hold and that

(C.1)' $Ph(\cdot, 0) = 0$,
(C.2)' $H(\theta) := Ph(\cdot, \theta)$ is differentiable at zero with a non-degenerate first derivative $H'(0)$, which we assume without loss of generality to be I,
(C.3)' the classes of functions $\mathcal{F}_i = \{h_i(\cdot, \theta) : |\theta| \le M\}, i = 1, \ldots, d$, are P-Donsker,
(C.4)' $\theta \to Var_P h_i(\cdot, \theta)$ is continuous at 0, $i = 1, \ldots, d$,
(C.5)' for all $Q \in \mathcal{P}_{\mathcal{F}}$ there exists $T_n(Q)$ such that $|Qh(\cdot, T_n(Q))| < \varepsilon_n / n^{1/2}$ for some $\varepsilon_n \to 0$,

the conclusions of Theorem 4.6 hold for $T_0(P) = 0$, $T_n(\hat{P}_n)$ and $T_N(\hat{P}_{N,n})$. We note that the same conclusions follow if (C.3)' and (C.4)' are replaced by asymptotic equicontinuity at zero of $\{\nu_n(h_i) : h_i \in \mathcal{F}_i\}$, $i = 1, \ldots, d$ (Romo (1990); see Remark 3.3).

Acknowledgements. A seminar talk by Prof. R. Dudley at CUNY stimulated our interest in the bootstrap of M-estimators and differentiable statistical functionals. A. van der Vaart and J. Wellner pointed out to us a gap in the first version of the proof of Theorem 2.6. We also acknowledge several interesting conversations with J. Romo.

References

Alexander, K.S. (1984). Probability inequalities for empirical processes and a law of the iterated logarithm. Ann. Probability 12, 1041-1067.

Alexander, K. S. (1985). Rates of growth for weighted empirical processes. In *Proceedings of the Berkeley Converence in Honor of Jerzy Neyman and Jack Kiefer* (L.M. LeCam and R. A. Olshen, eds.) 2, 475-493. Wadsworth, Monterey, CA.

Alexander, K. S. (1987). The central limit theorem for empirical processes on Vapnilk-Cĕrvonekis classes. Ann. Probability 15, 178-203.

Alexander, K. S. and Talagrand, M. (1989). The law of the iterated logarithm for empirical processes on Vapnik-Cĕrvonenkis classes. J. Multivariate Analysis 30, 155-166.

Araujo, A. and Giné, E. (1980). *The central limit theorem for real and Banach valued random variables.* Wiley, New York.

Arcones, M. and Giné, E. (1990). The bootstrap of the mean with arbitrary bootstrap sample size. Ann. Inst. H. Poincaré 25, 457-481.

Beran, R. (1984). Bootstrap methods in statistics. Jber. d. Dt. Math. Verein 86, 14-30.

Bickel, P. J. and Freedman, D. A. (1981). Some asymptotic theory for the bootstrap. Ann. Statist. 9, 1196-1217.

Cohn, D. L. (1980). *Measure theory.* Birkhauser, Boston.

Cuesta, J. A. and Matrán, C. (1988). The strong law of large numbers for k-means and best possible nets of Banach valued variables. Probability Theory and Rel. Fields 78, 523-534.

Dudley, R. M. (1984). A course on empirical processes. Lect. Notes in Math. 1097, 1-142. Springer, Berlin.

Dudley, R. M. (1985). An extended Wichura theorem, definition of Donsker class and weighted empirical distributions. Lect. Notes in Math. 1153, 141-178. Springer, Berlin.

Dudley, R. M. (1990). Nonlinear functionals of empirical measures and the bootstrap. In *Probability in Banach spaces VII*, 63-82. Progress in Probability Series. Birkhauser, Boston.

Dudley, R. M. (1989). *Real Analysis and Probability.* Wadsworth, Pacific Grove, California.

Dudley, R. M. and Philipp, W. (1983). Invariance principles for sums of Banach space valued random elements and empirical processes. Z. Wahrsch. v. Geb. 62, 509-552.

Gill, R. R. (1989). Non- and semi-parametric maximum likelihood estimators and the von Mises method (Part II). Scand. J. Statist. 16, 97-128.

Giné E. and Zinn, J. (1984). Some limit theorems for empirical processes. Ann. Probability 12, 929-989.

Giné E. and Zinn, J. (1986). Lectures on the central limit theorem for empirical processes. Lect. Notes in Math. 1221, 50-113. Springer, Berlin.

Giné E. and Zinn, J. (1990). Bootstrapping general empirical measures. Ann. Probability 18, 851-869.

Giné E. and Zinn, J. (1991). Gaussian characterization of uniform Donsker classes of functions. Ann. Probability 19, No. 3, to appear.

Hartigan, J. A. (1978). Asymptotic distributions for clustering criteria. Ann. Statistics 6, 117-131.

Huber, P. J. (1964). Robust estimation of a location parameter. Ann. Math. Statist. 35, 73-101.

Huber, P. J. (1967). The behavior of maximum likelihood estimates under non-standard conditions. *Proceedings Fifth Berkeley Symposium on Mathematical Statistics and Probability*, Vol. 1, 221-233. University of California Press, Berkeley.

LeCam, L. (1986). *Asymptotic methods in statistical decision theory.* Springer, Berlin.

Pollard, D. (1981). Strong consistency of k-means clustering. Ann. Statist. 9, 135-140.

Pollard, D. (1982). A central limit theorem for k-means clustering. Ann. Probability 10, 919-926.

Pollard. D. (1984). *Convergence of stochastic processes.* Springer, Berlin.

Pollard, D. (1985). New ways to prove central limit theorems. *Econometric Theory* 1, 259-314.

Pollard, D. (1989). Empirical processes: theory and applications. CBMS/NSF Regional Conference Series in Probability and Statistics, to appear.

Reeds, J. A. (1976). On the definition of von Mises functionals. Research Report # S-44, Dept. Statist., Harvard University.

Romo, L. (1990). Manuscript.

Sheehy, A. and Wellner, J. A. (1988). Uniformity in P of some limit theorems for empirical measures and processes. Technical Report #134, Department of Statistics, University of Washington, Seattle.

van der Vaart, A. W. and Wellner, J. (1989). Prohorov and continuous mapping theorems in the Hoffmann-Jørgensen weak convergence theory, with applications to convolution and asymptotic minimax theorems. Preprint.

Wellner, J. (1989). Discussion of: "Non- and semi-parametric maximum likelihood estimators and the von Mises method, Part I" by R. Gill, Scand. J. Statist. 16, 124-127.

Yang, S. S. (1985). On Bootstrapping a class of differentiable statistical functionals with applications to L- and M-estimators. Stat. Neerland. 39, 375-385.

BOOTSTRAPPING MARKOV CHAINS

K. B. Athreya[1]
Departments of Mathematics
and Statistics
Iowa State University
Ames, Iowa, USA

C. D. Fuh[2]
Institute of Statistical Science
Academia Sinica
Taipei, Taiwan, ROC

Abstract

We present a brief summary of some recent work on the application of the bootstrap method to estimate the distribution of the estimator of the transition probability matrix, and that of the hitting time distribution for finite and countable state space Markov chains that are ergodic. Some open problems are also indicated.

Introduction

Let $\{X_n;\ n \geq 0\}$ be a homogeneous ergodic (positive recurrent, irreducible and aperiodic) Markov chain countable with state space S and transition probability matrix $P = (p_{ij})$. The problem of estimating P and the distribution of the hitting time T_Δ of a state Δ arises in several areas of applied probability. The application of the bootstrap method of Efron (1979) to the finite state Markov chain case was considered by Kulperger and Prakasa Rao (1987)and Basawa et al (1989). Athreya and Fuh (1989) discussed the countable state space case. The general state space case is an important open problem.

The goal of the present paper is to give a brief survey of the results of the the above mentioned papers and also some related work of Datta and McCormick (1990) on second order correction of a method of bootstrap proposed by Basawa et al (1989) for the finite state space case. The latter paper also considers parametric bootstrap for finite state Markov chains. No proofs are given in the present paper.

Keywords: Primary 62G05; Secondary 60F05, 60J10
1980 ***Mathematics subject classifications:*** bootstrap estimation, central limit theorem, hitting times, Markov chains, stationary distributions, transition probabilities
[1] Research partially supported by NSF Grant 8803639.
[2] Research partially supported by NSC of ROC Grant 79-0208-M001-63.

Exploring the Limits of Bootstrap. Edited by Raoul LePage and Lynne Billard.
 ISBN: 0471-53631-8.

Bootstrapping a finite state space Markov chain

Let $\{X_n;\ n \geq 0\}$ be a time homogeneous irreducible aperiodic Markov chain with finite state space $S = \{1, 2, \cdots, k\}$ and transition probability matrix $P = (p_{ij})$. This implies that there exists an invariant probability measure $\Pi = (\pi_1, \cdots, \pi_k)$ such that $\pi_j > 0$, $\sum_{j=1}^{k} \pi_j = 1$, $\pi_j = \sum_\alpha \pi_\alpha p_{\alpha j}$, $j = 1, \cdots, k$, and $p_{ij}^{(n)} \to \pi_j$, as $n \to \infty$ for all $i \in S$.

Suppose $\underset{\sim}{x} = \{x_0, x_1, \cdots, x_n\}$ is a realization of the process $\{X_j;\ j = 0, \cdots, n\}$ observed up to time n. We estimate P by its maximum likelihood estimator $\hat{P}_n \equiv (\hat{p}_n(i, j))$, where

$$\hat{p}_n(i,j) = \begin{cases} n_{ij}/n_i, & \text{if } n_i > 0; \\ \delta_{ij}, & \text{otherwise,} \end{cases}$$

and estimate Π by $\hat{\Pi}_n \equiv (\hat{\pi}_n(i))$, where

$$\begin{aligned} \hat{\pi}_n(i) &= \frac{n_i}{n}, \\ n_{ij} &= \text{observed number of } ij \text{ transitions in } \{x_0, \cdots, x_n\}, \\ n_i &= \text{observed number of visits to state } i \text{ in } \{x_0, \cdots, x_n\}. \end{aligned}$$

Since the state space S is finite, we can consider the non-parametric case as a special case of the parametric case. So, the consistency and asymptotic normality of the maximum likelihood estimators can be deduced using the analogy with the multinomial distribution. This idea also can be used to prove the consistency of the bootstrap estimators of $\hat{P}_n$ given $\underset{\sim}{x}$

The consistency of $\widehat{\Pi}_n$ for Π follows from the strong law applied to the renewal sequence of return times to state i.

THEOREM 1. *For all i,*

$$\hat{\pi}_n(i) = n_i/n \longrightarrow \pi_i \qquad \textit{with probability } 1 \qquad \textit{as } n \to \infty.$$

The following theorem is a central limit theorem for the maximum likelihood estimator $\widehat{P}_n$ of P and is in Billingsley's book (1961).

THEOREM 2.

$$\sqrt{n}(\widehat{P}_n - P) \longrightarrow N(\underset{\sim}{0}, \Sigma_P) \qquad \textit{in distribution,}$$

where Σ_P, the variance-covariance matrix, is given by

$$(\Sigma_P)_{(ij,i'j')} = \delta_{ii'} p_{ij} \left(\delta_{jj'} - p_{ij'}\right).$$

Let $x^* = \{x_0^*, x_1^*, \ldots, x_n^*\}$ be a realisation of a Markov chain with transition probability matrix $\widehat{P}_n$ and let $\widetilde{P}_n$ be the $\widehat{P}_n$ function evaluated at this x^*. For this bootstrap method Kulperger and Prakasa Rao (1989) established the following central limit theorem.

THEOREM 3. *Under the notations given above, we have for almost all realizations of the Markov chain* $\{X_n;\ n \geq 0\}$,

$$\sqrt{N_n}(\widetilde{P}_n - \widehat{P}_n) \longrightarrow N(\underset{\sim}{0}, \Sigma_P) \qquad \text{in distribution}$$

as $n \to \infty$ *and* $N_n \to \infty$, *where* Σ_P *is the same variance-covariance matrix in* Theorem 2.

Let T_k be the first hitting time of state k. That is, we let

$$T_k = \begin{cases} \inf\{n \geq 0;\ X_n = k\}; & \\ \infty, & \text{if no such } n \text{ exist.} \end{cases}$$

Let $Pr(t; P) \equiv P(T_k \leq t \mid X_0 = 1; P)$ denote the probability that $T_k \leq t$ for $t \in \{1, 2, 3, \cdots\}$, where P is the transition probability matrix of a Markov chain $X = \{X_n;\ n \geq 0\}$ with initial state $X_0 = 1$.

For any $k \times k$ stochastic matrix P, let $A \equiv A(P)$ be the stochastic matrix which is the same as P except that the kth row is replaced by $(0, \cdots, 0, 1)$ with 1 in the kth position. Note that

$$Pr(t; P) = (A^t)_{1,k} \qquad (*)$$

The bootstrap estimate of the distribution $Pr(\cdot; P)$ of the hitting time T_k is $Pr\left(\cdot; \hat{P}_n\right)$. From $(*)$ and the fact that $\widehat{P}_n \to P$ with probability 1, we have for each t,

$$Pr(t; \hat{P}_n) - Pr(t; P) \longrightarrow 0 \quad \text{w.p.1} \quad \text{as } n \to \infty.$$

Here the problem is to estimate the distribution of

$$G_n(t; P) \equiv \sqrt{n}\left(Pr(t; \hat{P}_n) - Pr(t; P)\right).$$

The bootstrap approximation to the above distribution is the conditional distribution of

$$G_n(t; \hat{P}_n) \equiv \sqrt{n}\left(Pr(t; \tilde{P}_n) - Pr(t; \hat{P}_n)\right).$$

The problem here is to verify that these two distributions are asymptotically close. Kulperger and Prakasa Rao (1987) obtained the next two theorems.

THEOREM 4. *Let* $\hat{A}_n \equiv A\left(\hat{P}_n\right)$. *Then for all* $t = 1, 2, 3, \cdots$, *we have*

$$\sqrt{n}\left(\hat{A}_n^t - A^t\right) \longrightarrow N(\underset{\sim}{0}, Z_P^t) \quad \text{in distribution,}$$

where the variance-covariance matrix Z_P^t *is a function of* P.

REMARK 1: It turns out that Z_P^t is fairly complicated to compute for general t. For $t = 1, 2$ we have

$$Z_P^1 = \Sigma_A,$$
$$Z_P^2 = \text{Var}(AU + UA).$$

where $U \sim N(\underset{\sim}{0}$ and $\Sigma_A)$, for any $k \times k$ stochastic P, Σ_P is as in Theorem 2.

THEOREM 5. *Let $\tilde{A}_n \equiv A(\tilde{P}_n)$ where $\tilde{P}_n$ is defined as above. Then, for almost all realizations of the process, and for $t = 1, 2, 3, \cdots$, we have*

$$\sqrt{N_n}\left(\tilde{A}_n^t - \hat{A}_n^t\right) \longrightarrow N\left(\underset{\sim}{0}, Z_P^t\right) \quad \text{in distribution,}$$

where Z_P^t is the same variance-covariance matrix in Theorem 4.

Adapting the same type of analysis, one can estimate $m(P) \equiv E(T_k \mid X_0 = 1; P)$ the expected value of the hitting time T_k by $m(\hat{P}_n) \equiv E(T_k \mid X_0 = 1; \hat{P}_n)$. It can be shown below that $m(\hat{P}_n)$ is a consistent estimator of $m(P)$, and further

$$Q_n(t; P) \equiv \sqrt{n}\left(m(\hat{P}_n) - m(P)\right) \text{ is asymptotically normal.}$$

The bootstrap approximation to the above distribution is the conditional distribution of

$$Q_n(t; \hat{P}_n) \equiv \sqrt{N_n}\left(m(\tilde{P}_n) - m(\hat{P}_n)\right).$$

That these two distributions are asymptotically close can be shown using the δ method, Theorems 3 and 4, and the following

LEMMA. *Let P be a stochastic $k \times k$ matrix and $\{X_n;\ n \geq 0\}$ be a Markov chain with transition probability matrix P. Let*

$$m_{ijk}(P) \equiv E(\text{visits to } j \text{ before to } k \mid X_0 = i; P).$$

Then

$$m_{ijk}(P) = \sum_{r=0}^{\infty} (P_k^r)_{ij},$$

where P_k is the matrix given by P in which the kth column is replaced by $\underset{\sim}{0}$. Note that $m_{ijk}(P)$ is continuously differentiable with respect to the entries of P at all P such that $\inf_{i,j} p_{ij} > 0$.

Bootstrapping Markov chains: countable case

Athreya and Fuh (1989) have considered the case of a countable state space Markov chain. They propose three bootstrap methods.

Method I.

Let $\underset{\sim}{x} = (x_0, x_1, \cdots, x_n)$ be a realization of the Markov chain $\{X_n;\ n \geq 0\}$ with transition probability P. Let $\hat{P}_n \equiv P(n, \underset{\sim}{x})$ be an estimator of P based on the observed data $\underset{\sim}{x}$. A bootstrap method for estimating the sampling distribution H_n of $R(\underset{\sim}{x}, P) \equiv (\hat{P}_n - P)$ can be described as follows (same as in the finite state space case except that the resample size is changed from n to N_n

1) With $\hat{P}_n$ as its transition probability, generate a Markov chain realization of N_n steps $\underset{\sim}{x}^* = (x_0^*, x_1^*, \cdots, x_{N_n}^*)$. Call this the bootstrap sample, and let $\tilde{P}_n \equiv P(N_n, \underset{\sim}{x}^*)$. Note that $\tilde{P}_n$ bears the same relation to $\underset{\sim}{x}^*$ as $\hat{P}_n$ does to $\underset{\sim}{x}$.
2) Approximate the sampling distribution H_n of $R(\underset{\sim}{x}, P)$ by the conditional distribution H_n^* of $R(\underset{\sim}{x}^*, \hat{P}_n) \equiv (\tilde{P}_n - \hat{P}_n)$ given $\underset{\sim}{x}$.

Method II.

The existence of a recurrent state Δ which is visited infinitely often (i.o.) for a recurrent Markov chain is well-known. A well-known approach to its limit theory is via the embedded renewal process of returns to Δ. This is the so-called regeneration method. For a fixed state Δ, by the strong Markov property, the cycles $\{X_j;\ j = T_\Delta^{(n)}, \cdots, T_\Delta^{(n+1)} - 1\}$ are i.i.d. for $n = 1, 2, \cdots$, where $T_\Delta^{(n)}$ is the time of the nth return to Δ.

Fix an integer k and observe the chain up to the random time $n = T_\Delta^{(k+1)}$. Let

$$\{X_0, X_1, \cdots, X_n\}$$

be a realization of the process. Note that in this situation, $X_n = \Delta$. Fix i, j. Let $\eta_\alpha \equiv \{X_j;\ j = T_\Delta^{(\alpha)}, \cdots, T_\Delta^{(\alpha+1)} - 1\}$ denote the αth cycle, $g(\eta_\alpha)$ indicate the number of visits to state i during the cycle η_α, and $h(\eta_\alpha)$ indicate the number of ij transitions during the cycle η_α. Now, define

$$\hat{\pi}_k(i) \equiv \frac{\sum_{\alpha=1}^k g(\eta_\alpha)}{\sum_{\alpha=1}^k T_\alpha}, \qquad \hat{p}_k(i,j) \equiv \frac{\sum_{\alpha=1}^k h(\eta_\alpha)}{\sum_{\alpha=1}^k g(\eta_\alpha)}$$

be the estimators of π and P, where T_α is the length of η_α.

By the strong law both $\hat{\pi}_k$ and $\hat{P}_k$ are consistent. To estimate the distributions of $(\hat{\pi}_k - \pi)$ and $(\hat{P}_k - P)$, Athreya and Fuh (1989) propose the following bootstrap algorithm.

1) Decompose the original sample in the following fashion:

$$\{\eta_0, \eta_1, \eta_2, \cdots, \eta_k\}, \quad \text{where } \eta_0 \equiv \{X_0, X_1, \cdots, X_{T_\Delta^{(1)}-1}\}.$$

Let $\hat{F}_k$ denote the uniform probability measure on the cycles $\{\eta_\alpha;\ \alpha = 1, 2, \cdots, k\}$. If $X_0 = \Delta$ w.p.1., then one could take $n = T_\Delta^{(k)}$ and $\hat{F}_k$ to be the uniform probability measure on $\{\eta_\alpha;\ \alpha = 0, 1, 2, \cdots, k-1\}$.

2) With the original sample fixed, draw a "bootstrap sample" of size k according to $\hat{F}_k$. Denote this sample by $\eta_1^*, \eta_2^*, \cdots, \eta_k^*$. Then, the bootstrap

analogous of $\hat{\pi}_k(i)$, $\hat{p}_k(i,j)$ can be defined as follows:

$$\tilde{\pi}_k(i) \equiv \frac{\sum_{\alpha=1}^k g(\eta_\alpha^*)}{\sum_{\alpha=1}^k T_\alpha^*}, \qquad \tilde{p}_k(i,j) \equiv \frac{\sum_{\alpha=1}^k h(\eta_\alpha^*)}{\sum_{\alpha=1}^k g(\eta_\alpha^*)},$$

where T_α^* is the length of η_α^*.

3) Approximate the distribution of $\sqrt{k}(\hat{p}_k(i,j) - p_{ij})$ by the conditional distribution of $\sqrt{k}(\tilde{p}_k(i,j) - \hat{p}_k(i,j))$ given $\underline{x}$. Similarly for $\sqrt{k}(\tilde{\pi}_k(i) - \pi_i)$.

REMARK 2: A slightly different approach is to fix the original sample size n. Let k_n be the (random) number of full cycles included in the observation $\{X_0, X_1, \cdots, X_n\}$. Now generate the bootstrap sample by making i.i.d. selection uniformly from these k_n cycles. Since as $n \to \infty$, $k_n \to \infty$, this method should work even though the k_n cycles are no longer independent. Further, we may even choose Δ to depend on the observation $\{X_0, X_2, \cdots, X_n\}$. A natural choice is that Δ for which we get the largest number of cycles in $\{X_0, X_1, \cdots, X_n\}$.

Method III.

This method is especially suited for estimating the distribution of hitting times.

Let $\underline{x} \equiv \{x_0, x_1, \cdots, x_n\}$ be a realization of the Markov chain $\{X_n\}$ up to time n. Fix any two states l and k. Let $T_{r,i}$; $i = 1, 2, \cdots$ be the successive return time to state r. Now arrange $\{T_{l,i}\}$, $\{T_{k,i}\}$ such that

$$T_{k,i_1-1} < T_{l,1} < T_{k,i_1} \le T_{k,i_2-1} < T_{l,2} < T_{k,i_2} < \cdots .$$

(That is, for each m, there is an i_m such that $T_{k,i_m-1} < T_{l,m} < T_{k,i_m}$.) Define

$$\begin{aligned} Y_1 &= T_{k,i_1} - T_{l,1} \\ Y_2 &= T_{k,i_2} - T_{l,2} \\ &\vdots \\ Y_m &= T_{k,i_m} - T_{l,m} \end{aligned}$$

It is clear that by the Markov property, the $\{Y_i\}_1^\infty$ are i.i.d.

Let $M_n \equiv \sup\{m;\ T_{k,i_m} \le n\}$. Then, for each positive integer t, a natural estimator for the distribution

$$F_{lk}(t) = P(T_{k,1} \le t | X_0 = l)$$

is the empirical c.d.f.

$$\hat{F}_{lk}^{(n)}(t) = \frac{1}{M_n} \sum_{j=1}^{M_n} I(Y_j \le t).$$

The naive bootstrap will consist of fixing $Y_1, Y_2, \cdots, Y_{M_n}$ and drawing i.i.d. samples $Y_1^*, Y_2^*, \cdots, Y_{N_n}^*$ distributed as $\hat{F}_{lk}(\cdot)$ and defining

$$\tilde{F}_{lk}^{(n)}(t) \equiv \frac{1}{N_n} \sum_{j=1}^{N_n} I(Y_j^* \leq t).$$

For each fixed t, $\tilde{F}_{lk}^{(n)}(t) - \hat{F}_{lk}^{(n)}(t) \to 0$ in probability as $N_n \to \infty$ and w.p.1. if $\sum 1/N_n^2 < \infty$. Since $\hat{F}_{lk}^{(n)}(t) \to F_{lk}(t)$ w.p.1 (by the strong law) it is clear that $\tilde{F}_{lk}^{(n)}(\cdot)$ and $F_{lk}(\cdot)$ are close as $n \to \infty$.

This method is yet to be investigated fully and will not be treated further in this paper.

In order to justify the bootstrap Method I, the following weak law and central limit theorem for a double array of Markov chains was developed by Athreya and Fuh (1989).

Let $X_n = \{X_{nt};\ t \geq 0\}$ $n = 1, 2, \cdots$ be a sequence of Markov chains on a countable state space S with transition probability matrices $P_n = (p_{nij})$. Let $\pi_n = (\pi_{ni})$ be a stationary probability distribution for P_n. Suppose $\{X_{nt};\ t = 0, 1, 2, \cdots, N_n\}$ is a realization of the nth chain observed up to N_n steps. Let

$$\tilde{\pi}_n(i) = \frac{m_i^{(n)}}{N_n},$$

where

$$m_i^{(n)} \equiv \sum_{t=0}^{N_n} I(X_{nt} = i),$$

is the number of visits to i by X_n during $\{0, 1, \cdots, N_n\}$.

The following is a weak law for the *empirical distribution* $\tilde{\pi}_n(\cdot)$.

THEOREM 6. *Assume that for each n, there exists a probability distribution $\{\pi_{ni}, i \in S\}$ such that for each $\rho > 0$ and for fixed l, i, as $n, N_n \to \infty$, we have*

$$\delta_n(\rho) \equiv \sup_{m \geq \rho N_n} |\frac{1}{m} \sum_{s=1}^{m} p_{nli}^{(s)} - \pi_{ni}| \longrightarrow 0. \tag{2}$$

Then

$$\tilde{\pi}_n(i) - \pi_{ni} \longrightarrow 0 \text{ in probability.}$$

The hypothesis of Theorem 1, namely that $\delta_n(\rho) \to 0$ as $n \to \infty$ for each $0 < \rho < 1$ relates the length of observation N_n to the rate at which the nth chain approaches (in Cesaro mean) its stationary distribution. If P_n converges to a P that is nice, then it is reasonable to expect that the hypothesis of Theorem 6 holds. In fact, we have the following

THEOREM 7. *Let $P_n \equiv (p_{nij})$ and $P \equiv (p_{ij})$ be stochastic matrices such that*

i) *P is ergodic,*
ii) *$p_{nij} \longrightarrow p_{ij}$ for all i, j,*
iii) *$\pi_{ni} \longrightarrow \pi_i$ for all i,*

where π_n and π are stationary probabilities for P_n and P respectively. Then (2) holds and hence the weak law also holds, and for each i,

$$\tilde{\pi}_n(i) \longrightarrow \pi_i \text{ in probability.}$$

The weak law of Theorem 1 suggests the possibility of a central limit theorem for $\tilde{\pi}_n(\cdot)$. This turns out to be somewhat intractable and Athreya and Fuh (1989) address instead a related question motivated by the bootstrapping of P. Let $\tilde{p}_n(i,j)$ denotes the proportion of (i,j) transition to the number of visits to i in $\{X_{nt};\ 0 \leq t \leq N_n\}$. That is, let

$$m_i^{(n)} \equiv \sum_{t=1}^{N_n} I(X_{nt} = i),$$

$$m_{ij}^{(n)} \equiv \sum_{t=0}^{N_n - 1} I(X_{nt} = i) I(X_{n(t+1)} = j),$$

and

$$\tilde{p}_n(i,j) = \begin{cases} \frac{m_{ij}^{(n)}}{m_i^{(n)}}, & \text{if } m_i^{(n)} > 0; \\ \delta_{ij}, & \text{otherwise.} \end{cases}$$

Since $\tilde{\pi}_n(i) - \pi_i \to 0$ in probability, $\pi_i > 0$ and $N_n \to \infty$, we have that $m_i^{(n)} \to \infty$ in probability and hence

$$\tilde{p}_n(i,j) - p_{nij} \longrightarrow 0 \text{ in probability.}$$

The following is a central limit theorem for $(\tilde{P}_n - P_n)$.

THEOREM 8. *Let $P_n \equiv (p_{nij})$ and $P \equiv (p_{ij})$ be stochastic matrices such that*

i) *$p_{nij} \longrightarrow p_{ij}$ for all i, j,*
ii) *$\pi_{ni} \longrightarrow \pi_i$ for all i,*
iii) *$\pi_i > 0$, for all i and (2) holds.*

Then

$$\sqrt{N_n}(\tilde{P}_n - P_n) \longrightarrow N(\underset{\sim}{0}, \Sigma_P) \text{ in distribution,}$$

as $n \to \infty$ *and* $N_n \to \infty$ *independently of each other, where* Σ_P *is the same variance-covariance matrix in* Theorem 2.

Here the convergence in distribution means that for any finite set A of pairs (i,j),

$$\left\{\sqrt{N_n}\,(\tilde{p}_n(i,j) - p_{nij});\ (i,j) \in A\right\} \longrightarrow N\left(\underset{\sim}{0}, (\Sigma_P)_A\right),$$

where $(\Sigma_P)_A$ is a block diagonal matrix involving the states in A. In particular, if $A = \{1, 2, \cdots, k\}$, then

$$(\Sigma_P)_A = \begin{pmatrix} \frac{1}{\pi_1}\Gamma_1(P) & \underset{\sim}{0} & \cdots & \underset{\sim}{0} \\ \underset{\sim}{0} & \frac{1}{\pi_2}\Gamma_2(P) & \cdots & \underset{\sim}{0} \\ \vdots & \vdots & \ddots & \vdots \\ \underset{\sim}{0} & \underset{\sim}{0} & \cdots & \frac{1}{\pi_k}\Gamma_k(P) \end{pmatrix}$$

and $\Gamma_i(P) = (p_{ij}(\delta_{jl} - p_{il})), \quad i, j, l = 1, 2, \cdots, k.$

REMARK 3: By Theorem 7, condition *iii)* in the above theorem can be replaced by P being irreducible, positive recurrent and aperiodic.

Let $X \equiv \{X_t;\ t \geq 0\}$ be a homogeneous ergodic Markov chain with a countably infinite state space $S = \{1, 2, \cdots\}$ and transition probability matrix $P = (p_{ij})$. Let $\pi = (\pi_1, \pi_2, \cdots,)$ such that $\pi_j > 0$, $\sum_{j=1}^{\infty} \pi_j = 1$ and $\pi_j = \sum_{\alpha} \pi_\alpha p_{\alpha j}$, $j = 1, 2, \cdots$. It is known that the nth step transition probability $p_{ij}^{(n)} \mapsto \pi_j$, as $n \to \infty$ for all $i \in S$.

We now show that the Method I works for estimating the distribution of $(\hat{P}_n - P)$ when $\hat{P}_n$ is the maximum likelihood estimator. It is clear that a finite number of observations will not provide an estimator for all states of the transition probabilities.

Nevertheless, the *a.s.* convergence of $\hat{\pi}_n$ and $\hat{P}_n$ can be established using the recurrence of $\{X_n\}$ and the strong law of large numbers. The following are extensions of Theorem 1 and 2 to the countable state space case.

PROPOSITION 1. *Let* $X = \{X_n;\ n \geq 0\}$ *be a homogeneous irreducible recurrent Markov chain. Then for any initial distribution and any* (i,j), *we have*

$$\hat{p}_n(i,j) \longrightarrow p_{ij} \text{ with probability 1} \quad \text{as } n \to \infty.$$

Suppose in addition, the chain is also positive recurrent. Then, for any initial distribution and state i, *we have*

$$\hat{\pi}_n(i) \longmapsto \pi_i \quad \text{with probability 1} \quad \text{as } n \to \infty.$$

This suggests using $\hat{P}_n$ and $\hat{\pi}_n$ as estimators of P and π respectively and in order to obtain confidence intervals, we need to look at the distributions of $(\hat{P}_n - P)$ and $(\hat{\pi}_n - \pi)$.

PROPOSITION 2. (Derman (1956))
Let $X \equiv \{X_n;\ n \geq 0\}$ be a homogeneous irreducible, positive recurrent Markov chain. Then

$$\sqrt{n}(\hat{P}_n - P) \longrightarrow N(\underset{\sim}{0}, \Sigma_P) \quad \textit{in distribution,}$$

where Σ_P is the variance-covariance matrix.

The convergence is in the same sense as in Theorem 8 and so is Σ_P. It should be noted that Proposition 2 is a special case of Theorem 8.

PROPOSITION 3. (Derman (1956))
Let $X \equiv \{X_n;\ n \geq 0\}$ be a homogeneous ergodic Markov chain with $E_k T_k^2 < \infty$ for some k. Then for any initial distribution, we have

$$\sqrt{n}(\hat{\pi}_n - \pi) \longrightarrow N(\underset{\sim}{0}, \Sigma_P^*) \quad \textit{in distribution,}$$

where Σ_P^* is the variance-covariance matrix.

See Derman (1956) for an explicit form of Σ_P^*. Here again the convergence has the same meaning as Theorem 8.

In order to obtain confidence intervals for P and π, one can use Propositions 3 and 4, but use $\Sigma_{\hat{P}_n}$ and $\Sigma_{\hat{P}_n}^*$ in place of Σ_P and Σ_P^* respectively.

An alternate approach to finding confidence intervals is to use the method of bootstrap. Here for the pivotal quantity

$$\sqrt{n}(\hat{P}_n - P),$$

the bootstrap version is

$$\sqrt{N_n}(\tilde{P}_n - \hat{P}_n).$$

The following theorem states that for almost all realizations of $\{X_n\}$, the conditional distribution of $\sqrt{N_n}(\tilde{P}_n - \hat{P}_n)$ given $\underset{\sim}{x}$ converges in distribution to $N(\underset{\sim}{0}, \Sigma_P)$ as $n \to \infty$. The proof follows from Theorem 8 as $\hat{P}_n$ is consistent for P. The reader is referred to Athreya and Fuh (1989) for the details.

THEOREM 9. *Under the notations given above, we have for almost all realizations of $\{X_n;\ n \geq 0\}$,*

$$\sqrt{N_n}(\tilde{P}_n - \hat{P}_n) \longrightarrow N(\underset{\sim}{0}, \Sigma_P) \quad \textit{in distribution,}$$

as $n \to \infty$ and $N_n \to \infty$ independently, where Σ_P is the same variance-covariance matrix in Proposition 3.

REMARK 4: Note that in the above the only properties of m.l.e. $\hat{P}_n$ that we used were the consistency of $\widehat{P}_n$ for P and $\hat{\pi}_n$ for π. So the bootstrap will work with any other estimator of P enjoying these properties. For the finite state space, irreducible, aperiodic Markov chain, Theorem 9 is deducible from the methods of Kulperger and Rao (1989), who use different techniques and also from Basawa et al (1989). Theorem 9 in its present form is in Fuh's thesis (1989).

We now proceed to justify Bootstrap Method II.

Recall that Method II is based on k full cycles. Our goal is to estimate the distribution of $\sqrt{k}(\hat{\pi}_k(i) - \pi_i)$ and $\sqrt{k}(\hat{p}_k(i,j) - p_{ij})$.

The strong consistency property of the bootstrap estimators $\tilde{\pi}_k(i)$ and $\tilde{p}_k(i,j)$ is given by the following theorem.

THEOREM 10. (Athreya and Fuh (1989))
If there exists $\delta > 0$ such that $E_\Delta T_\Delta^{1+\delta} < \infty$, then

$$\tilde{\pi}_k(i) \longrightarrow \pi_i \quad \text{and} \quad \tilde{p}_k(i,j) \longrightarrow p_{ij} \quad \text{w.p.1.} \quad \text{as } k \to \infty.$$

It is known that if $E_\Delta T_\Delta^2 < \infty$, then the distribution of

$$\sqrt{k}\,(\hat{\pi}_k(i) - \pi_i) \text{ is asymptotically normal,}$$

and the bootstrap estimator of this distribution is the conditional distribution of $\sqrt{k}(\tilde{\pi}_k(i) - \hat{\pi}_k(i))$ given $(\eta_1, \cdots, \eta_k)$.

Method II works well to estimate the distribution of $\sqrt{k}(\hat{\pi}_k(i) - \pi_i)$ for an ergodic Markov chain with infinite state space. The statement is as follows. The reader is referred to Athreya and Fuh (1989) for the proof.

THEOREM 11. *With the notations given above, if $E_\Delta T_\Delta^2 < \infty$, then, for almost all realizations of $\{X_n;\ n \geq 0\}$, we have for each i*

$$\sqrt{k}(\tilde{\pi}_k(i) - \hat{\pi}_k(i)) \longrightarrow N(0, \sigma_i^2) \quad \text{in distribution} \quad \text{as } k \to \infty,$$

where $0 < \sigma_i^2 < \infty$.

For any real number $l_1, l_2, \cdots, l_r$,

$$\sqrt{k} \sum_{i=1}^{r} l_i\,(\tilde{\pi}_k(i) - \hat{\pi}_k(i))$$

can be written as

$$\sqrt{k}\left(\frac{\sum_{t=1}^{k} f(\eta_t^*)}{\sum_{t=1}^{k} T_t^*} - \frac{\sum_{t=1}^{k} f(\eta_t)}{\sum_{t=1}^{k} T_t}\right)$$

where for any cycle η, $f(\eta) = \sum_{i=1}^{r} l_i g_i(\eta)$, and $g_i(\eta)$ is the number of visits to i. Since

$$|f(\eta)| \leq (\sum_{1}^{r} |l_i|) T(\eta)$$

the hypothesis $E_\Delta T_\Delta^2 < \infty$ implies $E_\Delta (f(\eta))^2 < \infty$ yielding the following extension of the above theorem.

THEOREM 12. *With the above notations, if $E_\Delta T_\Delta^2 < \infty$, then for any finite subset A of the state space S, and for almost all realizations of $\{X_n;\ n \geq 0\}$,*

$$\left\{\sqrt{k}\,(\tilde{\pi}_k(i) - \hat{\pi}_k(i))\,;\ i \in A\right\} \longrightarrow N(\underset{\sim}{0}, \Sigma) \quad \text{in distribution} \quad \text{as } k \to \infty,$$

where Σ is the variance covariance matrix whose $(i,j)^{th}$ element is

$$\sigma_{ij} = a^2 Cov\,(g_i(\eta), g_j(\eta)) + a^4 V\,(T(\eta))\, Eg_i(\eta) Eg_j(\eta) \\ - 2a^3 Eg_i(\eta) Cov\,(T(\eta), g_j(\eta))\,,$$

where for any cycle η, $g_i(\eta)$ is the number of visits to i during η, and $T(\eta)$ is its length.

Turning now to $\tilde{p}_k(i,j) - \hat{p}_k(i,j)$, it is easily proved that mere positive recurrence (*i.e.* $E_\Delta T_\Delta < \infty$) is enough to yield the asymptotic normality of $\sqrt{k}(\hat{p}_k(i,j) - p_{ij})$). However, for the conditional distribution of $\sqrt{k}(\tilde{p}_k(i,j) - \hat{p}_k(i,j))$ to converge to normal for almost all realizations of $\{X_n;\ n \geq 0\}$, we need $E_\Delta T_\Delta^2 < \infty$.

THEOREM 13. *With the notations given above, if $E_\Delta T_\Delta^2 < \infty$, then, for almost all realizations of the process $\{X_n;\ n \geq 0\}$, we have for each i, j,*

$$\sqrt{k}(\tilde{p}_k(i,j) - \hat{p}_k(i,j)) \longrightarrow N(0, \sigma^2) \quad \text{in distribution,}$$

where $0 < \sigma^2 < \infty$.

With the same argument as in Theorem 12, we have the following extension of Theorem 13.

THEOREM 14. *With the above notations, if $E_\Delta T_\Delta^2 < \infty$, then for any finite subset A of $S \times S$, and for almost all realizations of $\{X_n;\ n \geq 0\}$*

$$\left\{\sqrt{k}\,(\tilde{p}_k(i,j) - \hat{p}_k(i,j))\,;\ (i,j) \in A\right\} \longmapsto N(\underset{\sim}{0}, \Sigma^*) \quad \text{in distribution,}$$

where Σ^* is the variance covariance matrix with the $(i,j),(i',j')^{th}$ element is

$$\begin{aligned}
&\sigma_{(i,j),(i',j')} \\
&= Cov\left(\frac{h_{ij}(\eta)}{Eg_i(\eta)} - \frac{Eh_{ij}(\eta)}{(Eg_i(\eta))^2}g_i(\eta), \frac{h_{i'j'}(\eta)}{Eg_{i'}(\eta)} - \frac{Eh_{i'j'}(\eta)}{(Eg_{i'}(\eta))^2}g_{i'}(\eta)\right) \\
&= \frac{1}{Eg_i(\eta)Eg_{i'}(\eta)}Cov\left(h_{ij}(\eta), h_{i'j'}(\eta)\right) \\
&+ \frac{Eh_{ij}(\eta)E_{i'j'}(\eta)}{(Eg_i(\eta)Eg_{i'}(\eta))^2}Cov\left(g_i(\eta), g_{i'}(\eta)\right) \\
&- 2\frac{Eh_{i'j'}(\eta)}{Eg_i(\eta)(Eg_{i'}(\eta))^2}Cov\left(h_{ij}(\eta), g_{i'}(\eta)\right) \\
&- 2\frac{Eh_{ij}(\eta)}{(Eg_i(\eta))^2Eg_{i'}(\eta)}Cov\left(h_{i'j'}(\eta), g_i(\eta)\right),
\end{aligned}$$

where $h_{ij}(\eta)$ is the number of ij transitions during the cycle η.

REMARK 5: If the bootstrap resample size k' is different from k, Theorem 13 still holds, as long as both k and k' go to ∞. The rate of growth is irrelevant, but the finite second moment for the hitting time T_Δ is crucial. (see Athreya (1987) for the details.)

REMARK 6: It was suggested earlier that a variation of method II is to fix the length n of observations and resample from the k_n (random) full cycles in $\{X_0, X_2, \cdots, X_n\}$, and that a natural choice for Δ is the state that yields the largest number of cycles in $\{X_0, X_1, \cdots, X_n\}$. This will make the cycles $\{\eta_1, \eta_2, \cdots, \eta_{k_n}\}$ dependent and hence the use of Proposition 6 in justifying the CLT for $\sqrt{k_n}(\tilde{\pi}_{k_n}(i) - \hat{\pi}_{k_n}(i))$ would not be valid. However, an examination of Singh's (1981) or Athreya's (1987) proof shows that even here the conditional Lindeberg's condition holds for almost all realizations of $\{X_t;\ t \geq 0\}$ and hence the CLT does hold for the bootstrap. Also by Renyi's CLT for random sums $\sqrt{k_n}(\hat{\pi}_{k_n}(i) - \pi_i)$ obeys a CLT. Thus this modified bootstrap also works. Similar remarks apply for bootstrapping $(\hat{P}_n - P)$.

Accuracy of the bootstrap in the finite state space case

Basawa et al (1989) have also considered the finite state space case. They propose the following (conditional) bootstrap method. Recall the definitions of n_i and n_{ij} from (1). Then, generate independent random variables W^*_{it}, $t = 1, 2, \cdots$, $i = 1, 2, \cdots$, such that

$$P\left(W^*_{it} = j \mid \underset{\sim}{x}\right) = \hat{p}_{ij}, \qquad 1 \le i, j \le k,\ t = 1, 2, \cdots .$$

Define

$$n^*_{ij} = \sum_{t=1}^{n_i} I\left(W^*_{it} = j\right),$$

and

$$\hat{p}^*_{ij} = \frac{n^*_{ij}}{n_i}, \qquad 1 \le i, j \le k.$$

Then, the bootstrap distribution H^*_n of

$$\left\{\sqrt{n}\left(\hat{p}^*_{ij} - \hat{p}_{ij}\right); 1 \le i, j \le k\right\}$$

is suggested as approximation to H_n, the distribution of

$$\left\{\sqrt{n}\left(\hat{p}_{ij} - p_{ij}\right); 1 \le i, j \le k\right\}.$$

Basawa et al show (using the fact that multinomial goes to multivariate normal) that H^*_n converges w.p.1 to the same multivariate normal as the limit of H_n (as claimed in Theorem 2), thus showing that this bootstrap method is consistent.

In a recent paper Datta and McCormick (1990) have investigated the accuracy of the above method and have established the following theorems.

THEOREM 15. *Fix i, j. Let $0 < p_{ij} < 1$. Then*

(a) $\sup_x \left|P\left(\sqrt{n_i}(\hat{p}_{ij} - p_{ij}) \le x\right) - P^*(\sqrt{n_i}(\hat{p}^*_{ij} - \hat{p}_{ij}) \le x)\right|$
$= O\left(\sqrt{\frac{\log n}{n}}\right)$,

(b) $\sup_x \left|P\left(\frac{n_i}{\sqrt{n}}(\hat{p}_{ij} - p_{ij}) \le x\right) - P^*\left(\frac{n_i}{\sqrt{n}}(\hat{p}^*_{ij} - \hat{p}_{ij}) \le x\right)\right|$
$= O\left(\sqrt{\frac{\log\log n}{n}}\right)$,

(c) $\sup_x \left|P\left(\frac{n^{-1/2} n_i(\hat{p}_{ij} - p_{ij})}{(\pi_i p_{ij}(1-p_{ij}))^{1/2}} \le x\right) - P^*\left(\frac{n^{-1/2} n_i(\hat{p}^*_{ij} - \hat{p}_{ij})}{(\hat{\pi}_i \hat{p}_{ij}(1-\hat{p}_{ij}))^{1/2}} \le x\right)\right|$
$= O\left(\frac{1}{\sqrt{n}}\right)$.

They also obtain an Edgeworth type expansion for the case when p_{ij} is irrational.

The asymptotic accuracy of the method proposed by Basawa et al is almost of order $O(n^{-1/2})$. Datta and McCormick (1990) propose three modified bootstrap schemes and obtain Edgeworth expansions for each of them. See Datta and McCormick (1990) for details.

Some Open Problems

The following are further research topics related to this paper.

1. The investigation of the accuracy of the bootstrap method proposed by Kulperger and Prakasa Rao (1989) for the finite state space and those proposed by Athreya and Fuh (1989) for the countable state Markov chain.
2. Doing numerical simulation for the comparison of the bootstrap methods proposed in the above papers, with a view to determine resample sizes k, N_n etc..
3. Bootstrapping the transition kernel $P(\cdot,\cdot)$ of an ergodic Harris chain, is an extension of bootstrapping the transition probability matrix of an ergodic Markov chain with discrete state space. The histogram estimator $\hat{P}_n(\cdot,\cdot)$ of $P(\cdot,\cdot)$ has asymptotic normality under Doeblin's condition and regularity hypotheses. An interesting problem here is to find the asymptotic behavior of the bootstrap estimator as well as to investigate more general chains that are Harris recurrent.

Acknowledgement: We would like to thank Raoul LePage for inviting us to contribute this paper to this Proceedings. We also would like to thank S. Datta for making his paper with McCormick available to us.

References

[1] K. B. Athreya. *Bootstrap of the mean in the infinite variance case.* Ann. Statist. **15** (1987): 724-731.

[2] K. B. Athreya and C. D. Fuh. *Bootstrapping Markov chains: countable case.* Technical Report: B-89-7, Institute of Statistical Science, Academia Sinica, Taipei, Taiwan, ROC, 1989.

[3] I. Basawa, T. Green, W. McCormick and R. Taylor. *Asymptotic bootstrap validity for finite Markov chains.* Submitted, 1989.

[4] P. Billingsley. *Statistical Inference for Markov Processes.* The University of Chicago Press, Chicago, 1961.

[5] S. Datta and W. P. McCormick. *Bootstrap for a finite Markov chain based on i.i.d. sampling.* Technical Report 139, University of Georgia, Athens, GA 30602, 1990.

[6] C. Derman. *Some asymptotic distribution theory for Markov chains with a denumerable number of states.* Biometrika **43** (1956): 285-294.

[7] B. Efron. *Bootstrap method: another look at the jackknife.* Ann. Statist. **7** (1979): 1-26.

[8] C. D. Fuh. *The bootstrap method for Markov chains.* Ph. D. dissertation, Iowa State University, 1989.

[9] R. J. Kulperger and B. L. S. Prakasa Rao. *Bootstrapping a finite state Markov chain.* University of Western Ontario, Statistics Dept. Preprint (1987). To appear in Sankhya.

Theoretical Comparison of Different Bootstrap t Confidence Bounds

By

P.J. Bickel
Univ. of California, Berkeley

Summary. We compare in a formal way the behaviour to second order of bootstrap confidence bounds for a parameter θ based on t statistics. We investigate the effect of:

1) Varying the estimate of scale in the denominator
2) Varying the estimate $\hat{\theta}$ of θ used in the numerator
3) Varying the bootstrap method, parametric or nonparametric in terms of
   a) Equivalence of the resulting procedures
   b) Correctness of the probability of coverage
   c) Minimization of the amount of undershoot
   d) Robustness to failure of parametric assumptions

Key Words: (AMS 1980 Classification) Bootstrap, Confidence bounds, Second order properties.

Department of Statistics, University of California, Berkeley, California 94720

Exploring the Limits of Bootstrap. Edited by Raoul LePage and Lynne Billard.
 ISBN: 0471-53631-8.

1. *Introduction.* Recent years have seen the development of a large number of approximate confidence bounds for a population parameter $\theta(F)$ on the basis of an i.i.d. sample $X_1, \ldots, X_n$ from a population F. These bounds, all based on resampling ideas, include the original Efron (1982) percentile method and BC (1985) and BCA (1987) modifications as well as the bootstrap t discussed extensively by Hall (1988), Beran's (1987) prepivoting approach as well as many others — see Hall (1988) and diCiccio and Romano (1988) for recent surveys.

Methods leading to bounds $\underline{\theta}_{n1}$, $\underline{\theta}_{n2}$ respectively have been considered equivalent to first order if,

$$\underline{\theta}_{n1} - \underline{\theta}_{n2} = O_p(n^{-1}) \tag{1.1}$$

and to second order if

$$\underline{\theta}_{n1} - \theta_{n2} = O_p(n^{-3/2}). \tag{1.2}$$

The principal criterion for a good method has been correctness of the coverage probability,

$$P[\underline{\theta}_n \leq \theta(F)] = 1 - \alpha + O(n^{-1/2}) \text{ (1st order)} \tag{1.3}$$

$$P[\underline{\theta}_n \leq \theta(F)] = 1 - \alpha + O(n^{-1}) \text{ (2nd order)}. \tag{1.4}$$

It is plausible and under assumptions explored in the literature true that bounds equivalent to a given order have coverage probabilities correct to the same order.

Hall (1988), (see also my discussion), diCiccio and Romano (1988), Beran (1987), diCiccio and Efron (1990) have shown that the various methods of construction of bootstrap confidence bounds are equivalent to first or second order to bootstrap t methods.

In this note we limit ourselves to different bootstrap t methods and investigate the effect of:

1) Varying the estimate of scale in the denominator

2) Varying the estimate $\hat{\theta}$ of θ used in the numerator

3) Varying the bootstrap method (nonparametric or parametric)

in terms of,

a) Equivalence of the resulting procedures

b) Correctness of probability of coverage and two aspects which have not been studied as extensively,

c) Minimization of the amount of undershoot

d) Robustness to failure of parametric assumption.

Our observations in c) are a simple application of the results of Pfanzagl (1981, 1985) — see also Bickel, Cibisov, van Zwet (1981) that first order efficiency implies second order efficiency.

The robustness remarks are trivial but we believe worth noting — see also Parr (1985).

2. *Second order correctness and equivalence.*

As we indicated our observations are $X_1, \ldots, X_n$ i.i.d. F, with empirical distribution which we denote by F_n. We are interested in a parameter θ: $\mathbf{F} \to R$ where $\mathbf{F}$ is a nonparametric family of distributions containing all distributions with finite support, for example $\mathbf{F} = \{$all distributions$\}$ or $\{$all distributions with finite variance$\}$. We consider the natural estimate $\hat{\theta} \equiv \theta(F_n)$. We assume as a model $\mathbf{F}_0 \subset \mathbf{F}$ and we are interested in lower confidence bounds for θ restricted to $\mathbf{F}_0$ which are based on $\hat{\theta}$. The estimate $\hat{\theta}$ may or may not be efficient for estimating θ on $\mathbf{F}_0$ but we can essentially always think of efficient estimates in this way. For instance, if $\mathbf{F}_0$ is parametric with densities $p(\cdot, \theta)$ then (under regularity conditions) $\theta(F)$ formally solving

$$\int \nabla \log p(x, \theta)\, dF(x) = 0$$

where ∇ is the gradient, corresponds to $\hat{\theta}$ = M.L.E. We also assume that $\hat{\theta}$ is asymptotically linear with influence function $\psi(\cdot, F)$ see Hampel (1986). That is,

$$\hat{\theta} = \theta(F) + n^{-1} \Sigma_{i=1}^{n} \psi(X_i, F) + O_P(n^{-1}) \tag{2.1}$$

where

$$\int \psi(x, F)\, dF(x) = 0$$

and

$$\sigma^2(F) \equiv \int \psi^2(x, F)\, dF(x) < \infty,$$

for all $F \in \mathbf{F}$. Thus if $\mathbf{L}(\cdot | F)$ denotes the distribution of a function of $X_1, \ldots, X_n$ under F then

$$\mathbf{L}\left[\frac{\sqrt{n}(\hat{\theta} - \theta(F))}{\sigma(F)} \,|\, F\right] \to \mathbf{N}(0, 1).$$

We further suppose that we are given an asymptotically linear estimate $\hat{\sigma}$ of $\sigma(F)$ on $\mathbf{F}_0$ with influence function r,

$$\hat{\sigma} = \sigma(F) + n^{-1} \Sigma_{i=1}^{n} r(X_i, F) + O_P(n^{-1}). \tag{2.2}$$

for $F \in \mathbf{F}_0$. If,

$$T_n(F_n, F) \equiv \sqrt{n}(\hat{\theta} - \theta(F)) / \hat{\sigma}$$

and we know the exact distribution $\mathbf{L}(T_n(F_n, F) \mid F)$ we are led to the exact $1 - \alpha$ LCB

$$\underline{\theta}_{EX} \equiv \hat{\theta} - \frac{\hat{\sigma}}{\sqrt{n}} c_n(F)$$

where

$$P_F[T_n(F_n, F) \leq c_n(F)] = 1 - \alpha. \tag{2.3}$$

The bootstrap bound(s) corresponding to $\underline{\theta}_{EX}$ are as usual obtained by estimating F by $\tilde{F}_n$ and replacing the, in general, unknown $c_n(F)$ by $c_n(\tilde{F}_n)$ where,

$$P_{\tilde{F}_n}[T_n(F_n^*, \tilde{F}_n) \leq e_n(\tilde{F}_n)] = 1 - \alpha \tag{2.4}$$

and F_n^* is the empirical df. of a sample $X_1^*, \ldots, X_n^*$ i.i.d. $\tilde{F}_n$. That is

$$\underline{\theta}_{BOOT} = \hat{\theta} - \frac{\hat{\sigma}}{\sqrt{n}} c_n(\tilde{F}_n). \tag{2.5}$$

If $\tilde{F}_n = F_n$ we are dealing with the usual nonparametric bootstrap. If $\mathbf{F}_0 = \{F_\gamma\}$ where $\gamma = (\theta, \eta)$ is a Euclidean parameter, and $\hat{\gamma} = (\hat{\theta}, \hat{\eta})$ is the M.L.E. of γ, then $\tilde{F}_n \equiv F_{\hat{\gamma}}$ is the parametric bootstrap. How do different bootstrap bounds compare in terms of second order correctness and equivalence? The following ''theorem'', stated under unspecified regularity conditions gives the answer. Parts A and B of the theorem have been noted by Hall, Beran, and others. We give a heuristic proof and then indicate what kind of regularity conditions are needed.

We let $\underline{\theta}_{BOOT}$, $\underline{\theta}_{EXACT}$ denote the nonparametric bootstrap bounds corresponding to $\tilde{F}_n = F_n$ and the corresponding exact bound. Superscripts 1,2 will indicate different choices of numerator $\hat{\theta}^{(1)}$ or $\hat{\theta}^{(2)}$ while subscripts will similarly correspond to different choices of $\hat{\sigma}^{(1)}$ or $\hat{\sigma}^{(2)}$. $\underline{\theta}_{PBOOT}$ with or without indices will indicate $\tilde{F}_n = F_{(\hat{\theta}, \hat{\eta})}$ when $\mathbf{F} = \{F_\gamma\}$, $\gamma = (\theta, \eta)$.

Theorem: Under suitable regularity conditions, for $F \in \mathbf{F}_0$,

A. $\underline{\theta}_{BOOT}$, $\underline{\theta}_{PBOOT}$ are second order correct.

B. $\underline{\theta}_{BOOT} = \underline{\theta}_{PBOOT} + O_p(n^{-3/2}) = \underline{\theta}_{EXACT} + O_p(n^{-3/2})$

C. If $\hat{\sigma}_1 = \hat{\sigma}_2$ but $\hat{\theta}^{(2)} = \hat{\theta}^{(n)} + \frac{\Delta_n}{n}$ where $\mathbf{L}_F(\Delta_n) \to \mathbf{L}_F(\Delta)$ then

$$\mathbf{L}_F\{n(\underline{\theta}^{(2)}_{BOOT} - \underline{\theta}^{(1)}_{BOOT}\} \;\to\; \mathbf{L}_F(\Delta - d(F)) \tag{2.6}$$

where $d(F)$ is a constant. If Δ is constant, $\Delta = d(F)$. Thus, unless $\hat{\theta}^{(1)}$ and $\hat{\theta}^{(2)}$ differ to second order only in bias i.e. $\hat{\theta}^{(2)} = \hat{\theta}^{(1)} + \frac{\Delta(F)}{n} + O_p(n^{-3/2})$ for Δ a constant, then $\underline{\theta}^{(2)}_{BOOT}$ and $\underline{\theta}^{(1)}_{BOOT}$ are not equivalent to second order.

D. Suppose $\hat{\theta}^{(1)} = \hat{\theta}^{(2)} = \hat{\theta}_{eff}$, an efficient estimate for θ on $\mathbf{F}_0$, but $r_1 \neq r_2$ where r_j correspond to $\hat{\sigma}_j$ via (2.2). Then,

$$\mathbf{L}_F\{n(\underline{\theta}_{BOOT1} - \underline{\theta}_{BOOT2})\} \;\to\; \mathbf{L}_F(\Delta) \;\neq\; 0$$

Comments: A,B: If $F \in \mathbf{F}_0$ there is nothing to choose between the parametric and nonparametric bootstrap t bounds to second order

(C) If $\hat{\theta} \equiv$ MLE, $b_n \equiv E_F(\hat{\theta}) - \theta$ then typically $b_n = \frac{b(F)}{n} + O(n^{-2})$ so that if $\hat{\theta}^{(2)} = \hat{\theta} - \frac{b(F_n)}{n}$, the debiased MLE, we expect (under regularity conditions!) bootstrap t bounds based on $\hat{\theta}$, $\hat{\theta}^{(2)}$ both to be second order correct and second order equivalent even though the parent estimates differ to second order. This continues to hold for estimates $\hat{\theta} + d(F_n)/n$ for d smooth. However, if $\hat{\theta}^{(2)}$ is an efficient estimate produced by an alternative method such as modified minimum χ^2 where we can expect $\mathbf{L}_F(\Delta)$ to be nondegenerate then the bounds are no longer second order equivalent. Admittedly such estimates $\hat{\theta}^{(2)}$ can always be improved to second order by procedures of the form $\hat{\theta} + \frac{d(F_n)}{n}$. For a discussion see Berkson (1980), Pfanzagl (1981)

(D) If $\hat{\theta}$ is the MLE of θ for a parametric model $\mathbf{F}_0$ suppose

$$\hat{\sigma}_1^2 \;\equiv\; \int \psi^2(x, F_n)\, dF_n(x),$$

the nonparametric estimate of $\sigma^2(F_\gamma)$ while

$$\hat{\sigma}_2^2 \;=\; \sigma^2(F_{\hat{\gamma}})$$

where $\hat{\gamma}$ is the MLE of γ. Then $r_1 \neq r_2$ unless $\hat{\sigma}_1^2$ is also efficient for $\mathbf{F}_0$. Thus the bounds are not equivalent to second order. As a consequence we note the following phenomenon. Hall (1988) shows that the Efron parametric and nonparametric BCA bounds are second order equivalent to bootstrap t bounds based on $\frac{\hat{\theta}-\theta}{\hat{\sigma}_i}$, $i = 1, 2$, respectively. We conclude that, in general, the parametric and nonparametric BCA bounds are not second order equivalent despite the equivalence in A. Using the parametric or nonparametric bootstrap for the distribution of $\hat{\theta}$ which is a starting point in this method does make a difference.

Formal proof of theorem:

Our heuristic argument supposes that, for each studentized statistic we consider $T(F_n, F)$ both $\mathbf{L}(T(F_n, F) | F)$ and $\mathbf{L}(T(F_n^*, \tilde{F}_n) | \tilde{F}_n)$ admit (Edgeworth) expansions to order $n^{-3/2}$. That is,

$$P_F[T(F_n, F) \le x] = \Phi(x) - n^{-1/2}\phi(x)A(x, F) + O(n^{-1}) \tag{2.7}$$

where $A(\cdot, F)$ is a polynomial of degree 2 and $O(n^{-1})$ is uniform in x. Similarly, we require,

$$P_{\tilde{F}_n}[T(F_n^*, \tilde{F}_n) \le x] = \Phi(x) - n^{-1/2}\phi(x)A(x, \tilde{F}_n) + O_p(n^{-1}). \tag{2.8}$$

Agreement of $\underline{\theta}_{BOOT}$ and $\underline{\theta}_{EXACT}$ (part B) follows from asymptotic inversion of (2.7), (2.8) and

$$A(\cdot, \tilde{F}_n) = A(\cdot, F) + O_p(n^{-1/2}). \tag{2.9}$$

If we further suppose that we can substitute the random $x = c_n(\tilde{F}_n)$ into (2.7) with $O(n^{-1})$ changing to $O_p(n^{-1})$ then evidently part A follows. We note in passing that the same type of heuristics indicate that all $\underline{\theta}_{BOOT}$ are first order equivalent to each other as well as first order correct.

The heuristics for C and D are based on the following lemma from Bai, Bickel, Olshen (1989).

Lemma: Suppose, for $j = 1, 2$, statistics T_{nj},

1) T_{nj} have Edgeworth expansions to order n^{-1}.

$$P[T_{nj} \le x] = \Phi(x) + n^{-1/2}A_j(x) + O(n^{-1}) \tag{2.10}$$

with $O(n^{-1})$ uniform in x.

2) If $T_{n2} \equiv T_{n1} + \frac{\Delta_n}{\sqrt{n}}$

$$\mathbf{L}(T_{n1}, \Delta_n) \rightarrow \mathbf{L}(U, V) \quad \text{with} \quad E\,|V| < \infty .$$

Then,

$$A_1(x) = A_2(x) \quad \text{for all} \quad x \quad \text{iff} \quad E(V\,|\,U) = 0 .$$

The proof is obtained by considering the Edgeworth expansions of $Ee^{itT_{nj}}$ which are valid under (2.10).

To prove C, D we note first that by A we can equivalently consider $\underline{\theta}^{(1)}_{EXACT}$ and $\underline{\theta}^{(2)}_{EXACT}$.

By (2.7) we expect

$$c_n^{(1)}(F) = c_n^{(2)}(F) + \frac{p(F)}{\sqrt{n}} + O(n^{-1}) .$$

Since

$$\hat{\sigma} = \sigma(F) + O_p(n^{-1/2})$$

we obtain that (2.6) holds with

$$d(F) = \sigma(F) + (F) .$$

If Δ is constant we need only show that

$$r(F) = \frac{\Delta}{\sigma(F)} \tag{2.11}$$

since then,

$$\underline{\theta}^{(2)}_{EXACT} - \underline{\theta}^{(1)}_{EXACT} = \frac{\hat{\sigma}}{n}\left[\frac{\Delta_n}{\hat{\sigma}} - \frac{\Delta}{\sigma(F)}\right]$$

$$= o_p(n^{-1}) .$$

But if,

$$T_{n1} \equiv \sqrt{n}\,\frac{(\hat{\theta}^{(1)} - \theta)}{\hat{\sigma}}, \qquad T_{n2} = \sqrt{n}\,\frac{(\hat{\theta}^{(2)} - \theta)}{\hat{\sigma}} - \frac{\Delta n^{-1/2}}{\sigma(F)}$$

evidently $n^{1/2}(T_{n1} - T_{n2}) = \frac{\Delta_n}{\hat{\sigma}} - \frac{\Delta}{\sigma(F)} = o_p(1)$. The lemma applies with $V = 0$ and (2.11) follows. For D we also need a result from the theory of efficient estimation, (see Pfanzagl (1981), Bickel, Klaassen, Ritov, and Wellner

(1991)) which we again state without explicit regularity conditions. These may however be found in the works cited above.

Proposition: Let $\mathbf{F}_0 = \{F_\gamma : \gamma \in \Gamma\}$, $\Gamma \subset R^k$ be a regular parametric model and $\mu : \mathbf{F}_0 \to R$ given by $\mu(F_\gamma) \equiv q(\gamma)$ where q is smooth. Let $\hat{\mu}_{eff}$ be an efficient (BAN) estimate of μ and $\hat{\mu}$ be another regular, asymptotically linear estimate of μ. That is, for all $F \in \mathbf{F}_0$, in a uniform sense,

$$\hat{\mu} = \mu(F) + n^{-1}\Sigma_{i=1}^n w(X_i, F) + o_p(n^{-1/2}) \tag{2.12}$$

where $\int w(x,F)\,dF(x) = 0$, $\int w^2(x,F)\,dF(x) < \infty$ and the same holds for w_{eff}. Let $\frac{\partial l}{\partial \gamma_j}(X_1, \gamma)$, $j = 1, \ldots, k$ be the score function (derivatives of the loglikelihood of X_1). Then if $F = F_\gamma$, for $j = 1, \ldots, k$,

$$\text{cov}_\gamma(w(X_1, F_\gamma) - w_{eff}(X_1, F_\gamma),\ \frac{\partial l}{\partial \gamma_j}(X_1, \gamma)) = 0. \tag{2.13}$$

That is, $w(X_1, F_\gamma) - w_{eff}(X_1, F_\gamma)$ is orthogonal to the tangent space of $\mathbf{F}_0$ at F_0. This is essentially a consequence of the Hájek-Le Cam convolution theorem. Alternatively it can be viewed as a consequence of the differential geometry of statistical models, see for example Efron (1975). If the linear span of $\{\frac{\partial l}{\partial \gamma_j}(X_1, \gamma)\}$ is interpreted as the tangent space of $\mathbf{F}_0$ at F_γ claim (2.14) is true for semiparametric models as well see Bickel et al (1991). To prove D it is enough to establish that

$$c_{n1}(F) = c_{n2}(F) + O(n^{-1}). \tag{2.14}$$

For then, by A, the same applies to $c_{n1}(F_n)$ and $c_{n2}(F_n)$ and hence,

$$n(\underline{\theta}_{BOOT2} - \underline{\theta}_{BOOT1}) = n^{-1/2} c_{n1}(F)\Sigma_{i=1}^n (r_1(X_i, F) - r_2(X_i, F)) + o_p(1),$$

and D follows. But (2.15) follows from the lemma and proposition. Since $\hat{\theta}$ is efficient, $\psi(X_1, F_\gamma)$ is in the linear span of $\frac{\partial l}{\partial \gamma_j}(X_i, \gamma)$ $j = 1, \ldots, k$. If, without loss of generality we take, $\hat{\sigma}_1 = \hat{\sigma}_{eff}$ then, by the proposition, for all $F_\gamma \in \mathbf{F}_0$,

$$\text{cov}_\gamma(\psi(X_1, F_\gamma),\ r_2(X_1, F) - r_1(X_1, F)) = 0. \tag{2.15}$$

Write,

$$U_n = \sqrt{n}\,\frac{(\hat\theta - \theta)}{\hat\sigma_1}, \qquad V_n = \sqrt{n}\,\frac{(\hat\theta - \theta)}{\hat\sigma_2}.$$

Then,

$$V_n = U_n + U_n \frac{(\hat\sigma_1 - \hat\sigma_2)}{\hat\sigma_2} \tag{2.16}$$

$$= U_n + \frac{\Delta_n}{\sqrt{n}}$$

where

$$\Delta_n = U_n (n^{-1/2} \Sigma_{i=1}^n (r_1(X_i, F) - r_2(X_i, F))\sigma^{-1}(F)) + o_p(1).$$

Evidently, $L_F(U_n, \Delta_n)$ tends to a $\mathbf{L}(Z, WZ)$ where $Z \sim \mathbf{N}(0, \sigma^{-2}(F) \int \psi^2(x, F)\, dF(x))$, W is independent of Z and $\mathbf{N}(0, \sigma^{-2}(F) \int (r_1 - r_2)^2(x, F)\, dF(x))$. Evidently $E(WZ|Z) = 0$ and we can apply the lemma. □

To make these results rigorous we need to justify (2.7), (2.8), (2.9) and substitution of the random $c_n(F_n)$ into (2.7). When the estimates are smooth functions of vector means $n^{-1} \Sigma_{i=1}^n M(X_i)$, $M_{p\times 1}$, the argument is due to Cibisov (1973) and Pfanzagl (1981), see also Hall (1986). In general the idea is:

a) To expand $T(F_n, F)$ in a von Mises or Hoeffding expansion and show that the remainder after three terms can be neglected in the Edgeworth expansion. That is, write for suitable a_j

$$\begin{aligned} T(F_n, F) = \sqrt{n}\{&\textstyle\int a_1(x)\, d(F_n - F)(x) \\ &+ \textstyle\int a_2(x, y)\, d(F_n - F)(x)\, d(F_n - F)(y) \\ &+ \textstyle\int a_3(x, y, z)\, d(F_n - F)(x)\, d(F_n - F)(y)\, d(F_n - F)(z) \\ &+ r_n\} \end{aligned} \tag{2.17}$$

where $P[|r_n| \geq n^{-3/2-\delta}] = O(n^{-1})$ for some $\delta > 0$. This is to be expected since we expect $r_n = O_p(n^{-2})$ Conditions such that the sum of the first three terms has an Edgeworth expansion of the from (2.7) may be gleaned from Bickel, Götze, van Zwet (1989) for example. Of course (2.9) can, in principle, be justified in the same way save that techniques such as those of Singh (1981) and Bickel and Freedman (1980) have to be employed to get by the failure of Cramer's condition due to the discreteness of F_n. Substitution of $c_n(F_n)$ in (2.7) can be justified once we express $T(F_n, F) - c_n(F_n)$ in the form (2.17) by using the inversion of (2.8) for $c_n(F_n)$.

3. *Second order optimality and robustness.*

It is natural to define second order efficiency in terms of undershoot for a lower confidence bound $\underline{\theta}^*$ by: For all $F \in \mathbf{F}_0$,

i) $\underline{\theta}^*$ is second order correct

ii) If $\underline{\theta}$ is second order correct and $\delta_n > 0$, $n^{1/2}\delta_n = O(1)$ then

$$P_F[\underline{\theta}^* \leq \theta(F) - \delta_n] \leq P_F[\underline{\theta} \leq \theta(F) - \delta_n] + o(n^{-1/2}). \tag{3.1}$$

In fact to avoid superefficiency phenomena we essentially have to require second order correctness to hold uniformly on shrinking neighbourhoods of every fixed F and then require (3.1) to similarly hold uniformly. If $o(n^{-1/2})$ is replaced in (3.1) by $o(1)$ then $\underline{\theta}^*$ is first order efficient. It is shown in Pfanzagl (1981) and Bickel, Cibisov, and van Zwet (1981) that first order efficiency implies second order efficiency. But $\underline{\theta}_{BOOT} = \hat{\theta} - \frac{\sigma(F)}{\sqrt{n}} z_{1-\alpha} + O_p(n^{-1})$. If $\hat{\theta}$ is efficient as an estimate it follows that $\underline{\theta}_{BOOT}$ is first order efficient.

We conclude that second order efficiency of bootstrap t confidence bounds depends only on the first order efficiency of the estimate $\hat{\theta}$ defining them and not on the choice of $\hat{\sigma}$ beyond its consistency. If $\hat{\theta}$ is not first order efficient then $\underline{\theta}_{BOOT}$ is not first order and a fortiori not second order efficient.

Robustness: Suppose $\theta(F)$ is the parameter we wish to estimate for all $F \in \mathbf{F}$. If $\mathbf{F}_0 = \{F_\gamma : \gamma \in \Gamma\}$, $\Gamma \subset R^k$ we may wish to estimate $\sigma(F_\gamma)$ by $\sigma(F_{\hat{\gamma}})$ in forming $\underline{\theta}_{BOOT}$. If $F \notin \mathbf{F}_0$ and $\hat{\gamma} = \gamma(F) + o_p(1)$ we expect $\sigma(F_{\hat{\gamma}}) = \sigma(F_{\gamma(F)}) + o_p(1)$. Unless $F_{\gamma(F)} = F$, $\underline{\theta}_{BOOT}$ will not even be first order correct. This does not happen if we use $\hat{\sigma}^2 = \int \psi^2(x, F_n)\, dF_n(x)$ and the nonparametric bootstrap.

On the other hand if we use $\hat{\sigma}^2 = \int \psi^2(x, F_n)\, dF_n(x)$ but use the parametric bootstrap, when $F \notin \mathbf{F}_0$, we are, in general, first order correct. The reason is that,

$$1 - \alpha = P_{\tilde{F}_n}\left[\frac{\sqrt{n}(\hat{\theta}^* - \hat{\theta})}{\hat{\sigma}^*} \leq c_n(\tilde{F}_n)\right] = P_{F_{\gamma(F)}}\left[\frac{\sqrt{n}(\hat{\theta} - \theta)}{\sigma(F_{\gamma(F)})} \leq z_{1-\alpha}\right] + o_p(1).$$

But $\underline{\theta}_{PBOOT}$ is not second order correct since that depends on $A(\cdot, F_{\gamma(F)}) = A(\cdot, F)$, which, in general, is false. Thus, robustness considerations strongly mandate a robust estimate of variance and more weakly mandate use of the nonparametric bootstrap.

References

Bai, C., Bickel, P.J. and Olshen, R. (1989). The bootstrap for prediction. Proceedings of an Oberwolfach Conference, Springer-Verlag.

Beran, R. (1987). Prepivoting to reduce level error of confidence sets. *Biometrika* **74**, 457-468.

Berkson, J. (1980). Minimum chi-square not maximum likelihood (with discussion). *Ann. Statist.* **8**, 457-487.

Bickel, P.J. (1974). Edgeworth expansion in nonparametric statistics. *Ann. Statist.* **2**, 1-20.

Bickel, P.J. and Freedman, D.A. (1981). Some asymptotics on the bootstrap. *Ann. Statist.* **9**, 1196-1217.

Bickel, P.J., Chibisov, D.M. and van Zwet W.R. (1981). On efficiency of first and second order. *International Statistical Review* **49**, 169-175.

Bickel, P.J., Götze, F., and van Zwet, W.R. (1983). A simple analysis of third-order efficiency of estimates. Proceedings of the Berkeley Conference in Honor of Jerzy Neyman and Jack Kiefer. Wadsworth. Belmont.

Bickel, P.J., Götze, F. and van Zwet, W.R. (1989). The Edgeworth expansion for U statistics of degree two. *Ann. Statist.* **14**, 1463-1484.

Chibisov, D.M. (1973). An asymptotic expansion for a class of estimators containing maximum likelihood estimators. *Theory Probab. Appl.* **18**, 295-303.

diCiccio, T. and Romano, J. (1988). A review of bootstrap confidence intervals. *JRSS* **B 50**, 338-355.

diCiccio, T. and Efron, B. (1990). Better approxiamte confidence intervals in exponential families. *Tech Report, Stanford University.*

Efron, B. (1975). Defining the curvature of a statistical problem (with applications to second order efficiency). *Ann. Statist.* **3**, 1189-1242.

Efron, B. (1982). *The Jackknife, the Bootstrap and Other Resampling Plans.* SIAM. Philadelphia.

Efron, B. (1985). Bootstrap confidence intervals for a class of parametric problems. *Biometrika* **72**, 45-58.

Efron, B. (1987). Better bootstrap confidence intervals (with discussion). *J. Amer. Statist. Assoc.* **82**, 171-200.

Hall, P.J. (1986). On the bootstrap and confidence intervals. *Ann. Statist.* **14**, 1431-1452.

Hall, P.J. (1988). Theoretical comparison of bootstrap confidence intervals (with discussion). *Ann. Statist.* **16**, 927-985.

Hampel, F., Renchotti, E., Rousseuw, P., Stahel, W. (1986). Robust statistics: the approach based on influence functions. J. Wiley. New York.

Singh, K. (1981). On the asymptotic accuracy of Efron's bootstrap. *Ann. Statist.* **9**, 1187-1195.

Parr, W. (1985). The bootstrap: Some large sample theory and connections with robustness. *Stat. Prob. Letters* **3**, 97-100.

Pfanzagl, J. (1981). Contributions to a general asymptotic statistical theory. *Lecture Notes in Statistics* **13**, Springer Verlag.

Pfanzagl, J. (1985). Asymptotic expansions for general statistical models. *Lecture Notes in Statistics* **31**, Springer Verlag.

BOOTSTRAP FOR A FINITE STATE MARKOV CHAIN BASED ON I.I.D. RESAMPLING

Somnath Datta and William P. McCormick
University of Georgia

ABSTRACT

This paper investigates the scope of bootstrap schemes based on i.i.d. resampling for estimating the sampling distribution of the m.l.e. $\hat{p}_{ij}$ *of a transition probability* p_{ij} *of a finite state Markov chain.*

The asymptotic accuracy of a bootstrap method proposed by Basawa et al. is studied. It is shown that the best rate possible with this method is $O(n^{-1/2})$, *where* n *is the sample size. Three modified bootstrap schemes are proposed for the above problem. It is shown that an Edgeworth correction is possible with each of these new methods when estimating the sampling distribution of standardized* $\hat{p}_{ij}$, *if* p_{ij} *is irrational.*

1. Introduction

A number of bootstrap methods for estimating the sampling distribution of hitting time, transition counts and proportions of a Markov chain have recently been proposed by various authors. Basawa et al. (1989) and Kulperger and Prakasa Rao (1989) considered Markov chains with finite state spaces; the countable state space case has been investigated by Athreya and Fuh (1989).

Whereas the asymptotic validity of these methods has been established by the respective authors, nothing is known about their rates of approximation. In this paper, we study the asymptotic rates of a method proposed by Basawa et al. (1989). This method is easy to implement in practice because, given the original data, the bootstrap distribution is constructed using i.i.d. resampling.

We show that the best rate possible with this method is $O(n^{-1/2})$, where n is the sample size, which is the same as that of the classical normal approximation. This method is referred to as "conditional bootstrap" by the previous authors. The main reason it fails to be any better is that this approximation is based on a rather naive i.i.d. sampling which cannot account for a part of the skewness term arising from the dependent structure in the original sample.

Department of Statistics, Univ. of Georgia, Athens, GA 30605, USA.
The first author's research was partially supported by a grant from the Univ. of Georgia.

Exploring the Limits of Bootstrap. Edited by Raoul LePage and Lynne Billard.
 ISBN: 0471-53631-8.

In this paper we propose several new methods to overcome the above problem. These new bootstrap schemes are also based on i.i.d. sampling and therefore easy to use; yet they achieve a rate of $o(n^{-1/2})$, which is better than that of the normal approximation.

Throughout this paper, $\{X_n: n \geq 1\}$ will denote an aperiodic, irreducible Markov chain with finite state space $S = \{1, 2, ..., N\}$. It is well known that in this case a unique stationary distribution exists. The one-step transition matrix will be denoted by $P = (p_{ij})$ and the stationary distribution by $\underline{p} = (p_i)$.

The maximum likelihood estimator $\hat{P} = (\hat{p}_{ij})$ of P based on $X_1, ..., X_{n+1}$ is given by

$$\hat{p}_{ij} = \begin{cases} \dfrac{n_{ij}}{n_i}, \text{ if } n_i > 0, \\ 0, \text{ otherwise,} \end{cases} \tag{1.1}$$

with

$$n_{ij} = \sum_{t=1}^{n}[X_t = i,\ X_{t+1} = j] \text{ and } n_i = \sum_{t=1}^{n}[X_t = i],$$

where, for a set A, [A] stands for its indicator function. The problem is to bootstrap the sampling distribution of $\hat{p}_{ij}$ on the basis of the original data $X_1, ..., X_{n+1}$. The following is a method proposed by Basawa et al. with motivation deriving from a multinomial type representation of the Markov chain $\{X_n\}$. See Basawa et al. for the details.

Given the original data $X_1, ..., X_{n+1}$, obtain $\hat{p}_{ij}$ by (1.1). Then generate independent random variables W_{it}^* with probability mass function

$$P^*(W_{it}^* = j) = \hat{p}_{ij},\ 1 \leq i, j \leq N;\ t \geq 1. \tag{1.2}$$

Define

$$n_{ij}^* = \sum_{t=1}^{n_i} [W_{it}^* = j], \tag{1.3}$$

and

$$p_{ij}^* = n_{ij}^*/n_i,\ \ 1 \leq i, j \leq N. \tag{1.4}$$

Then the distribution of $n_i^{1/2}(p_{ij}^* - \hat{p}_{ij})$, given $X_1, ..., X_{n+1}$, is a bootstrap approximation to the distribution of $n_i^{1/2}(\hat{p}_{ij} - p_{ij})$.

Basawa et al. showed that this approximation is asymptotically valid, in sup norm on $\mathbf{R}$, along almost all sample paths $X_1, ..., X_{n+1}$. In this paper, we obtain a one-term Edgeworth expansion for the bootstrap distribution. Comparisons of this with the corresponding expansion for the sampling distribution yield the rate results mentioned earlier. These results are presented in the next section.

Section 3 describes some modified bootstrap methods and the corresponding rate results. As mentioned, earlier, the main result of these section is that a better rate of approximation is possible with these methods if p_{ij} is irrational.

All the proofs are presented in Section 4.

2. Asymptotics for the Conditional Bootstrap

Let P denote the probability governing the Markov chain $\{X_n\}$ and P* the (conditional) probability governing the bootstrap distribution. All the results in this section involving P* are valid for almost all sample paths $X_1, X_2, \cdots$. All the proofs are given in Section 4. In what follows, $1 \le i, j \le N$.

The first theorem establishes a rate for the bootstrap approximation to the distribution of centered $\hat{p}_{ij}$. Throughout this section, p_{ij}^* will stand for the bootstrap estimate defined by (1.4). Let $\hat{p}_i = n_i/n$.

Theorem 2.1

Let $0 < p_{ij} < 1$. Then

(a) $$\sup_x |P\{n_i^{1/2}(\hat{p}_{ij} - p_{ij}) \le x\} - P^*\{n_i^{1/2}(p_{ij}^* - \hat{p}_{ij}) \le x\}| = O((\frac{\log n}{n})^{1/2}),$$

(b) $$\sup_x |P\{n^{-1/2}n_i(\hat{p}_{ij} - p_{ij}) \le x\} - P^*\{n^{-1/2}n_i(p_{ij}^* - \hat{p}_{ij}) \le x\}| = O((\frac{\log\log n}{n})^{1/2}).$$

The next theorem shows that a better rate of convergence is possible for the bootstrap approximation to the sampling distribution of the standardized $\hat{p}_{ij}$.

Theorem 2.2

Let $0 < p_{ij} < 1$. Then

$$\sup_x |P\{\frac{n^{-1/2}n_i(\hat{p}_{ij}-p_{ij})}{(p_i p_{ij}(1-p_{ij}))^{1/2}} \le x\} - P^*\{\frac{n^{-1/2}n_i(p_{ij}^*-\hat{p}_{ij})}{(\hat{p}_i\hat{p}_{ij}(1-\hat{p}_{ij}))^{1/2}} \le x\}| = O(n^{-1/2}).$$

A comparison of the one term Edgeworth expansions would reveal that the above rate cannot be improved. The Edgeworth expansion for the distribution of standardized $\hat{p}_{ij}$ has been obtained by Datta and McCormick (1990). We only present the case when p_{ij} is irrational. In the other case, the expansion will have an additional discontinuous $n^{-1/2}$ term. The Edgeworth expansion for the bootstrap approximation follows from Theorem 2.2 of Datta (1989).

Theorem 2.3

Suppose that for some $1 \le k \le N$, $p_{kk} > 0$. If p_{ij} is irrational then

$$P\{\frac{n^{-1/2} n_i(\hat{p}_{ij}-p_{ij})}{(p_i p_{ij}(1-p_{ij}))^{1/2}} \le x\} = \Phi(x) + \frac{n^{-1/2}\phi(x)(1-x^2)}{6(p_i p_{ij}(1-p_{ij}))^{1/2}}(1-2p_{ij} + 3p_{ij}S_{ij}) + o(n^{-1/2}),$$

uniformly in $x \in \mathbf{R}$, where Φ is the standard normal c.d.f., ϕ is the standard normal p.d.f. and

$$S_{ij} = \sum_{k=1}^{\infty} (p_{ji}^{(k-1)} - p_{ii}^{(k)}).$$

Remark 2.1 The above theorem can be obtained under a more general condition. See Datta and McCormick (1990) for details.

Theorem 2.4

Let $0 < p_{ij} < 1$. Then

$$P^*\{\frac{n^{-1/2} n_i(p_{ij}^* - \hat{p}_{ij})}{(\hat{p}_i\hat{p}_{ij}(1-\hat{p}_{ij}))^{1/2}} \le x\} = \Phi(x) + \frac{n^{-1/2}\phi(x)(1-x^2)(1-2p_{ij})}{6(p_i p_{ij}(1-p_{ij}))^{1/2}}$$

$$+ \frac{n^{-1/2}\phi(x)Q(n^{1/2}x(\hat{p}_i\hat{p}_{ij}(1-\hat{p}_{ij}))^{1/2})}{(p_i p_{ij}(1-p_{ij}))^{1/2}} + o(n^{-1/2}),$$

uniformly in x, where $Q(x) = [x] - x + 1/2$.

It can be seen from the above two theorems that the $n^{-1/2}$ terms in the two expansions do not match. Hence the closeness of the bootstrap approximation to the true sampling distribution cannot be any smaller than $O(n^{-1/2})$.

Roughly speaking, this bootstrap technique is ignoring certain dependency present in the representation of n_{ij} while mimicing it; and therefore over-simplifying the matter by generating n_{ij}^* from a binomial distribution. Consequently the third moments of the two distributions do not match.

Note that, because of the latticeness of the summands, the Edgeworth expansion for the bootstrap has a discontinuous term which contributes to further mismatch. This term can, however, be removed by adding a small correction term $n_i^{-3/8}\eta$ to the bootstraped pivot $n_i^{1/2}(p_{ij}^* - \hat{p}_{ij})/(\hat{p}_{ij}(1-\hat{p}_{ij}))^{1/2}$, where η is a zero mean random variable independent of W^*'s having a finite third absolute moment and a characteristic function with a compact support. This would follow from a continuous version of a special case of the Theorem in Babu and Singh (1989).

In the next section, we propose a few bootstrap techniques which will take care of both the above mentioned problems and thereby improve the rate of approximation.

3. Some New Bootstrap Methods

We now propose three bootstrap methods based on i.i.d. sampling, more or less along the same line, with the intention of making the first order Edgeworth expansion match with that of the sampling distribution. Consequently, the bootstrap schemes may appear to be somewhat artificial. However, because of the i.i.d. nature of the samples, these should be easier to implement than some other natural methods, e.g., parametric bootstrap, proposed by Basawa et al. and Athreya and Fuh.

As observed in Theorem 2.3 the dependence in the data shows up in the one term Edgeworth expansion. In order to match the Edgeworth expansion in Theorem 2.3 we choose in Theorem 3.1 to use random weightings applied to the transition counts. To motivate our bootstrap scheme let F_{ij} be a non-lattice distribution with $\int x dF_{ij} = 0$, $\sigma_{ij}^2 = \int x^2 dF_{ij} = p_i p_{ij}(1-p_{ij})$ and $\int x^3 dF_{ij} = r_{ij}\sigma_{ij}^2(1-2p_{ij})$ where $r_{ij} = (1-2p_{ij}+3p_{ij}S_{ij})/(1-2p_{ij})$. Then the one-term Edgeworth expansion for a standardized sum of i.i.d. random variables with distribution F_{ij} is exactly the expansion appearing in the right hand side of Theorem 2.3. Our objective then is to mimic the maximum likelihood estimate $\hat{p}_{ij} = n_{ij}/n_i$ by a bootstrap estimate $\hat{p}_{ij}^*$ whose standardized conditional distribution is appropriately close to that corresponding to F_{ij} and then invoke a continuous Edgeworth expansion result.

Given the data $X_1, \ldots, X_{n+1}$, we compute $\hat{p}_{ij}$ as before. For $k \geq 1$ let $(\hat{p}_{ij}^{(k)}) = \hat{P}^k$ be the estimated k-step transition matrix. We estimate S_{ij} by

$$\hat{S}_{ij} = \begin{cases} \sum_{k=1}^{K_n} (\hat{p}_{ji}^{(k-1)} - \hat{p}_{ii}^{(k)}), & \text{if } i \neq j \\ 1 - \hat{p}_i, & \text{if } i = j, \end{cases} \tag{3.1}$$

where $K_n \uparrow \infty$, and r_{ij} by

$$\hat{r}_{ij} = \begin{cases} 1 + \frac{3\hat{p}_{ij}\hat{S}_{ij}}{1-2\hat{p}_{ij}}, & \text{if } \hat{p}_{ij} \neq \frac{1}{2} \\ 0, & \text{otherwise.} \end{cases} \tag{3.2}$$

It will be shown that $\hat{r}_{ij}$ is strongly consistent for r_{ij}.

The rest of the plan indicated before can be carried out in several ways. Here we propose two most natural schemes and a slight modification of one of them.

The first bootstrap scheme can be described as follows. Generate i.i.d. pairs (X_{1t}^*, X_{2t}^*), $1 \leq t \leq n$ such that

$$P^*(X_{1t}^* = i) = \hat{p}_i, \quad 1 \leq i \leq N, \tag{3.3}$$

and

$$P^*(X_{2t}^* = j \mid X_{1t}^* = i) = \hat{p}_{ij}, \; 1 \leq i, j \leq N. \tag{3.4}$$

Let $[X_{1t}^* = i, X_{2t}^* = j]$ be a transition count. In place of the naive bootstrap estimate $\Sigma[X_{1t}^* = i, X_{2t}^* = i] \,/\, \Sigma[X_{1t}^* = i]$ we will consider a bootstrap estimate of the form

$$\sum_{t=1}^{n} I_{ijt}^*[X_{1t}^* = i, X_{2t}^* = j] \,/\, \sum_{t=1}^{n} I_{ijt}^*[X_{1t}^* = i]$$

where the I_{ijt}^* are i.i.d weights, with $E^* I_{ijt}^{*2} = 1$ and $E^* I_{ijt}^{*3} = \hat{r}_{ij}$.

More precisely, given i and j, generate i.i.d. I_{ijt}^*, $1 \leq t \leq n$ such that

$$P^*(I_{ijt}^* = A_{ij}) = p_{1ij},$$

$$P^*(I^*_{ijt} = B_{ij}) = p_{2ij}, \tag{3.5}$$

$$P^*(I^*_{ijt} = 0) = 1 - p_{1ij} - p_{2ij},$$

and I^*_{ij}'s are independent of (X^*_1, X^*_2)'s, where

$$A_{ij} = \begin{cases} -1, & \text{if } \hat{r}_{ij} > -1, \\ \hat{r}_{ij}, & \text{if } \hat{r}_{ij} \le -1, \end{cases} \tag{3.6}$$

and

$$B_{ij} = \begin{cases} \hat{r}_{ij}, & \text{if } \hat{r}_{ij} > 1, \\ 1, & \text{if } \hat{r}_{ij} \le 1, \end{cases} \tag{3.7}$$

$$p_{1ij} = \frac{B_{ij} - \hat{r}_{ij}}{A^2_{ij}(B_{ij} - A_{ij})}, \tag{3.8}$$

$$p_{2ij} = \frac{\hat{r}_{ij} - A_{ij}}{B^2_{ij}(B_{ij} - A_{ij})}. \tag{3.9}$$

Let

$$n^*_{i(j)} = \sum_{t=1}^{n} I^*_{ijt}[X^*_{1t} = i] \tag{3.10}$$

and

$$n^*_{ij} = \sum_{t=1}^{n} I^*_{ijt}\,[X^*_{1t} = i,\, X^*_{2t} = j]. \tag{3.11}$$

Then

$$p^*_{ij} = \begin{cases} n^*_{ij}/n^*_{i(j)}\,, & \text{if } n^*_{i(j)} \ne 0, \\ 0\,, & \text{otherwise,} \end{cases}$$

is a bootstrap estimate of $\hat{p}_{ij}$ and we have the following result:

Theorem 3.1

Along almost all sample paths $X_1, X_2, \cdots,$

(a) $\sup_x |P\{n_i^{1/2}(\hat{p}_{ij}-p_{ij}) \le x\} - P^*\{n_i^{-1/2}n_{i(j)}^*(p_{ij}^*-\hat{p}_{ij}) \le x\}| = O((\frac{\log n}{n})^{1/2}),$

if $0 < p_{ij} < 1$,

(a′) $\sup_x |P\{n^{-1/2}n_i(\hat{p}_{ij}-p_{ij}) \le x\} - P^*\{n^{-1/2}n_{i(j)}^*(p_{ij}^*-\hat{p}_{ij}) \le x\}| = O((\frac{\log\log n}{n})^{1/2}),$

if $0 < p_{ij} < 1$;

moreover, if the conditions of Theorem 2.3 hold then

(b) $$\sup_x |P\{\frac{n^{-1/2}n_i(\hat{p}_{ij}-p_{ij})}{(p_i p_{ij}(1-p_{ij}))^{1/2}} \le x\} - P^*\{\frac{n^{-1/2}n_{i(j)}^*(p_{ij}^*-\hat{p}_{ij})}{(\hat{p}_i\hat{p}_{ij}(1-\hat{p}_{ij}))^{1/2}} \le x\}| = o(n^{-1/2}).$$

Next, we describe another scheme which is slightly simpler than the above. The resulting method will be a direct modification of the conditional bootstrap of Basawa et al.(1989).

Given the data $X_1, \ldots, X_{n+1}$, generate W^*'s as in Section 1, and independently of W^*'s, generate i.i.d. I_{ijt}^*, $1 \le t \le n_i$, as in the first method of this section. Let

$$n_{i(j)}^{*\prime} = \sum_{t=1}^{n_i} I_{ijt}^*, \quad n_{ij}^{*\prime} = \sum_{t=1}^{n_i} I_{ijt}^*[W_{it}^* = j],$$

and

$$p_{ij}^{*\prime} = \begin{cases} n_{ij}^{*\prime}/n_{i(j)}^{*\prime}, & \text{if } n_{i(j)}^{*\prime} \ne 0, \\ 0, & \text{otherwise.} \end{cases}$$

Then $p_{ij}^{*\prime}$ is our bootstrap estimate of $\hat{p}_{ij}$ under this method. In order to remove the effect of latticeness and obtain $o(n^{-1/2})$ rate, we need to add a correction term to the the standardized $p_{ij}^{*\prime}$.

Theorem 3.2

Along almost all sample paths $X_1, X_2, \ldots,$

(a) $\sup_x |P\{n_i^{1/2}(\hat{p}_{ij} - p_{ij}) \le x\} - P^*\{n_i^{-1/2} n_{i(j)}^{*\prime}(p_{ij}^{*\prime} - \hat{p}_{ij}) \le x\}| = O((\frac{\log n}{n})^{1/2})$

if $0 < p_{ij} < 1$,

(a′) $\sup_x |P\{n^{-1/2} n_i(\hat{p}_{ij} - p_{ij}) \le x\} - P^*\{n^{-1/2} n_{i(j)}^{*\prime}(p_{ij}^{*\prime} - \hat{p}_{ij}) \le x\}| = O((\frac{\log\log n}{n})^{1/2})$

if $0 < p_{ij} < 1$;

moreover if the conditions of Theorem 2.3 hold then

(b)

$$\sup_x |P\{\frac{n^{-1/2} n_i(\hat{p}_{ij} - p_{ij})}{(p_i p_{ij}(1 - p_{ij}))^{1/2}} \le x\} - P^*\{\frac{n^{-1/2} n_{i(j)}^{*\prime}(p_{ij}^{*\prime} - \hat{p}_{ij})}{(\hat{p}_i \hat{p}_{ij}(1 - \hat{p}_{ij}))^{1/2}} + n_i^{-3/8}\eta \le x\}| = o(n^{-1/2}),$$

where η is a zero mean random variable (independent of W^*'s and I_{ij}^*'s) having finite absolute third moment and compactly supported characteristic function.

It is possible to remove the correction term in the last theorem by adjusting the random weights appropriately. The idea is to modify the distribution of the weights suitably that would guarantee the non-latticeness of the summands in the limit.

Given the data $X_1, X_2, \ldots, X_{n+1}$, generate W^*'s as in Section 1. Let $\hat{r}_{ij}$ be as in (3.2).

Define

$$A_{ij}' = \begin{cases} -1 - 1/2, & \text{if } \hat{r}_{ij} > -1 \\ [\hat{r}_{ij}] - 1/2, & \text{if } \hat{r}_{ij} \le -1, \end{cases} \tag{3.12}$$

and

$$B_{ij}' = \begin{cases} [\hat{r}_{ij}+1]+\varepsilon_n, & \text{if } \hat{r}_{ij} > 1 \\ 1 + \varepsilon_n, & \text{if } \hat{r}_{ij} \le 1 \end{cases} \tag{3.13}$$

where $0 < \varepsilon_n \to \varepsilon$, and ε is irrational. Define p_{1ij}' and p_{2ij}' as p_{1ij} and p_{2ij} with

A_{ij} and B_{ij} replaced by A'_{ij} and B'_{ij} respectively. Generate $I^{*'}_{ijt}$, $1 \le t \le n_i$, independently of W^*'s such that $I^{*'}_{ij}$'s are i.i.d. having probability mass function given by (3.5) with the p, A and B replaced by the corresponding p′, A′ and B′. Letting

$$n^{*''}_{i(j)} = \sum_{t=1}^{n_i} I^{*'}_{ijt}$$

and

$$n^{*''}_{ij} = \sum_{t=1}^{n_i} I^{*'}_{ijt}\,[W^*_{it} = j],$$

$$p^{*''}_{ij} = \begin{cases} n^{*''}_{ij} \,/\, n^{*''}_{i(j)}, & \text{if } n^{*''}_{i(j)} \ne 0, \\ 0, & \text{otherwise,} \end{cases}$$

defines a bootstrap estimate of $\hat{p}_{ij}$. For this third method we can conclude the following:

Theorem 3.3

Along almost all sample paths $X_1, X_2, \ldots,$

(a) $\sup_x |P\{n_i^{½}(\hat{p}_{ij} - p_{ij}) \le x\} - P^*\{n_i^{-½} n^{*''}_{i(j)}(p^{*''}_{ij} - \hat{p}_{ij}) \le x\}| = O((\frac{\log n}{n})^{½})$

if $0 < p_{ij} < 1$,

(a′) $\sup_x |P\{n^{-½} n_i(\hat{p}_{ij} - p_{ij}) \le x\} - P^*\{n^{-½} n^{*''}_{i(j)}(p^{*''}_{ij} - \hat{p}_{ij}) \le x\}| = O((\frac{\log\log n}{n})^{½})$

if $0 < p_{ij} < 1$;

moreover, if the conditions of Theorem 2.3 hold then

(b) $\sup_x |P\{\frac{n^{-½} n_i(\hat{p}_{ij} - p_{ij})}{(p_i p_{ij}(1 - p_{ij}))^{½}} \le x\} - P^*\{\frac{n^{-½} n^{*''}_{i(j)}(p^{*''}_{ij} - \hat{p}_{ij})}{(\hat{p}_i \hat{p}_{ij}(1 - \hat{p}_{ij}))^{½}} \le x\}| = o(n^{-½}).$

To the best of our knowledge, the present paper is the only work so far studying the rates of bootstrap approximations for Markov chains. It is hoped that this paper will initiate further studies in this direction. In particular, the following questions deserve investigations:

(1) How does the natural parametric bootstrap (see, e.g., Basawa et al., Athreya and Fuh (method I)) perform in terms of rates of approximations? To answer this question, one essentially needs to establish a continuous version of the Edgeworth expansion for $\hat{p}_{ij}$.

(2) How to bootstrap the joint distribution of $(\hat{p}_{ij})$ with a rate better (if possible!) than that of normal approximation? The parametric bootstrap may be a candidate for this. Clearly, the method in this section can handle only one $\hat{p}_{ij}$ at a time.

(3) Is it possible to bootstrap a studentized version of $\hat{p}_{ij}$, e.g., $n_i^{1/2}(\hat{p}_{ij} - p_{ij})/(\hat{p}_{ij}(1-\hat{p}_{ij}))^{1/2}$, with a rate of $o(n^{-1/2})$? Certainly, this will be more useful, e.g., in the construction of confidence interval for p_{ij}, than the present results.

4. Proofs

We are going to use the following terms and notations throughout this section. For any sequence $V_1, ..., V_n$ of i.i.d., non-degenerate random variables with finite second moment, $n^{1/2}(\overline{V}-\mu)/\sigma$ will be called the standardized sum of the V's, where $\overline{V} = n^{-1}\Sigma V_i$, $\mu = EV_1$, $\sigma^2 = E(V_1-\mu)^2$. A sequence of distributions $\{F_n\}$ on $\mathbf{R}$ is said converge in d_3 to a distribution F on $\mathbf{R}$, written as $F_n \xrightarrow{(3)} F$, if $\{F_n\}$ converges to F weakly, and $\int |x|^3 dF_n \to \int |x^3| dF$. Φ and ϕ will denote the standard normal c.d.f. and p.d.f. respectively.

Let $Y_t = (X_t, X_{t+1})$, $t \geq 1$, where $\{X_t\}$ is the original Markov chain. Then $\{Y_t\}$ is also a Markov chain with state space $S_2 = \{(u,v){:}\ 1 \leq u,v \leq N, p_{uv} > 0\}$, which has at most N^2 states. Moreover the stationary distribution for Y_t is given by $p_{(u,v)} = p_u\, p_{uv}$.

Since $\{X_t\}$ is irreducible, aperiodic, and has a finite state space, there exist $M < \infty$ and $\rho < 1$ such that

$$\max_{1\leq i,j\leq N} |p_{ij}^{(k)} - p_j| \leq M\rho^k, \text{ for all } k \geq 1. \tag{4.1}$$

Consequently,

$$\max_{(u,v),(x,y)\in S_2} |p^{(k+2)}_{(u,v)(x,y)} - p_{(x,y)}|$$

$$\max_{(u,v),(x,y)\in S_2} p_{xy} |p^{(k)}_{vx} - p_x| \le M\rho^k, \text{ for all } k \ge 1,$$

where $p^{(k)}_{(u,v)(x,y)}$ denote the k-step transition probabilities for $\{Y_t\}$. Therefore $\{Y_t\}$ satisfies the basic condition (0.1) of Nagaev (1961), for some k_0 large enough.

Fix $1 \le i, j \le N$. Define

$$f(Y_t) = [X_t = i,\ X_{t+1} = j] - p_{ij}[X_t = i].$$

Note that $Ef(Y_t) = 0$, for all t and

$$\sum_1^n f(Y_t) = n_i(\hat{p}_{ij} - p_{ij}). \tag{4.2}$$

Moreover, under the stationary initial distribution,

$$\sigma^2 = \lim_{n\to\infty} E_{\underline{p}}(\frac{1}{\sqrt{n}} \sum_1^n f(Y_t))^2 = p_i\, p_{ij}(1 - p_{ij}) > 0, \tag{4.3}$$

if $0 < p_{ij} < 1$. Therefore,

$$\frac{n^{-1/2} n_i(\hat{p}_{ij}-p_{ij})}{(p_i p_{ij}(1-p_{ij}))^{1/2}} = \frac{\sum_1^n f(Y_t)}{n^{1/2}\sigma}. \tag{4.4}$$

Theorem 1 of Nagaev (1961) will be used to prove Theorems 2.1 and 2.2.

Proof of Theorem 2.2

The conditions (0.5) of Theorem 1 of Nagaev are satisfied by $g(x) = x$ and $f(Y_t)$. Therefore it follows by that theorem and the identification (4.4) above that

$$\sup_x |P\{\frac{n^{-1/2} n_i(\hat{p}_{ij}-p_{ij})}{(p_i p_{ij}(1-p_{ij}))^{1/2}} \le x\} - \Phi(x)| = O(n^{-1/2}). \tag{4.5}$$

On the other hand, since $n_{ij}^* \sim$ Binomial $(n_i, \hat{p}_{ij})$ under the P* and $p_{ij}^* = n_{ij}^*/n_i$, by the ordinary Berry Esseen Theorem for i.i.d. summands it follows that

$$\sup_x | P^*\{\frac{n_i^{1/2}(p_{ij}^*-\hat{p}_{ij})}{(\hat{p}_{ij}(1-\hat{p}_{ij}))^{1/2}} \le x\} - \Phi(x) | \le \frac{\text{constant}}{(\hat{p}_{ij}(1-\hat{p}_{ij}))^{3/2}} n_i^{-1/2},$$

$$= O(n^{-1/2}), \tag{4.6}$$

a.s. (P), because $\liminf_{n\to\infty} \hat{p}_{ij}(1-\hat{p}_{ij}) = p_{ij}(1-p_{ij}) > 0$ and $n_i/n \to p_i > 0$, a.s. (P).

Theorem 2.2 follows, by the triangle inequality, from (4.5) and (4.6). □

Proof of Theorem 2.1

(a) From (4.5) and (4.6) we get that

$$\sup_x | P\{(\frac{\hat{p}_i}{p_i})^{1/2} n_i^{1/2}(\hat{p}_{ij} - p_{ij}) \le x\} - P^*\{n_i^{1/2}(p_{ij}^* - \hat{p}_{ij}) \le x\} |$$

$$\le O(n^{-1/2}) + \left| \frac{1}{(p_{ij}(1-p_{ij}))^{1/2}} - \frac{1}{(\hat{p}_{ij}(1-\hat{p}_{ij}))^{1/2}} \right| M, \tag{4.7}$$

a.s. (P), where $M = \sup |t\, \phi(t)|$.

By the law of iterated logarithm (Chung, I16, Theorem 5), the second term in the above bound is $O((\log\log n/n)^{1/2})$ a.s. (P).

We now complete the proof by showing that

$$\sup_x |P\{(\frac{\hat{p}_i}{p_i})^{1/2} n_i^{1/2}(\hat{p}_{ij}-p_{ij}) \le x\} - P\{n_i^{1/2}(\hat{p}_{ij}-p_{ij}) \le x\}| = O((\frac{\log n}{n})^{1/2}). \tag{4.8}$$

For any $0 < \varepsilon < 1$, the LHS above is bounded by

$$\sup_x |P\{Z_n \le x\} - P\{Z_n \le x(1+\varepsilon)^{1/2}\}|$$

$$+ \sup_x |P\{Z_n \le x\} - P\{Z_n \le x(1-\varepsilon)^{1/2}\}| \tag{4.9}$$

$$+ P\{|\hat{p}_i - p_i| > \varepsilon p_i\},$$

where

$$Z_n = \left(\frac{\hat{p}_i}{p_i}\right)^{1/2} \frac{n_i^{1/2}(\hat{p}_{ij} - p_{ij})}{(p_{ij}(1-p_{ij}))^{1/2}}.$$

By (4.5) and the mean value theorem, the first two terms in (4.9) are $O(\varepsilon) + O(n^{-1/2})$. The third term is no more than

$$2(1 - \Phi(\frac{p_i n^{1/2}\varepsilon}{\sigma_i})) + O(n^{-1/2}),$$

where $\sigma_i^2 = \sigma^2(\hat{p}_i) = p_i(1-p_i) + 2p_i \sum_{t=1}^{\infty} (p_{ii}^{(t)} - p_i)$, by the Berry-Esseen bound (Nagaev, Theorem 1) with $f(X_t) = [X_t = i]$,

$$= O(e^{-\frac{p_i^2 n\varepsilon^2}{4\sigma_i^2}}) + O(n^{-1/2}),$$

$$= O(n^{-1/2}), \text{ if } \varepsilon = \left(\frac{4\sigma_i^2 \log n}{np_i^2}\right)^{1/2}.$$

This completes the proof. □

(b) We once again get from (4.5) and (4.6) that

$$\sup_x | P\{n^{-1/2}n_i(\hat{p}_{ij} - p_{ij}) \le x\} - P^*\{n^{-1/2}n_i(p_{ij}^* - \hat{p}_{ij}) \le x\} |$$

$$\le O(n^{-1/2}) + \left| \frac{1}{(p_i p_{ij}(1-p_{ij}))^{1/2}} - \frac{1}{(\hat{p}_i \hat{p}_{ij}(1-\hat{p}_{ij}))^{1/2}} \right| M,$$

a.s. (P), where $M = \sup |t\,\phi(t)|$.

By the law of iterated logarithm applied to $\hat{p}_i$ and $\hat{p}_{ij}$, the second term in the above bound is $O((\log \log n/n)^{1/2})$ a.s. (P). This completes the proof. □

Proof of Theorem 2.3

See Datta and McCormick (1990). □

Proof of Theorem 2.4

It can be verified that $n^{-½}n_i(p_{ij}^* - \hat{p}_{ij})/(\hat{p}_i\hat{p}_{ij}(1-\hat{p}_{ij}))^{½}$ is the standardized sum corresponding to the summands $[W_{it}^* = j]$, $t = 1, 2, ..., n_i$. Under P*, $[W_{it}^* = j] \sim$ Binomial $(1,\hat{p}_{ij}) \overset{(3)}{\rightarrow}$ Binomial $(1,p_{ij})$, a.s. (P). Moreover, since $0 < p_{ij} < 1$, Binomial $(1,p_{ij})$ and, a.s. for large n, Binomial $(1,\hat{p}_{ij})$ are lattice with span one. Hence the expansion for p_{ij}^* follows from Theorem 2.2 of Datta (1989), with $p = \infty$ and $x_G = 0$, and the fact that $\hat{p}_i = n_i/n \rightarrow p_i$ (>0) a.s. (P). □

The following lemma is needed for proving the theorems in Section 3.

Lemma 4.1

For any $1 \leq i, j \leq N$,

$$\hat{S}_{ij} \rightarrow S_{ij} \quad \text{a.s. (P), as } n \rightarrow \infty.$$

Proof

For $i = j$, $S_{ij} = S_{ii} = 1 - \lim_{k\rightarrow\infty} p_{ii}^{(k)} = 1 - p_i$. Hence the conclusion is immediate from the fact that $\hat{p}_i \rightarrow p_i$ a.s. (P).

The conclusion in the $i \neq j$ case will follow if we prove that

$$\sum_1^{\infty} | \hat{p}_{ji}^{(k-1)} - \hat{p}_{ii}^{(k)} - p_{ji}^{(k-1)} + p_{ii}^{(k)} | \rightarrow 0, \text{ a.s. (P)}. \tag{4.10}$$

Since $\hat{P} \rightarrow P$ a.s., it follows that each summand in the above sum converges a.s. to zero, as $n \rightarrow \infty$. It can be seen from (4.1) that for k_0 large enough,

$$\sup_{i,j,A} | p_i^{(k_0)}(A) - p_j^{(k_0)}(A) | < 1 , \tag{4.11}$$

where the supremum is taken over all $1 \leq i,j \leq N$ and $A \subset S$; and $p_i^{(k_0)}(A) = \sum_{j\in A} p_{ij}^{(k_0)}$ for any $1 \leq i \leq N$, $A \subset S$.

Fix any $r_1 \in (r,1)$ where r denotes the LHS of (4.11). Since $\hat{P} \overset{a.s.}{\rightarrow} P$,

$$\sup_{i,j,A} | \hat{p}_i^{(k_0)}(A) - \hat{p}_j^{(k_0)}(A) | < r_1$$

for almosts all sample paths, for large enough $\underline{n}$ (depending upon the path). Therefore (see Nagaev), a stationary distribution $\hat{\underline{p}}' = (\hat{p}_i)$ for $\hat{P}$ exists and

$$\sup_{i,j} |\hat{p}_{ij}^{(k)} - \hat{p}_j'| \leq r_1^{(k/k_0)-1}, \quad \text{for all } k > k_0.$$

Thus, for almost all sample paths, for large enough n,

$$|\hat{p}_{ji}^{(k-1)} - \hat{p}_{ii}^{(k)} - p_{ji}^{(k-1)} + p_{ii}^{(k)}| \leq 2M\rho^k + 2r_1^{(k/k_0)-1}$$

for all $k > k_0$, where M and ρ are given by (4.1).

The proof now ends by D.C.T. □

Proof of Theorem 3.1

(a) and (a′). Fix $1 \leq i,j \leq N$. Note that $n^{-1/2} n_{i(j)}^* (p_{ij}^* - \hat{p}_{ij})/(\hat{p}_i\hat{p}_{ij}(1-\hat{p}_{ij}))^{1/2}$ is the standardized sum of the i.i.d summands

$$Y_t^* = I_{ijt}^*([X_{1t}^* = i, X_{2t}^* = j] - \hat{p}_{ij}[X_{1t}^* = i]), \tag{4.12}$$

$1 \leq t \leq n$. It is easy to check that,

$$E^*Y_t^* = 0, \quad \sigma_n^2 = E^*Y_t^{*2} = \hat{p}_i\hat{p}_{ij}(1-\hat{p}_{ij}),$$

$$\mu_{3,n} = E^*Y_t^{*3} = \begin{cases} (1-2\hat{p}_{ij}+3\hat{p}_{ij}\hat{S}_{ij})\hat{p}_i\hat{p}_{ij}(1-\hat{p}_{ij}), & \text{if } \hat{p}_{ij} \neq \frac{1}{2}, \\ 0, & \text{if } \hat{p}_{ij} = \frac{1}{2}. \end{cases}$$

Since $\hat{p}_i \xrightarrow{a.s.} p_i > 0$, $\hat{p}_{ij} \xrightarrow{a.s.} p_{ij} \in (0,1)$ and $\hat{S}_{ij} \xrightarrow{a.s.} S_{ij}$, we get that $\limsup_{n\to\infty} \frac{|\mu_{3,n}|}{\sigma_n^3} < \infty$, a.s. (P). Hence by the Berry-Esseen Theorem for i.i.d. summands,

$$\sup_x |P^*\{\frac{n^{-1/2} n_{i(j)}^*(p_{ij}^*-\hat{p}_{ij})}{(\hat{p}_i\hat{p}_{ij}(1-\hat{p}_{ij}))^{1/2}} \leq x\} - \Phi(x)| = O(n^{-1/2}), \tag{4.13}$$

a.s. (P). Rest of the proof is essentially the same as that of Theorem 2.1.

(b) Let Y_∞ be a random variable generated in the same way as Y_1^* with $\hat{p}_i$ replaced by p_i, $\hat{p}_{ij}$ by p_{ij}, $\hat{r}_{ij}$ by $r_{ij} = 1 + 3p_{ij}\, S_{ij}/(1-2p_{ij})$. Then Y_∞ has a non-lattice distribution F_{ij} (say) since p_{ij} is irrational, with

$$\int x dF_{ij} = 0, \quad \int x^2 dF_{ij} = p_i p_{ij}(1-p_{ij}),$$

$$\int x^3 dF_{ij} = r_{ij} p_i p_{ij}(1-p_{ij})(1-2p_{ij}) = p_i p_{ij}(1-p_{ij})(1-2p_{ij}+3p_{ij}S_{ij})$$

By the Ergodic Theorem and Lemma 4.1, $\hat{p}_i \xrightarrow{a.s.} p_i$, $\hat{p}_{ij} \xrightarrow{a.s.} p_{ij}$ and $\hat{r}_{ij} \xrightarrow{a.s.} r_{ij}$. Hence $F_{ij}^* \xrightarrow{(3)} F_{ij}$, a.s. (P), where F_{ij}^* is the distribution of Y_t^*. Therefore, Theorem 2.1 of Datta (1989) applies with $p = \infty$, $F = F_{ij}$ and $G = F_{ij}^*$ to yield

$$P^*\left\{\frac{n^{-½} n_{i(j)}^* (p_{ij}^* - \hat{p}_{ij})}{(\hat{p}_i \hat{p}_{ij}(1-\hat{p}_{ij}))^{½}} \le x\right\} = \Phi(x) + \frac{n^{-½}\phi(x)(1-x^2)}{6(p_i p_{ij}(1-p_{ij}))^{½}}(1-2p_{ij}+ 3p_{ij}\, S_{ij}) + o(n^{-½}),$$

uniformly in x, a.s. (P). The proof now ends by Theorem 2.3 and the triangle inequality. □

Next we prove a general result for i.i.d. summands which will be used to prove Theorem 3.2 (b). This result is a continuous version of special case of the Corollary in Babu and Singh (1989).

Theorem 4.1

Let $Y_1, \ldots, Y_n$ i.i.d. with distribution $G = G_n$ such that $\int y dG = 0$. If $G \xrightarrow{(3)} F$ such that $\int y^2 dF > 0$, $\int |y|^3 dF < \infty$, then

$$P\left\{\frac{n^{½}\overline{Y}}{\sigma_n} + n^{-3/8}\,\eta \le x\right\} = \Phi(x) + \frac{n^{-½}\mu_3(1-x^2)\phi(x)}{6\sigma^3} + o(n^{-½}) \qquad (4.14)$$

uniformly in x, where $\sigma_n^2 = \int y^2 dG$, $\sigma^2 = \int y^2 dF$, $\mu_3 = \int y^3 dF$ and η is a zero mean random variable which is independent of $Y_1, Y_2, \ldots, Y_n$, has finite third absolute moment and a characteristic function with compact support.

Proof

Let ϕ_G and ψ be the characteristic functions of G and η respectively. Let

$$\gamma_n(t) = e^{-t^2/2}(1 + \frac{n^{-1/2}\mu_{3,n}}{6\sigma_n^3}(it)^3),$$

where, $\mu_{3,n} = \int y^3 dG$.

For any $\delta > 0$, by Esseen's Lemma (Lemma XVI.3.2 of Feller), the LHS of (4.14) is bounded by

$$\int_{|t|<n^{1/2}\delta\sigma_n} |(\phi_G^n(n^{-1/2}t/\sigma_n)\,\psi\,(n^{-3/8}t) - \gamma_n(t))/t\,|\,dt \tag{4.15}$$

$$+ \int_{|t|>n^{1/2}\delta\sigma_n} |\gamma_n(t)/t\,|\,dt,$$

for large enough n such that $\{\,|t|>n^{1/2}\,\delta\,\sigma_n\} \cap$ support $(\psi) = \phi$.

Since, $\mu_{3,n} \to \mu_3 < \infty$, $\sigma_n \to \sigma > 0$, and $e^{-t^2/2} \le e^{-n\delta^2\sigma_n^2/4}\,e^{-t^2/4}$, for $|t| > n^{1/2}\delta\sigma_n$, it follows that the second term in (4.15) goes to zero exponentially fast, as $n \to \infty$.

The first term in (4.15) can be further bound by

$$\int_{|t|<n^{1/2}\delta\sigma_n} |(\phi_G^n(n^{-1/2}t/\sigma_n) - \gamma_n(t))/t\,|\,dt + \int_{|t|<n^{1/2}\delta\sigma_n} |\gamma_n(t)|\,|1-\psi(n^{-3/8}t)|/|t|\,dt. \tag{4.16}$$

The first term above is handled in exactly the same way as in the proofs of Lemma 4.2 and Theorem 2.1 of Datta (1989). To bound the second term we use the Taylor expansion for ψ:

$$\psi(t) = 1 + t\,\psi'(0) + \frac{t^2}{2}\,\psi''(t'),$$

$0 < t' < t$. Since η has mean zero, $\psi'(0) = 0$. Also, ψ'' is continuous, because Y has finite third absolute moment, and hence is bounded on the compact support (ψ). Let $\sup|\psi''| = 2K$. Then the second term in (4.16) is no more than

$$K\,n^{-3/4} \int_{|t|<n^{1/2}\delta\sigma_n} e^{-t^2/2}\,(1 + \frac{n^{-1/2}|\mu_{3,n}|\,|t|^3}{6\sigma_n^3})\,|t|\,dt,$$

$$= O(n^{-3/4}).$$

This completes the proof. □

Proof of Theorem 3.2

Since $n^{-1/2}n^{*\prime}_{i(j)}(p^{*\prime}_{ij}-\hat{p}_{ij})/(\hat{p}_i\hat{p}_{ij}(1-\hat{p}_{ij}))^{1/2}$ is the standardized sum of the i.i.d. summands $Y_t^* = I^*_{ijt}([W^*_{it} = j] - \hat{p}_{ij})$, $1 \le t \le n_i$, parts (a) and (a′) follows in the same way as those of Theorem 3.1,since $n_i/n \overset{a.s.}{\rightarrow} p_i$.

By similar arguments as in the proof of Theorem 3.1(b), $F^*_{ij} \overset{(3)}{\rightarrow} F_{ij}$, a.s. (P), where F^*_{ij} is the distribution of Y_1^*, F_{ij} is the distribution of Y_∞ which is generated in the same way as Y_1^* with $\hat{p}_{ij}$ replaced by p_{ij}, and $\hat{r}_{ij}$ by r_{ij}. Clearly, $\int x dF_{ij} = 0$, $\int x^2 dF_{ij} = p_{ij}(1-p_{ij})$, $\int x^3 dF_{ij} = p_{ij}(1-p_{ij})(1-2p_{ij}+3p_{ij}S_{ij})$. Hence, by Theorem 4.1, along almost all sample paths,

$$P^*\{\frac{n^{-1/2}n^{*\prime}_{i(j)}(p^{*\prime}_{ij}-\hat{p}_{ij})}{(\hat{p}_i\hat{p}_{ij}(1-\hat{p}_{ij}))^{1/2}} + n_i^{-3/8}\eta \le x\} = \Phi(x) + \frac{n_i^{-1/2}\phi(x)(1-x^2)}{6(p_{ij}(1-p_{ij}))^{1/2}}(1-2p_{ij}+3p_{ij}S_{ij})$$

$$+ o(n_i^{-1/2}), \text{ uniformly in } x,$$

$$= \Phi(x) + \frac{n^{-1/2}\phi(x)(1-x^2)(1-2p_{ij} + 3p_{ij}S_{ij})}{6(p_i\, p_{ij}(1-p_{ij}))^{1/2}} + o(n^{-1/2}),$$

uniformly in x. The proof once again ends by Theorem 2.3. □

Proof of Theorem 3.3

It can easily be checked that $n^{-1/2}\, n^{*\prime\prime}_{i(j)}\, (p^{*\prime\prime}_{ij} - \hat{p}_{ij})/(\hat{p}_i\hat{p}_{ij}(1-\hat{p}_{ij}))^{1/2}$ is the standardized sum corresponding to the summands

$$Y_t^* = I^{*\prime}_{ijt}([W^*_{it} = j] - \hat{p}_{ij}), \quad 1 \le t \le n_i, \tag{4.17}$$

which are i.i.d. under P* , with

$$E^*Y_t^* = 0, \;\; E^*Y_t^{*2} = \hat{p}_{ij}(1-\hat{p}_{ij}), \;\; EY_t^{*3} = \hat{r}_{ij}\hat{p}_{ij}(1-\hat{p}_{ij})(1-2\hat{p}_{ij}).$$

Thus, parts (a) and (a′) of the theorem follow as before (i.e., parts (a) and (a′) of Theorem 3.1) since $n_i/n \overset{a.s.}{\rightarrow} p_i$.

For part (b), define Y_∞ in the same way as Y_1^* with $\hat{p}_{ij}$ replaced by p_{ij}, $\hat{r}_{ij}$ by r_{ij} and ε_n by ε. Also, define Y'_∞ similarly with $\hat{p}_{ij}$ replaced by p_{ij} and A'_{ij}, B'_{ij} by A''_{ij}, B''_{ij} given by

$$A''_{ij} = \begin{cases} -1-½, & \text{if } r_{ij} > -1, \\ [r_{ij} - 1] - ½, & \text{if } r_{ij} \le -1, \end{cases}$$

and

$$B''_{ij} = \begin{cases} [r_{ij}] + \varepsilon, & \text{if } r_{ij} > 1, \\ 1 + \varepsilon, & \text{if } r_{ij} \le 1. \end{cases}$$

Then Y_∞ and Y'_∞ are both non-lattice random variables because ε is irrational. Furthermore,

$$E\, Z_\infty = 0, \quad E\, Z_\infty^2 = p_{ij}(1-p_{ij}),$$

$$E\, Z_\infty^3 = r_{ij}\, p_{ij}(1-p_{ij})(1-2p_{ij}) = p_{ij}(1-p_{ij})(1-2p_{ij}+3p_{ij}S_{ij}),$$

where $Z_\infty = Y_\infty$ or Y'_∞.

To obtain the desired expansion, first consider the case when r_{ij} is not an integer. Then, by similar arguments as in the proof of Theorem 3.1(b), $F^*_{ij} \xrightarrow{(3)} F_{ij}$; F^*_{ij} and F_{ij} being the distributions of Y^*_t and Y_∞ respectively. Consequently, by Theorem 2.1 of Datta (1989)

$$P^*\{\frac{n^{-½} n^{*''}_{i(j)}(p^{*''}_{ij} - \hat{p}_{ij})}{(\hat{p}_i \hat{p}_{ij}(1-\hat{p}_{ij}))^{½}} \le x\} \tag{4.18}$$

$$= \Phi(x) + \frac{n_i^{-½}\phi(x)(1-x^2)}{6(p_{ij}(1-p_{ij}))^{½}}(1-2p_{ij} + 3p_{ij}S_{ij}) + o(n_i^{-½}),$$

uniformly in x, a.s. (P),

$$= \Phi(x) + \frac{n^{-½}\phi(x)(1-x^2)}{6(p_i p_{ij}(1-p_{ij}))^{½}}(1-2p_{ij}+3p_{ij}S_{ij}) + o(n^{-½}), \tag{4.19}$$

uniformly in x, a.s. (P) since $n_i/n \xrightarrow{a.s.} p_i > 0$.

For the case when r_{ij} is an integer, consider two subsequences of n_i, viz., $\{n_i'\} = \{n_i: \hat{r}_{ij} \geq r_{ij}\}$ and $\{n_i''\} = \{n_i: \hat{r}_{ij} < r_{ij}\}$. Then, a.s. (P), along n_i', $F_{ij}^{*(3)} \rightarrow F_{ij}$, and along n_i'', $F_{ij}^{*(3)} \rightarrow F_{ij}'$ where F_{ij}' is the distribution of Y_∞'. Since Y_∞ and Y_∞' have the same first three moments, we get the same expansions (4.18) and (4.19) along both the subsequences.

Finally, we get the conclusion of the theorem by comparing (4.19) with Theorem 2.3. □

REFERENCES

Athreya, K. B. and Fuh, Cheng-Der (1989). Bootstrapping Markov chains: countable case. Preprint #21, Iowa State University, Ames, Iowa.

Babu, G. J. and Singh, K. (1989). On Edgeworth expansions in the mixture case. Ann. Statist. **17,** 443-447.

Basawa, I. V., Green, T. A., McCormick, W. P. and Taylor, R. L. (1989). Asymptotic bootstrap validity for finite Markov chains. To appear in Comm. Statist.

Chung, K. L. (1967). Markov chains with stationary transition probabilities, Second Edn., Springer-Verlag.

Datta, Somnath (1989). A note on continuous Edgeworth expansions and the bootstrap. To appear in Sankhya A.

Datta, Somnath and McCormick, W. P. (1990). On first order Edgeworth expansions for a Markov chain. Technical report, Dept. of Stat., Univ. of Georgia, Athens.

Feller, William (1966, 1971). An introduction to probability theory and its applications, Vol II., John Wiley & Sons.

Kulperger, R. J. and Prakasa Rao, B.L.S.P. (1989). Bootstrapping a finite state Markov chain. To appear in Sankhya A.

Nagaev, S. V. (1961). More exact statements of limit theorems for homogeneous Markov chains. Th. Prob. and Appl. **6,** 62-80.

SIX QUESTIONS RAISED BY THE BOOTSTRAP

B. Efron*

Abstract

Investigations of bootstrap methods often raise more general questions in statistical inference. This talk discusses six such questions: (1) Why do distributions estimated by maximum likeklihood tend to have too short tails? (2) Why does the delta method tend to underestimate standard errors? (3) Why are cross-validation estimates so variable? (4) What is a "correct" confidence interval? (5) What is a good nonparametric pivotal quantity? (6) Can we get bootstrap-like answers without Monte Carlo?

Introduction. Working on the bootstrap tends to raise broader questions of statistical theory. This paper considers six such questions. The first three worry about standard methods, and how they sometimes fail us. The second three questions concern matters not much examined by standard theory. Only the last question directly concerns the bootstrap, but bootstrap considerations appear in all six.

The title may give the impression that six answers will be provided. The actual number is closer to 1.5. My hope in presenting this paper is to attract more solutions, or at least more interest, in some questions that seem to me to be of considerable importance.

1. Why do maximum likelihood estimated distributions tend to be short-tailed? Estimates of probability distributions obtained by maximum likelihood tend to be more concentrated than the distributions themselves. Here is a familiar example: suppose that $y_1, y_2, \cdots, y_n$ is an independent and identically distributed (i.i.d.) sample from an unknown distribution F,

$$F \overset{\text{i.i.d.}}{\rightarrow} y_1, y_2, \cdots, y_n. \tag{1.1}$$

* Department of Statistics, Sequoia Hall, Stanford University, Stanford, CA 94305-4065

Exploring the Limits of Bootstrap. Edited by Raoul LePage and Lynne Billard.
 ISBN: 0471-53631-8.

The nonparametric maximum likelihood estimate (MLE) of F is the empirical distribution $\hat{F}$,

$$\hat{F}: \quad \text{probability } 1/n \text{ on } y_i, \qquad i = 1, 2, \cdots, n. \tag{1.2}$$

The empirical distribution assigns probability

$$\text{Prob}_{\hat{F}}\{A\} = \frac{\#\{y_i \in A\}}{n} \tag{1.3}$$

to any set A in the sample space of the $y's$. This is an unbiased estimate of the true probability $\text{Prob}_F\{A\}$,

$$E\{\text{Prob}_{\hat{F}}\{A\}\} = \text{Prob}_F\{A\}. \tag{1.4}$$

However the same unbiasedness does not apply to the variance functional: $\text{Var}_{\hat{F}}\{Y\} = \sum_{i=1}^{n}(y_i - \bar{y})^2/n$ has expectation

$$E \ \text{Var}_{\hat{F}}\{Y\} = \frac{n-1}{n}\text{Var}_F\{Y\}. \tag{1.5}$$

We see that the variance function is underestimated by maximum likelihood, albeit mildly so. Elementary statistics courses recommend estimating the variance by

$$\frac{n}{n-1}\text{Var}_{\hat{F}}\{Y\} = \sum_{i=1}^{n}(y_i - \bar{y})^2/(n-1), \tag{1.6}$$

rather than by $\text{Var}_{\hat{F}}\{Y\}$ itself. This removes the bias, but doesn't explain it.

The underestimation effect seen in (1.5) is quite small from the point of view of asymptotic theory. Most often we are estimating $\sigma^2 = \text{Var}_F\{Y\}$ in order to form a confidence interval for the expectation $\mu = E_F\{Y\}$, say of the form

$$\bar{y} \pm z^{(\alpha)}\widehat{\text{se}}, \tag{1.7}$$

where $z^{(\alpha)}$ is the $100\alpha^{\text{th}}$ normal percentile, e.g. $z^{(.95)} = 1.645$, and $\widehat{\text{se}} = \hat{\sigma}/\sqrt{n}$, the estimated standard error of $\bar{y}$.

The choice of the MLE versus the unbiased estimate of σ^2 makes a difference of magnitude only

$$O(\widehat{\text{se}}/n) \tag{1.8}$$

in (1.7). This is a third order correction in the usual parlance. By comparison, ignoring the skewness of F in forming interval (1.7) makes a difference of $O(\widehat{\text{se}}/\sqrt{n})$, a second order error; and using $\bar{y}$ instead of an efficient estimate $\hat{\mu}$, makes a difference of $O(\widehat{\text{se}})$, a first order error. (See Efron (1987).) Third order notwithstanding, the underestimation effect in (1.5) can cause substantial problems when n is small.

We can exacerbate the phenomenon seen in (1.5) by considering a standard linear model,

$$y_i = x_i\beta + e_i, \tag{1.9}$$

where x_i is an observed p-dimensional covariate vector, β is an unknown p-dimensional parameter vector, and the e_i are an i.i.d. sample from some unknown distribution F with mean 0 and variance σ^2. Suppose we assume that F is normal, $F \sim N(0, \sigma^2)$. Then the MLE for σ^2 is

$$\hat{\sigma}^2 = \sum_{i=1}^{n}(y_i - x_i\hat{\beta})^2/n, \tag{1.10}$$

having expectation

$$E\{\hat{\sigma}^2\} = \frac{n-p}{n}\sigma^2. \tag{1.11}$$

If p is a large function of n, then the MLE $\hat{\sigma}^2$ badly underestimates the dispersion parameter σ^2. Nevertheless, with p fixed and n going to infinity, (1.11) still represents a third-order error.

The underestimation seen in (1.5) gets worse as we consider measures of dispersion more extreme than the variance.

Table 1 shows the results of a Monte Carlo experiment involving 100 samples, each of which consisted of 10 independent observations from a standard exponential distribution. Four functionals of F were considered, the mean, the standard deviation, the skewness and the kurtosis. The sampling experiment shows that skew($\hat{F}$) badly underestimates skew(F), the situation being worse for the kurtosis. Fisher's theory of k-statistics, Kendall and Stuart, Chapter 12, (1958), reduces the bias of skew($\hat{F}$) and kurt($\hat{F}$), without explaining it.

My own concern with the dispersion-reducing tendencies of maximum likelihood estimation stems from its effect on bootstrap confidence intervals. The simplest way to form a bootstrap confidence interval is to

	Mean	Sd	Skew	Kurt
True F:	1	1	2	6
Ave $\hat{F}$:	1.01	.863	.910	.052
%<True:	55%	71%	93%	100%

Table 1. Sampling Experiment. 100 i.i.d. samples of size $n = 10$, from a standard exponential distribution. For each sample, the mean, standard deviation, skewness, and kurtosis of $\hat{F}$ were evaluated. Shown are the average values of these four functionals, and the proportion of the 100 samples for which the sample functional was less than the true functional.

use the order statistics of the bootstrap replications. Suppose $\theta = t(F)$ is a real-valued parameter of interest, such as a mean, a correlation, an eigenvalue, etc., estimated by $\hat{\theta} = t(\hat{F})$. We draw some large number "B" of independent bootstrap samples $\mathbf{x}^{*1}, \mathbf{x}^{*2}, \cdots, \mathbf{x}^{*B}$, where in the one-sample situation (1.1), each $\mathbf{x}^*$ is a random sample of size n from $\hat{F}$,

$$\hat{F} \stackrel{\text{i.i.d.}}{\rightarrow} (y_1^*, y_2^*, \cdots, y_n^*) \equiv \mathbf{x}^*. \tag{1.12}$$

The y_i do not have to be real-valued; F can be a distribution on an arbitrary probability space.

Each bootstrap sample gives a bootstrap replication of $\hat{\theta} = t(\hat{F})$, namely $\hat{\theta}^{*b} = t(\hat{F}^{*b})$, where $\hat{F}^{*b}$ is the empirical distribution corresponding to the bth bootstrap sample $\mathbf{x}^{*b}$. The <u>percentile confidence interval</u>, of approximate coverage 90%, is defined to be the central 90% range of the ordered bootstrap replication. In the usual order-statistic notation, the 90% percentile interval is $[\hat{\theta}^*_{(.05B)}, \hat{\theta}^*_{(.95B)}]$.

The length of the percentile interval,

$$\widehat{\text{length}} = \hat{\theta}^*_{(.95B)} - \hat{\theta}^*_{(.05B)} \tag{1.13}$$

is a dispersion functional of $\hat{F}$, similar to $\text{Var}_{\hat{F}}\{Y\}$ except that it is usually evaluated by a Monte Carlo process. Like $\text{Var}_{\hat{F}}$, $\widehat{\text{length}}$ often tends to be a little too small, leading to undercoverage for the percentile confidence interval.

Once again, this effect is tiny from the asymptotic point of view. The <u>standard confidence interval</u> for a parameter θ,

$$\hat{\theta} \pm z^{(\alpha)} \widehat{\text{se}}, \tag{1.14}$$

where $\hat{\theta} = \theta(\hat{F})$ and $\widehat{\text{se}}$ is any reasonably efficient estimate of standard error for $\hat{\theta}$, is typically first-order accurate: its non-coverage probability, in each tail, approaches the nominal value $1-\alpha$ at rate $1/\sqrt{n}$. The bootstrap confidence interval "BC_a", Efron (1987), which is an improved version of the percentile interval, is second-order accurate: its non-coverage probabilities approach $(1-\alpha)$ at the factor rate $1/n$. Its coverage errors are of the third order, $O(\frac{1}{n})$, corresponding to the third order shortening effect seen in (1.5). However, even third-order errors can be consequential in small samples. This is especially true if the confidence intervals are used for hypothesis testing purposes, where small errors in the end-points can change the verdict of the test.

There are other third-order effects that produce coverage errors of order $1/n$ in the BC_a intervals. Efron (1985) details the third-order errors for bootstrap confidence intervals in a certain class of parametric problems. Perhaps the moral here is that "plug-in" methods, like maximum likelihood and the bootstrap, give answers accurate to the second-order but not to the third; and that the tendency toward shortness of maximum likelihood estimated distributions accounts for part of the third-order error.

2. Why does the delta method tend to underestimate standard errors? The delta method is the oldest device for assessing the standard errors of complicated statistical estimates, predating the jackknife and bootstrap methods by at least 150 years. The delta method is also known as the method of statistical differentials, propagation of errors formula, and the Taylor series method. Despite its ancient pedigree, I have found the delta method to be less reliable than the jackknife or the bootstrap, with an occasional tendency to badly underestimate the true standard error.

Table 2 shows the results of all four sampling experiments in Efron (1982) that compare the delta method, bootstrap, and jackknife estimates of standard error with the true standard error. In the first experiment, each sample consisted of ten independent pairs (x_i, y_i), with x_i having a uniform distribution on $[0, 1]$, independent of y_i, which had a G_1 (one-sided exponential) distribution. The ratio statistic $\hat{\theta} = \bar{y}/\bar{x}$ then had true standard error .37, as verified by extensive Monte Carlo sampling. Each sample in the experiment gave a delta method, bootstrap, and jackknife estimate of standard error. In this case all three methods were nearly unbiased, giving average standard error estimates of .35, .37, and .37 respectively. Details of the four sampling experiments appear in Efron (1982), Section 3.

In the last three sampling experiments, the delta method has a noticeable downward bias. This is particularly evident in the last case, where the statistic of interest was the $\tanh^{-1}$ transformation of a simple correlation coefficient from 15 independent bivariate normal points.

A puzzling aspect of the differences seen in Table 2 is that the delta method is intimately related to both the jackknife and the bootstrap. In fact the delta method is identical to Jaeckel's (1972) infinitesimal jackknife. Suppose $\hat{\theta}$ is a functional statistic $\hat{\theta} = S(\hat{F})$, such as the ordinary mean $S(\hat{F}) = \int x d\hat{F} = \bar{x}$. Let $\hat{F}_i^\epsilon$ be a distorted version of the empirical distribution $\hat{F}$ that puts extra probability on the ith data point,

$$\hat{F}_i^\epsilon = \begin{cases} (1-\epsilon)/n + \epsilon & \text{probability on } x_i \\ (1-\epsilon)/n & \text{probability on } x_j, \quad j \neq i. \end{cases} \tag{2.1}$$

The empirical influence function is the derivative

statistic	se delta	estimate boot	jack	True se	n	Sample consists of n independent pairs (x_i, y_i), with distribution
$\bar{y}/\bar{x}$	.35	.37	.37	.37	10	$x_i \sim U(0,1)$ indep. of $y_i \sim G_1$
$\bar{y}/\bar{x}$	.53	.64	.70	.67	10	$x_i \sim U(0,1)$ indep. of $y_i \sim G_1^2/2$
sample correlation	.175	.206	.223	.218	14	$(x_i, y_i) \sim N_2(\mu, \Sigma)$ true corr = .5
$\tanh^{-1}$ (corr)	.244	.301	.314	.297	14	$(x_i, y_i) \sim N_2(\mu, \Sigma)$, true corr = .5

Table 2. Four sampling experiments comparing the delta method, bootstrap, and jackknife estimates of standard error, from Efron (1982). Number tabled is the average estimate in the sampling experiment. The delta method gives the smallest estimates. In the last three cases, the delta method estimates are considerably too small.

$$U_i \equiv \frac{\partial}{\partial \epsilon} S(\hat{F}_i^\epsilon)|_{\epsilon=0}. \tag{2.2}$$

Efron (1981) showed that the usual nonparametric delta function estimate of standard error using a linear Taylor series expansion of $\hat{\theta}$ is identical to the infinitesimal jackknife formula suggested by Jaeckel,

$$\text{se}_{\text{delta}}\{\hat{\theta}\} \equiv [\sum_{i=1}^{n} U^2/n^2]^{1/2}. \tag{2.3}$$

Notice that setting $\epsilon = -1/(n-1)$ in (2.1) gives the empirical distribution for the reduced data set with the ith point removed,

$$\hat{F}_i^{-(1/(n-1))} \equiv \hat{F}_{(i)} = \begin{cases} 0 & \text{probability on } x_i \\ \frac{1}{n-1} & \text{probability on } x_j, \quad j \neq i, \end{cases} \tag{2.4}$$

Let $\hat{\theta}_{(i)} = S(\hat{F}_{(i)})$ and $\hat{\theta}_{(\cdot)} = \sum_{i=1}^{n} \hat{\theta}_{(i)}/n$. Tukey's jackknife estimate of standard error is

$$\text{se}_{\text{jack}}\{\hat{\theta}\} = \{\frac{n-1}{n}\Sigma[\hat{\theta}_{(i)} - \hat{\theta}_{(\cdot)}]^2\}^{1/2}. \tag{2.5}$$

This is almost the same as (2.3), except that ϵ in (2.2) has been set equal to $-1/(n-1)$ instead of going to zero. (Tukey's formula also incorporates an extra factor of $n/(n-1)$, in order that $\text{se}_{\text{jack}}\{\bar{x}\}$ exactly equal the usual formula for the standard error of the mean, $[\Sigma(x_i-\bar{x})^2/n(n-1)]^{1/2}$.)

Efron and Stein (1981) showed that the jackknife estimates of variance, $\text{se}_{\text{jack}}\{\hat{\theta}\}^2$, tends to be biased upward as an estimate of the variance. A moderate upward bias is discernable in the jackknife column of Table 2. The close relationship of definitions (2.5) and (2.3) suggests that the jackknife and delta method should behave similarly, but such is not always the case.

The bootstrap method gives the nonparametric maximum likelihood estimate (MLE) of standard error: let $\text{se}\{t; F\}$ indicate the true standard error of a statistic $\hat{\theta} = t(\mathbf{x})$, where $\mathbf{x} = (x_1, x_2, \cdots, x_n)$ is an i.i.d. sample from F; then the bootstrap estimate of standard error is

$$\text{se}_{\text{boot}}\{\hat{\theta}\} = \text{se}\{t; \hat{F}\}, \tag{2.6}$$

where $\hat{F}$ is the nonparametric MLE of F, i.e. the empirical distribution of the sample $x_1, x_2, \cdots, x_n$. In other words, $\text{se}_{\text{boot}}\{\hat{\theta}\}$ is the nonparametric MLE of the true standard error.

We expect maximum likelihood estimates to be nearly unbiased, as demonstrated by the bootstrap column of Table 2. The next argument shows the close theoretical connection between all three estimators, the bootstrap, the delta method, and the jackknife. This deepens the puzzle of the delta method's poor performance.

A bootstrap sample is a random sample of size n from $\hat{F}$,

$$\hat{F} \xrightarrow{\text{i.i.d.}} (x_1^*, x_2^*, \cdots, x_n^*) \equiv \mathbf{x}^*. \tag{2.7}$$

Corresponding to $\mathbf{x}^*$ is the bootstrap replication $\hat{\theta}^* = t(\mathbf{x}^*)$. Definition (2.6) is an ideal version of $\text{se}_{\text{boot}}\{\hat{\theta}\}$, which is usually approximated by Monte Carlo: independent bootstrap samples $\mathbf{x}^{*1}, \mathbf{x}^{*2}, \cdots, \mathbf{x}^{*B}$ give independent bootstrap replications $\hat{\theta}^{*1}, \hat{\theta}^{*2}, \cdots, \hat{\theta}^{*B}$; then

$$\text{se}_{\text{boot}}\{\hat{\theta}\} \doteq [\sum_{b=1}^{B}(\hat{\theta}^{*b} - \hat{\theta}^{*\cdot})^2/(B-1)]^{1/2}, \tag{2.8}$$

where $\hat{\theta}^{*\cdot} \equiv \sum_b \hat{\theta}^{*b}/B$. Expression (2.8) approaches (2.6) as $B \to \infty$. Usually B in the range $50 - 200$ gives good results, see Efron (1987).

Each member of a bootstrap sample $\mathbf{x}^* = (x_1^*, x_2^*, \cdots, x_n^*)$ is randomly selected from the original data set $(x_1, x_2, \cdots, x_n)$. Let $\mathbf{P}^*$ be the resampling vector $(P_1^*, P_2^*, \cdots, P_n^*)$, where

$$P_i^* = \#\{x_j^* = x_i\}/n, \tag{2.9}$$

the proportion of the bootstrap sample equalling x_i. Then $\mathbf{P}^*$ has a rescaled multinomial distribution, of n draws on n categories each with probability $1/n$,

$$\mathbf{P}^* \sim \text{Mult}(n, \mathbf{P}^o)/n \qquad [\mathbf{P}^o \equiv (\frac{1}{n}, \frac{1}{n}, \cdots, \frac{1}{n})]. \tag{2.10}$$

With the original sample $\mathbf{x}$ fixed, we can think of $\hat{\theta}^* = t(\mathbf{x}^*)$ as a function of $\mathbf{P}^*$, say $\hat{\theta}(\mathbf{P}^*)$. Another way to state (2.6) is

$$\text{se}_{\text{boot}}\{\hat{\theta}\} = [\text{Var}_*\{\hat{\theta}(\mathbf{P}^*)\}]^{1/2}, \tag{2.11}$$

where Var_* indicates variance under the multinomial distribution (2.11).

Figure 1 schematically represents (2.11). The prone triangle is the simplex $\mathcal{S}$ in which $\mathbf{P}^*$ takes its values. The function $\hat{\theta}(\mathbf{P}^*)$ is represented by a curved surface over $\mathcal{S}$. At the center of $\mathcal{S}$ is $\mathbf{P}^o$, with $\hat{\theta}(\mathbf{P}^o) = t(\mathbf{x})$, the actual simple value of the statistic. We usually need to approximate

(2.11) by Monte Carlo because there is no closed formula for the variance of a non-linear function of a multinomial vector.

Both the jackknife and the delta method avoid Monte Carlo by approximating $\hat{\theta}(\mathbf{P}^*)$ with a linear function of $\mathbf{P}^*$, say $\hat{\theta}_{\text{LIN}}(\mathbf{P}^*)$; and then theoretically evaluating $[\text{Var}_*\{\hat{\theta}_{\text{LIN}}(\mathbf{P}^*)\}]^{1/2}$, instead of (2.11), from the usual formula for the variance of a linear function of a multinomial.

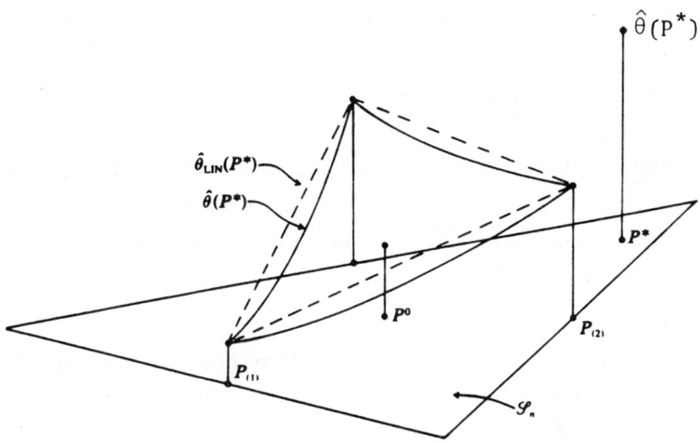

Figure 1. Schematic representation of bootstrap resampling; resampling vector $\mathbf{P}^*$ takes its values in $\mathcal{S}$; $\hat{\theta}(\mathbf{P}^*)$ is the curved surface over $\mathcal{S}$; bootstrap standard error is square root of variance of $\hat{\theta}(\mathbf{P}^*)$ where $\mathbf{P}^*$ has multinomial distribution (2.10). Dashed lines indicate the linear approximation to $\hat{\theta}(\mathbf{P}^*)$ used by the jackknife. The delta method uses the linear function tangent to the resampling surface at the central point $(\mathbf{P}^o, \hat{\theta}(\mathbf{P}^o))$.

The delta method standard error is based on the most obvious linear approximation to $\hat{\theta}(\mathbf{P}^*)$: the tangent hyperplane to the surface $\hat{\theta}(\mathbf{P}^*)$ through the central point $(\mathbf{P}^o, \hat{\theta})$. The jackknife uses the hyperplane which equals $\hat{\theta}(\mathbf{P}^*)$ at the n "jackknife points" $\mathbf{P}_{(1)}, \mathbf{P}_{(2)}, \cdots, \mathbf{P}_{(n)}$, where $\mathbf{P}_{(i)}$ is

$$(1, 1, \cdots, 1, 0, 1, \cdots, 1)/(n-1), \tag{2.12}$$

0 in the ith place. Once again, it is difficult to see why se_{jack} should be biased upwards and se_{delta} biased downwards.

We can set up an artificial estimation problem which gives some insight into the relationship between se_{delta} and se_{boot}. Consider the data $\mathbf{x}$ as fixed, and suppose that we observe a multinomial vector $\mathbf{P} \sim \text{Mult}(n, \underset{\sim}{\pi})/n$, where now $\underset{\sim}{\pi}$ can be any vector in the simplex $\mathcal{S}$, not just $\mathbf{P}^o$ as in (2.10). The artificial problem is to estimate $\hat{\theta}(\underset{\sim}{\pi})$ having observed $\mathbf{P}$.

The MLE of $\hat{\theta}(\underset{\sim}{\pi})$ in the artificial problem is $\hat{\theta}(\mathbf{P})$. It then turns out that $[\text{se}_{\text{boot}}\{\hat{\theta}\}]^2$ is the variance of the MLE when $\underset{\sim}{\pi} = \mathbf{P}^{\mathbf{o}}$, while $[\text{se}_{\text{delta}}\{\hat{\theta}\}]^2$ is the Cramer-Rao lower bound for the unbiased estimation of $\hat{\theta}(\underset{\sim}{\pi})$, at $\underset{\sim}{\pi} = \mathbf{P}^{\mathbf{o}}$. This argument makes it plausible that se_{delta} would usually be less than se_{boot}. "Plausible" isn't a proof though, and it isn't true that $\text{se}_{\text{delta}} < \text{se}_{\text{boot}}$ in every case.

The delta method is much too useful a tool to throw away. However, it's numerical results shouldn't be accepted uncritically, since they seem liable to underestimation. The jackknife and bootstrap standard errors are both more dependable.

3. Why are cross-validation estimators so variable? Cross-validation, like the delta method, is a time-honored technique for assessing statistical error. Modern computational equipment has greatly increased the use of cross-validation for choosing estimation rules and estimating their prediction errors. "Time-honored" is not the same as "tried and true". My own experience, as described in this section, is that cross-validation can be undependable in some situations, and that substantially better methods are available. The discussion here is taken from Efron (1983) and Chapter 7 of Efron (1982). Good references for cross-validation include Stone (1974), Geisser (1975), and Lachenbruch and Mickey (1968).

Figure 2 shows a sample obtained in a simulation experiment: we observe a <u>training set</u> of data (y_i, z_i), $i = 1, 2, \cdots, 14$, where

$$y_i = \begin{cases} 0 & \text{with probability } 1/2 \\ 1 & \text{with probability } 1/2, \end{cases} \tag{3.1}$$

and z_i is bivariate normal, with identity covariance, and mean depending on y_i,

$$z_i | y_i \sim N_2((y_i - \frac{1}{2}, 0), I) \tag{3.2}$$

The two means, $(\frac{1}{2}, 0)$ and $(-\frac{1}{2}, 0)$, are indicated by stars in Figure 2. In real practice, of course, we wouldn't know the probability mechanism generating the data.

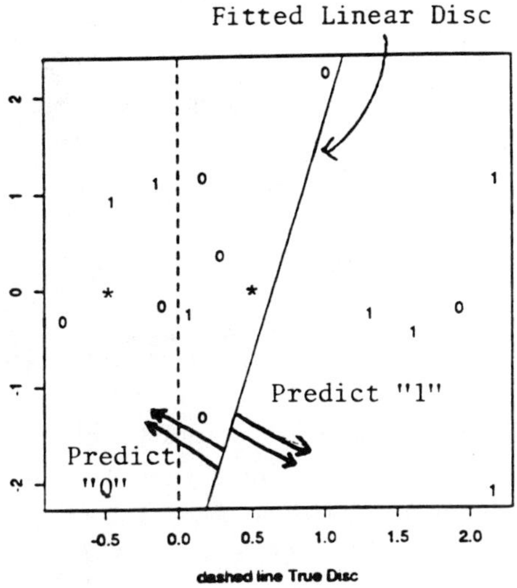

Figure 2. Fisher's linear discriminant (solid line) fit to 14 data points generated according to (3.1), (3.2); stars indicate mean values of the "0" and "1" distributions; the apparent error rate is 4/14 in this case.

Fisher's linear discriminant divides the prediction space, $\mathcal{R}^2$ in this case, into two regions, for the purpose of predicting the y value of a future pair (y, z), having observed only z. In practice, z might be some observable predictors, like age, sex, weight, etc. and y some dichotomous outcome that we want to predict, like the success or failure of a medical procedure.

The linear discriminant predicts 1 or 0 depending on the size of a certain linear function,

$$\widehat{\text{pred}}(z) = \begin{cases} 1 & if \quad a + b'z \geq 0 \\ 0 & if \quad a + b'z < 0. \end{cases} \tag{3.3}$$

The constant a and the vector b are functions of the training set $\{(y_i, z_i), i = 1, 2, \cdots, 14\}$, see (2.13) of Efron (1983). One of the main uses of cross-validation is to assess the prediction error of a data- based prediction rule like (3.3).

The True Error Rate of a prediction rule is the probability that it misclassifies a future observation,

$$\text{True Error} = \text{Prob}\{y \neq \widehat{\text{pred}}(z)\}. \tag{3.4}$$

The probability in (3.4) is conditional on the training set; $\widehat{\text{pred}}()$ is fixed, only y and z being random. For the sample in Figure 2, True Error = .344, as calculated from (3.1), (3.2). True Error rates cannot be calculated in practice since we usually don't know the probability mechanism generating (y, z).

An obvious estimate of True Error is the Apparent Error rate

$$\text{AppError} = \#\{y_i \neq \widehat{\text{pred}}(z_i)\}/14, \tag{3.5}$$

the proportion of the training set points misclassified by $\widehat{\text{pred}}$. Four of the 14 points are misclassified in Figure 2, three 1's and one 0, so AppError = 4/14 = .286. In this case the difference between True and Apparent rates,

$$\text{Diff} = \text{True} - \text{App}, \tag{3.6}$$

is positive. Diff is usually positive, at least in expectation, because $\widehat{\text{pred}}$ is fit to the training set, and so does better in "predicting" the training set than in predicting a genuinely new point.

Cross-validation gives a way of using the training set to obtain a nearly unbiased estimate of Diff. Let $\widehat{\text{pred}}_{(i)}$ be the prediction rule obtained from the reduced training set that excludes point (y_i, z_i). In our example

$$\widehat{\text{pred}}_{(i)}(z) = \begin{cases} 1 & \text{if} \quad a_i + b'_{(i)}z \geq 0, \\ 0 & \text{if} \quad a_i + b'_{(i)}z < 0, \end{cases} \tag{3.7}$$

where $a_{(i)}, b_{(i)}$ are the coefficients of Fisher's linear discriminant based on $(y_1, z_1), (y_2, z_2), \cdots,$
$(y_{i-1}, z_{i-1}), (y_{i+1}, z_{i+1}), \cdots, (y_{14}, z_{14})$. The cross-validation estimate of Diff is

$$\widehat{\text{Diff}}_{\text{CV}} = \#\{y_i \neq \widehat{\text{pred}}_{(i)}(z_i)\}/14 - \text{AppError}. \tag{3.8}$$

Intuitively, the first term of (3.8) should be a newly unbiased estimate of True Error, so $\widehat{\text{Diff}}$ should be nearly unbiased for Diff. We can illustrate this with a sampling experiment.

The sampling experiment comprised 100 independent replications of (3.1), (3.2). In other words, 100 independent realizations of Figure 2 were constructed. Table 3 shows the results. We see that the average of $\widehat{\text{Diff}}_{\text{CV}}$, the cross-validation estimate, averaged .091, compared to .096 for the true Diff. This corroborates the approximate unbiasedness of $\widehat{\text{Diff}}_{\text{CV}}$.

	Trial	True Err	App Err	Diff	$\widehat{\text{Diff}}$ CV	$\widehat{\text{Diff}}$ .632
	[1]	.458	.286	.172	.214	.083
	[2]	.312	.357	-.045	.000	.068
	[3]	.313	.357	-.044	.071	.095
	[4]	.351	.429	-.078	.071	.051
	[5]	.330	.357	-.027	.143	.094
	[6]	.318	.143	.175	.214	.086
	[7]	.310	.071	.239	.071	.083
	[8]	.380	.286	.094	.071	.130
	[9]	.360	.429	-.069	.071	.119
	[10]	.335	.143	.192	.000	.042
all	mean:	.360	.264	.096	.091	.076
100	sd:	.045	.123	.113	.073	.035
	RMSE:				.149	.117

Table 3. One hundred replications of (3.1), (3.2); shown are results of the first 10 replications, and summary statistics for all 100. The cross-validation estimte of Diff, "CV", is nearly unbiased for the true Diff, but has a large standard deviation.

Table 3 also shows that $\widehat{\text{Diff}}_{\text{CV}}$ is highly variable: its standard deviation over the 100 simulations was .073, about 80% as big as its mean .091. Unbiased or not, this makes $\widehat{\text{Diff}}_{\text{CV}}$ an undependable estimator of Diff.

Efron (1983) shows that cross-validation is closely related to the bootstrap, much as the delta method, jackknife, and bootstrap are related in Figure 1. This leads to several new estimators for Diff, based on variants of the underlying bootstrap argument. The most successful of these, "$\widehat{\text{Diff}}_{.632}$", also appears in Table 3. We see that it is moderately biased downward, but has much smaller standard deviation than $\widehat{\text{Diff}}_{\text{CV}}$. An objective way to compare the two procedures is in terms of their root mean square errors for estimating True Error = AppError + Diff,

$$\text{RMSE} = [E\{\text{True} - (\text{App} + \widehat{\text{Diff}})\}^2]^{1/2} \tag{3.9}$$

we see that RMSE was .149 for cross-validation, compared to .117 for the .632 rule.

Five sampling experiments are considered in Efron (1983), and eight estimators of Diff. $\widehat{\text{Diff}}_{.632}$ was clearly the winner in terms of RMSE, and cross-validation was even more clearly the loser. In two of the five experiments, using cross-validation was considerably worse than simply estimating True by App, i.e. taking $\widehat{\text{Diff}} = 0$.

We have another small mystery here: all of the Diff estimates, including cross-validation, are variants of the same bootstrap argument, and yet they perform quite differently in small-sample simulation experiments. My belief, or hope, is that future research will produce a dependable improvement over cross-validation. The problem of estimating the prediction error of a data-based prediction rule is important enough to discuss further study.

4. What is a correct confidence interval? Suppose we wish to construct a 90% central confidence interval for a real-valued parameter θ, having observed some data $\mathbf{x}$. A proposed interval $[\hat{\theta}_{\text{LO}}(\mathbf{x}), \hat{\theta}_{\text{UP}}(\mathbf{x})]$ is said to be <u>accurate</u> if it fails to cover θ exactly 5% of the time in each direction,

$$\text{Prob}\{\theta > \hat{\theta}_{\text{UP}}(\mathbf{x})\} = \text{Prob}\{\theta < \hat{\theta}_{\text{LO}}(\mathbf{x})\} = .05. \tag{4.1}$$

An accurate confidence interval is not necessarily a correct one, though. If $x_1, x_2, \cdots, x_{10}$ is a random sample from a $N(\theta, \gamma)$ distribution, with both the mean θ and the variance γ unknown, then the student's t interval for θ based on only $x_1, x_2, \cdots, x_5$ is obviously accurate in the sense of (4.1). Equally obvious, it is not the correct interval for θ. It is inferentially wrong, though probabilistically right.

The question of correctness arises forcibly in the construction of approximate confidence intervals. Various theories have been put forth to construct intervals accurate to a high degree of asymptotic approximation. We will discuss some of these theories in this section, and Sections 5 and 6 as well. Are the intervals they construct highly <u>correct</u> as well as highly accurate? Answering this question seems crucial if we are to avoid the pitfall of the student's t example above.

The notion of interval correctness is more difficult to pin down than interval accuracy. Correctness is clear-cut only in the simplest situations, where the data $\mathbf{x}$ can be reduced to a one-dimension sufficient statistic $\hat{\theta}$, and where the percentiles of $\hat{\theta}$ increase monotonically as a function of θ. Then the textbook method of confidence interval construction, taking $\hat{\theta}_{\text{LO}}$ to be that θ for which the observed value $\hat{\theta}$ is at the 95th percentile of possible outcomes, and similarly for $\hat{\theta}_{\text{HI}}$, gives what most statisticians would call the correct confidence interval for θ. Sometimes more complicated-looking situations can be reduced to the simple form. In a bivariate normal model for example, the maximum likelihood estimate $\hat{\rho}$ of the true correlation ρ has a distribution depending only on ρ.

Fieller's construction for the ratio of normal means gives an example of a correct confidence interval in a genuinely multiparametric situation. Suppose y is bivariate normal with identity covariance matrix, $y \sim N_2(\mu, I)$, and we want a confidence interval for the ratio $\theta(\mu) = \mu_2/\mu_1$. The level surface $\{\theta(\mu) = \theta_0\}$ is a straight line passing through the origin at angle $\tan^{-1}(\theta_0)$, as shown in the left panel of Figure 3.

There is an obvious way to test the hypothesis $H_0 : \theta(\mu) = \theta_0$. Let D_0 be the signed distance from y to the straight line $\{\theta(\mu) = \theta_0\}$. [Almost any sign convention will do, for instance $\text{sign}(D_0) = \text{sign}(y_2 - y_{02})$, where y_0 is the nearest point to y on $\{\theta(\mu) = \theta_0\}$.] Then D_0 has a standard $N(0,1)$ distribution if $\mu \in \{\theta(\mu) = \theta_0\}$. The obvious .90 two-sided test rejects H_0 if y lies outside the band $|D_0| < 1.645$. Fieller's confidence interval comprises those values θ_0 such that the test accepts the hypothesis $\theta(\mu) = \theta_0$. In other words, it is these values θ_0 such that the distance from y to $\{\theta(\mu) = \theta_0\}$ is less than 1.645, as shown in the right panel of Figure 3. See Section 5.14 of Miller (1986). Most, though not all, statisticians find the pivotality of D_0 an irresistable argument for the correctness of the Fieller intervals.

Fieller's construction depends on the level surface $\{\theta(\mu) = \theta_0\}$ being straight lines. Suppose we consider a parameter for which the level surface $C_{\theta_0} = \{\theta(\mu) = \theta_0\}$ are curved, for example $\theta(\mu) = \mu_1\mu_2$, where the C_{θ_0} are hyperbolae. Efron (1985) discusses an approximate version of Fieller's construction applying to the curved case.

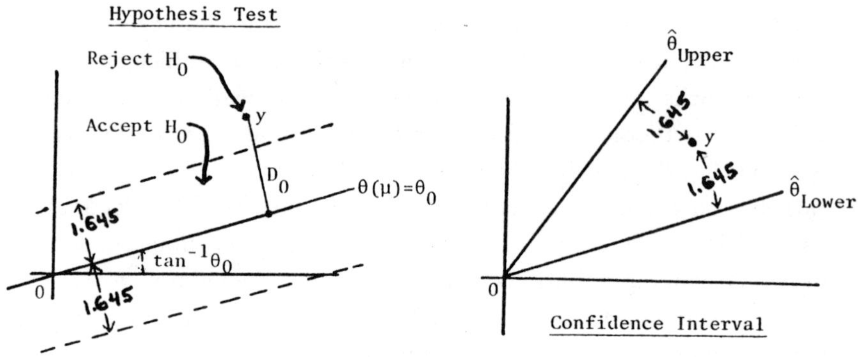

Figure 3. Fieller's construction of a 90% confidence interval for the rates of normal means; we observe $y \sim N_2(\mu, I)$ and want a confidence interval for $\theta(\mu) = \mu_2/\mu_1$; the level surface $\{\theta(\mu) = \theta_0\}$ is the straight line through the origin at angle $\tan^{-1}(\theta_0)$; hypothesis test for $H_0 : \theta = \theta_0$ accepts H_0 for $|D_0| < 1.645$, where D_0 is the signed distance from y to $\{\theta(\mu) = \theta_0\}$, left panel; the Fieller interval for θ is these values of θ_0 for which H_0 is accepted, right panel.

Let y_0 be the nearest point to the observed point $y \sim N_2(\mu, I)$ on $\mathcal{C}_{\theta_0}$, and let D_0 be the signed length of $y - y_0$ (using any reasonable sign convention). Also define curv_0 to be the curvature of $\mathcal{C}_{\theta_0}$ at the point y_0. Then the adjusted signed distance,

$$D_0' \equiv \frac{D_0 - \text{curv}_0}{1 - \text{curv}_0^2} \tag{4.2}$$

is approximately normal. This approximation is very good, the cdf of D_0' differing from the standard normal cdf by only $O(n^{-3/2})$ if y is actually a sufficient statistic obtained from n observations $y_1, y_2, \cdots, y_n \overset{\text{i.i.d.}}{\sim} N_2$ (μ, I). This is third-order accuracy, in the language of Section 1.

Inverting the approximate pivotal D_0' gives a third-order accurate approximate confidence interval for θ. Table 4 shows the results for the case $y \sim N_2(\mu, I)$, $\theta(\mu) = \|\mu\|$, when the observed vector y has length $\|y\| = 5$. (A version of (4.2) holds in higher dimensions.) In this case there is an exact confidence interval for θ based on inverting the non-central chi-square distribution of $\|y\|^2 \sim \chi_5^2(\theta^2)$.

We see that the approximate interval based on the adjusted signed distance gives excellent results in this case. The second order accurate bootstrap interval "BC_a" is not quite as good, while the first order accurate standard interval, (1.14), is far off.

	.05	.95	
Exact:	2.68	6.19	
D_0':	2.71	6.19	3rd order
BC_a:	2.94	6.06	2nd order
Standard:	3.36	6.64	1st order

Table 4. We observe $\|y\| = 5$, where $y \sim N_6(\mu, I)$, and want a confidence interval for $\theta(\mu) = \|\mu\|$; exact two-sided 90% interval $(2.68, 6.19)$ is based on inverting the noncentral χ^2 distribution of $\|y\|^2$; interval based on inverting the approximate pivotal D_0' (4.2), is third order accurate; bootstrap method BC_a is second order accurate; standard method is first order accurate. From Efron (1985).

In this example, most statisticians, or at least most frequentists, would consider the non-central chi-square intervals to be correct, as well as exactly accurate. The D_0' intervals are third order correct, as well as third order accurate, in the sense that their endpoints differ from the exact ones by only $O(n^{-3/2})$ if y is obtained from an i.i.d. sampling situation.

Barndorff-Nielsen (1986) extends (4.2) to general parametric families. The signed distance D_0 is replaced by the signed square root of the likelihood ratio statistic. "Bartlett corrections" are made to the mean and standard deviation of D_0, to get an adjusted statistic D_0' which has a standard normal distribution to the third order of asymptotic accuracy. Then D_0' is inverted to give third-order accurate approximate confidence intervals for θ.

Are these intervals third order correct as well as third order accurate? Efron (1985) argues for the correctness of the D_0' intervals based on the fact that they come from an approximate pivotal that is geometrically reasonable. But (4.2) lacks the immediacy of Fieller's construction, even in the normal case. Moreover, there are other third order intervals available, Hall (1988), Cox and Reid (1987), Welch and Peers (1963), which may or may not agree with Barndorff-Nielsen's intervals.

DiCiccio and Efron (1990) show that all of these methods give confidence interval agreeing at the second order, and in a certain sense they all are second order correct, as well as accurate. The best situation would be if all the methods continued to agree at the third order. This happy result might very well be false. If so, the question of correctness will be a pressing one. Highly accurate confidence statements are not worth pursuing if they lead to inferential errors.

5. What is a good nonparametric pivotal quantity? Pivotal quantities play a crucial role in the theory of confidence intervals, as we saw in the previous section. Much of the recent bootstrap literature concerns the construction of approximate confidence intervals in nonparametric settings. This raises the question of this section: what is a good approximate pivotal quantity in nonparametric estimation problems?

We consider the one-sample situation, where the observed data $\mathbf{x}$ is an i.i.d. sample from some unknown probability distribution F

$$F \overset{\text{i.i.d.}}{\rightarrow} (x_1, x_2, \cdots, x_n) = \mathbf{x}, \tag{5.1}$$

and where we want an approximate confidence interval for a real-valued parameter $\theta(F)$. This becomes a nonparametric problem if we assume that F can be any distribution at all on the sample space of the x_i.

We begin with an example of a bad guess at a nonparametric pivotal quantity. The left panel of Figure 4 shows the lawschool data, 15 pairs of points $x_i = (a_i, b_i)$, where a_i and b_i are measures of excellence for the entering 1975 class at lawschool i, $i = 1, 2, \cdots, 5$. See Section 2.5 of Efron (1982). Suppose we want a confidence interval for $\theta(F)$ the Pearson correlation coefficient. The nonparametric MLE of θ is $\hat{\theta} = \theta(\hat{F}) = .776$, the sample correlation coefficient based on the 15 points $\mathbf{x} = (x_1, x_2, \cdots, x_{15})$.

The right side of Figure 4 shows the histogram of $B = 1000$ bootstrap replications of $\hat{\theta}^* - \hat{\theta}$; $B = 1000$ independent bootstrap samples $\mathbf{x}^{*1}, \mathbf{x}^{*2}, \cdots, \mathbf{x}^{*B}$ were generated as in (2.7), and for each one the difference of correlation coefficients $\hat{\theta}^{*b} - \hat{\theta}$ (= sample correlation coefficient of bth bootstrap sample $\mathbf{x}^{*b}$, minus .776) were calculated. We see that the 5th and 95th percentiles of the 1000 $\hat{\theta}^{*b} - \hat{\theta}$ values were $-.254$ and .170 respectively.

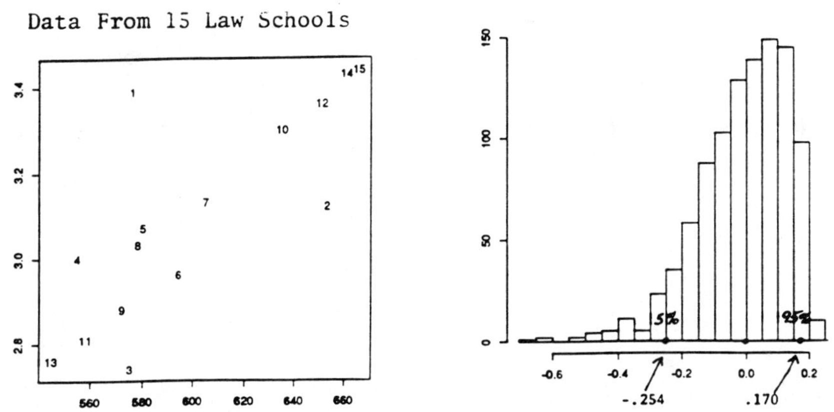

Figure 4. Lawschool data, $n = 15$ pairs of points, left panel. $B = 1000$ bootstrap replications of $\hat{\theta}^* - \hat{\theta}$, where $\hat{\theta}$ is the Pearson correlation, right panel. The quantity $\hat{\theta}^* - \hat{\theta}$ is a bad guess at a nonparametric pivotal quantity in this case.

Suppose we believe that $\hat{\theta} - \theta$ is an approximate nonparametric pivotal quantity. Then $\hat{\theta}^* - \hat{\theta}$ should have approximately the same percentiles as $\hat{\theta} - \theta$, so that

$$\text{Prob}\{-.254 < \hat{\theta} - \theta < .170\} \doteq .90. \tag{5.2}$$

Inverting (5.2) gives a bad guess at a confidence interval for θ,

$$\theta \in [\hat{\theta} - .170, \hat{\theta} + .254] = [.606, 1.03]. \tag{5.3}$$

The quantity $\hat{\theta} - \theta$ is a poor choice for an approximate pivotal statistic in most situations, either parametric or nonparametric. If $\hat{\theta} - \theta$ is long-tailed to the left, as in the correlation example, then usually the confidence interval for θ will extend further left of $\hat{\theta}$ than right. (The reader can check this by considering binomial or Poisson confidence intervals.) This is the opposite of what would happen if $\hat{\theta} - \theta$ were actually pivotal.

The second most obvious guess for an approximate nonparametric pivotal quantity is a t-like statistic, say

$$T = \frac{\hat{\theta} - \theta}{\hat{\sigma}}, \tag{5.4}$$

where $\hat{\sigma}(\mathbf{x})$ is some estimate of standard error for $\hat{\theta}(\mathbf{x})$, perhaps the jackknife or delta method estimates. If we believe in the pivotality of T, then we can use the bootstrap to construct a "bootstrap t" approximate confidence interval for θ; we generate some large number B of bootstrap replications of T,

$$T^* = \frac{\hat{\theta}(\mathbf{x}^*) - \hat{\theta}}{\hat{\sigma}(\mathbf{x}^*)}; \tag{5.5}$$

compute the 5th and 95th percentiles of the values T^{*b}, $b = 1, 2, \cdots, B$, say $T^{*(.05)}$ and $T^{*(.95)}$; and assign θ the approximate confidence interval

$$\theta \in [\hat{\theta} - T^{*(.95)}\hat{\sigma},\ \hat{\theta} - T^{*(.05)}\hat{\sigma}]. \tag{5.6}$$

A surprising and encouraging fact has emerged in the bootstrap literature. The bootstrap-t method gives second order accurate and correct intervals in a wide variety of situations. (See P. Hall (1988), Abramovitch and Singh (1985), and DiCiccio and Efron (1990).) There are good reasons to believe that the T statistic (5.4) is a nonparametric pivotal to the second order of asymptotic accuracy.

The favorable asymptotics of the bootstrap-t method are no guarantee of good small-sample behavior. In fact, there are practical difficulties connected with the choice of the denominator $\hat{\sigma}$ in (5.4). This can be seen in the lawschool correlation example. $B = 1000$ bootstrap replications of T were generated from the lawschool data, with

$$\hat{\sigma}(\mathbf{x}^*) = [(1 - \hat{\theta}(\mathbf{x}^*)^2)]/\sqrt{15} + .03. \tag{5.7}$$

Here $[1 - \hat{\theta}^2]/\sqrt{15}$ is an approximation to the standard error of $\hat{\theta}$, suggested by normal theory. The added quantity .03 was necessary to prevent occasional very large values of T^*. Section 5 of Efron (1990) gives a full explanation.

The 5th and 95th percentiles of the T^* values were $-.939$ and 2.93 respectively. Using (5.6), this gives the bootstrap-t confidence interval $[.40, .91]$ for θ. The 95th percentile 2.93 is suspiciously large, leading to a suspiciously low lower endpoint .40, see Table 5.

The same bootstrap replications that gave the bootstrap-t percentiles can be used to check the pivotality of T^*'s distribution. This relieves the statistician's reliance on asymptotic theory. Figure 5 shows the 5th, 10th,

16th, 50th, 84th, 90th, and 95th percentiles of the 1000 T^* bootstrap replications as dashed lines. So, for example, the 5% dashed line is at height $-.939$, and the 95% line at height 2.93.

		Nonparametric			Parametric		
	$\hat{\theta}^* - \hat{\theta}$	Standard	Boot-t	BC_a	Boot-t	BC_a	Exact
.05:	.61	.61	.40	.48	.53	.50	.49
.95:	1.03	.95	.91	.94	.93	.90	.90

Table 5. Approximate confidence intervals for the correlation coefficient, lawschool data; the bootstrap-t and BC_a methods are second order accurate and correct. Parametric intervals assume a bivariate normal model for the data. The lower endpoint of the nonparametric bootstrap-t interval is suspiciously low, compared with the BC_a answer and also with the parametric results.

The percentiles of the T distribution when sampling from F, as in (5.1), are functions of F. Bootstrap sampling gives these function values for $F = \hat{F}$, the empirical distribution (1.2). If T is approximately pivotal, then we expect the percentiles not to change much if we change F from $\hat{F}$ to some other nearby distribution. The jagged lines in Figure 5 trace the percentiles of T^* as we change $\hat{F}$ to the deleted-point distributions $\hat{F}_{(i)}$, (2.4). That is, we successively remove one point at a time from the original data set $\mathbf{x}$, and compute the percentiles of the bootstrap-t distribution when sampling from $F = \hat{F}_{(i)}$, the empirical distribution of the remaining 14 points. Efron (1990a) shows how this computation can be done without requiring any bootstrap samples beyond the original 1000.

The upper bootstrap-t percentiles are seen to be quite variable under small changes in $\hat{F}$. This argues against the pivotality of T, in this situation. The jackknife-after-bootstrap theory in Efron (1990a) uses the jagged line variability to assign a standard error of ± 1.8 to the estimated value 2.93 for the 95th percentile. We simply do not have enough data to estimate the nonparametric bootstrap-t percentiles very well in this problem.

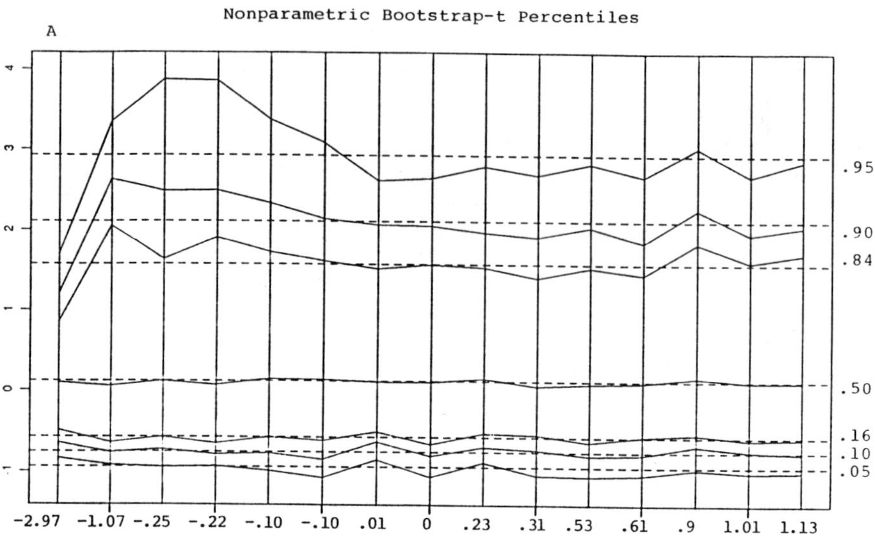

Figure 5. The percentiles of the 1000 bootstrap-t replications T^*, dashed lines; also the percentiles of T^*, successively removing one point at a time from the original data set, jagged lines. The upper percentiles are highly sensitive to removing single data points, indicating a lack of pivotality. From Section 5 of Efron (1990a). Bottom numbers show point removed, numbered as in Figure 4.

The T statistic (5.4) is a reasonable answer to the question "what is a good nonparametric pivotal quantity?", at least in theory. Further development will be needed to make the bootstrap-t a dependable method in practice. It is not yet clear when bootstrap-t methods, even with the bugs worked out, will be preferable to other nonparametric confidence interval methods like the BC_a.

6. What are computationally efficient ways to bootstrap? Typical problems require $50 - 200$ bootstrap replications to estimate a standard error, and $1000 - 2000$ replications to compute a bootstrap confidence interval, see Section 9 of Efron (1987). These numbers assume that the bootstrap estimation is done in the most obvious way. Various computational and probabilistic methods have been suggested to reduce the number of replications required. The promise of such methods is not only a reduction of the computational burden, but also a deeper understanding of the bootstrap. This section discusses just two of the methods, with references to many others.

Suppose again that we are in the one-sample nonparametric situation (5.1), and that we wish to assess the bias of the statistic $\hat{\theta}(\mathbf{x})$ as an estimate of the parameter $\theta(F)$. The straightforward bootstrap estimate of bias is calculated as follows: B bootstrap samples $\mathbf{x}^{*b}$ give replications $\hat{\theta}^{*b} = \hat{\theta}(\mathbf{x}^{*b})$, $b = 1, 2, \cdots, B$; then the bias of $\hat{\theta}$ is estimated by

$$\bar{\text{bias}}_B = \frac{1}{B}\sum_{b=1}^{B} \hat{\theta}^{*b} - \hat{\theta}(\mathbf{x}). \tag{6.1}$$

In the resampling vector notation of (2.9)-(2.11),

$$\bar{\text{bias}}_B = \frac{1}{B}\sum_{b=1}^{B} \hat{\theta}(\mathbf{P}^{*b}) - \hat{\theta}(\mathbf{P}^{o}), \tag{6.2}$$

where $\mathbf{P}^{*b}$ is the resampling vector corresponding to $\mathbf{x}^{*b}$, and $\mathbf{P}^{o}$ is $(1, 1, \cdots, 1)'/n$, the center point of the resampling simplex $\mathcal{S}$ in Figure 1. As B goes to infinity, $\bar{\text{bias}}_B$ approaches the ideal bootstrap bias estimate

$$\text{bias}_\infty = E_*\{\hat{\theta}(\mathbf{P}^*)\} - \hat{\theta}(\mathbf{P}^{o}), \tag{6.3}$$

E_* indicating expectation with respect to the multinomial distribution (2.10).

Section 2 of Efron (1990b) discusses an improvement on $\bar{\text{bias}}_B$,

$$\widehat{\text{bias}}_B = \frac{1}{B}\sum_{b=1}^{B} \hat{\theta}(\mathbf{P}^{*b}) - \hat{\theta}(\bar{\mathbf{P}}), \tag{6.4}$$

where $\bar{\mathbf{P}} = \sum_b \mathbf{P}^{*b}/B$, the average of the B resampling vectors. (This assumes that $\hat{\theta}(\mathbf{P}^*)$ is continuously defined as a function of $\mathbf{P}^*$, so that $\hat{\theta}(\bar{\mathbf{P}})$ is well specified.) As B increases, the improved estimate $\widehat{\text{bias}}_B$ approaches the ideal value bias_∞ more quickly than does $\bar{\text{bias}}_B$.

Table 6 refers to the following simulation experiment: ten samples were selected, each consisting of $n = 10$ independent bivariate normal points $x_i = (y_i, z_i)$, drawn from the distribution $F = N_2((8, 4)'/\sqrt{10}, I)$.

We are interested in the bias of $\hat{\theta}(\mathbf{x}) = \bar{z}/\bar{y}$ as an estimate of $\theta(F) = E\{y\}/E\{z\}$.

Table 6 compares $\widehat{\text{bias}}_B$ with $\overline{\text{bias}}_B$ for $B = 20$. The bias estimates are multiplied by 1000 for each reading. Also shown is $\overline{\text{bias}}_{8000}$, which we use in place of bias_{20} as the ideal bootstrap bias estimate. We see that $\widehat{\text{bias}}_{20}$ is much better than $\overline{\text{bias}}_{20}$ in matching this ideal, the ratio of squared errors for the 10 samples being

$$\Sigma[\overline{\text{bias}}_{20} - \overline{\text{bias}}_{8000}]^2/\Sigma[\widehat{\text{bias}}_{20} - \overline{\text{bias}}_{8000}] = 23.9.$$

In a much larger version of Table 6, this ratio was estimated to be 50.3. In effect, this means that $\widehat{\text{bias}}_{20}$ is about as effective as $\overline{\text{bias}}_{20\times 50}$!

Sample	$\overline{\textit{bias}}_{8,000}$	$\widehat{\textit{bias}}_{20}$	$\overline{\textit{bias}}_{20}$
1	7.35	4.08	−16.53
2	23.45	26.78	31.90
3	10.90	11.49	27.19
4	17.30	15.00	−31.12
5	.95	−.60	2.56
6	3.50	3.35	−10.00
7	14.55	13.99	.56
8	3.90	3.18	9.82
9	4.45	6.68	.70
10	12.85	23.90	3.55

NOTE: Entries are 1,000 × bias estimate. In this case $\widehat{\text{bias}}_{20}$ is about 50 times as good an estimator as $\overline{\text{bias}}_{20}$.

Table 6. Estimates of bias for $\hat{\theta} = \bar{z}/\bar{y}$; ten samples, each of size $n = 10$ drawn from $(y_i, z_i) \sim N_2/(8,4)'/\sqrt{10}, I)$; $\widehat{\text{bias}}_{20}$ compared with $\overline{\text{bias}}_{20}$ as an estimate of $\overline{\text{bias}}_{8000} \doteq \text{bias}_\infty$; $\widehat{\text{bias}}_{20}$ tends to be closer to $\overline{\text{bias}}_{8000}$. Entries are 1000 × bias estimate. From Section 2 of Efron (1990b).

The bias estimate $\widehat{\text{bias}}_B$ corrects $\overline{\text{bias}}_B$ by taking into account the discrepancy between $\bar{\mathbf{P}}$, the average of the observed resampling vectors, and $\mathbf{P}^o$, their theoretical expectation. Davison, Hinkley, and Schechtman (1987) make the correction another way: they draw the resampling vectors $\mathbf{P}^{*b}$, $b = 1, 2, \cdots, B$, in a manner which forces $\bar{\mathbf{P}}$ to equal $\mathbf{P}^o$. Then (6.2) performs very much like $\widehat{\text{bias}}_B$. Various methods of improved bootstrap sampling for reducing the number of bootstrap replications appear in Johns (1988), Graham et al. (1987), Ogbonmwan and Wynn (1986), Therneau (1983), and Hesterberg (1988).

In some situations it is possible to obtain bootstrap results without any Monte Carlo sampling at all. DiCiccio and Efron (1990) discuss a method for confidence interval construction within exponential families. Their method called "ABC", for "Approximate BC_a" or "Approximate Bootstrap Confidence" intervals, replaces the Monte Carlo bootstrap replications of the (parametric) BC_a method with simple analytic approximations. This is possible because exponential family problems tend to be very smooth. The ABC method requires only $4p + 6$ recomputations of the statistic of interest, compared to a few thousand for the full BC_a intervals, where p is the dimension of the exponential family.

Figure 6 shows two examples of the ABC method. The right panel concerns the lawschool data of Figure 4. Now we assume that the data is bivariate normal, $x_i \overset{\text{i.i.d.}}{\sim} N_2(\mu, \Sigma)$ for $i = 1, 2, \cdots, 15$, with the mean vector μ and covariance matrix Σ unknown. This is a $p = 5$ parameter exponential family. We want a confidence interval for the correlation coefficient. The first-order correct standard intervals, dashed lines, are compared with the second-order correct ABC intervals, solid lines. The ABC limits lie far below the standard limits. The solid points indicate the exact limits for the correlation coefficient, assuming bivariate normality. We see that the ABC limits are nearly, but not perfectly, correct.

The right side of Figure 6 concerns the student score data from Mardia, Kent, and Bibby (1979). This comprises 5 test scores from each of $n = 88$ students. We assume that the scores have a multivariate normal distribution, $x_i \overset{\text{i.i.d.}}{\sim} N_5(\mu, \Sigma)$, $i = 1, 2, \cdots, 88$ a $p = 20$ parameter exponential family. In this case the ABC limits are shifted upwards from the standard limits. The solid points show the BC_a limits from a full bootstrap analysis involving $B = 4000$ parametric bootstrap samples. We see that the ABC limits are indeed good approximations to those from the BC_a, even though they require only a few percent as much computation.

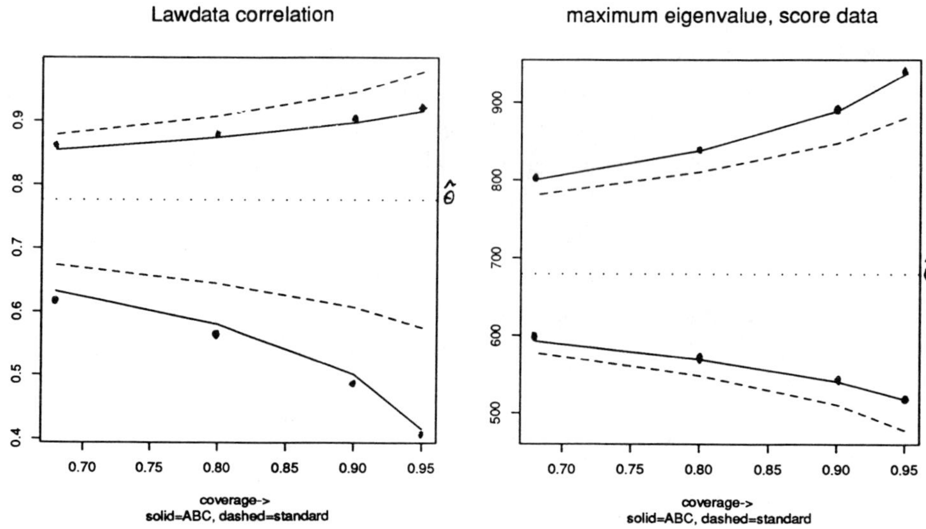

Figure 6. ABC second-order correct confidence intervals, solid lines, compared with the standard interval (1.14), dashed lines; horizontal axis is coverage probability of the two-sided interval; dotted line is MLE $\hat{\theta}$. Left panel lawschool data of Figure 4; assumes $x_i \overset{\text{i.i.d.}}{\sim} N_2(\mu, \Sigma)$, $i = 1, 2, \cdots, 15$; parameter of interest is correlation coefficient of Σ; solid points indicate exact confidence intervals for correlation. Right panel score data, Mardia, Kent, and Bibby (1979); assumes $x_i \overset{\text{i.i.d.}}{\sim} N_5(\mu, \Sigma)$, $i = 1, 2, \cdots, 88$; parameter of interest is maximum eigenvalue of Σ; solid points are BC_a limits, $B = 4000$ replications.

REFERENCES

Abramovitch, L. and Singh, K. (1985). Edgeworth corrected pivotal statistics and the bootstrap. *Ann. Stat.* **13**, 116-132.

Barnforff-Nielsen, O. E. (1986). Inference on full or partial parameters based on the standardized signed log likelihood ratio. *Biometrika* **73**, 307-322.

Cox D. R., and Reid, N. (1987). Parameter orthogonality and approximate conditional inference (with discussion). *J. Royal Stat. Soc., Ser. B* **49**, 1-39.

Davison, A. C., Hinkley, D. V., and Schechtman, E. (1987). Efficient bootstrap simulation. *Biometrika* **73**, 555-561.

DiCiccio, T., and Efron, B. (1990). Better approximate confidence intervals in exponential families. Submitted to *JASA*.

Efron, B. (1981). Nonparametric estimates of standard error: the jackknife, the bootstrap, and other resampling methods. *Biometrika* **68**, 589-599.

Efron, B. (1982). The jackknife, the bootstrap, and other resampling plans. *CBMS* **38**, *SIAM-NSF*.

Efron, B. (1983). Estimating the error rate of a prediction rule: improvements in cross-validation. *JASA* **78**, 316-331.

Efron, B. (1985). Bootstrap confidence intervals for a class of parametric problems. *Biometrika* **72**, 45-58.

Efron, B. (1987). Better bootstrap confidence intervals. *JASA* **82**, 171-185.

Efron (1990a). Jackknife-after-bootstrap standard errors and influence functions. Submitted to *J. Royal Stat. Soc.*.

Efron, B. (1990b). More efficient bootstrap computations. *JASA* **85**, 79-89.

Efron, B. and Stein, C. (1981). The jackknife estimate of variance. *Ann. Stat.* **9**, 586-596.

Geisser, S. (1975). The predictive sample reuse method with applications. *JASA* **70**, 320-328.

Graham, R. L., Hinkley, D. V., John, P. W. M., and Shi, S. (1987). Balanced design of bootstrap simulations. Technical Report 48, University of Texas at Austin, Mathematics Dept.

Hall, P. (1988). Theoretical comparison of bootstrap confidence intervals (with discussion). *Ann. Stat.* **16**, 927-985.

Hesterberg, T. (1988). Variance reduction techniques for bootstrap and other Monte Carlo simulations. Unpublished Ph.D. dissertation, Stanford University, Dept. of Statistics.

Jaeckel, L. (1972). The infinitesimal jackknife. *Memorandum MM72-1215-11*, Bell Labs, Murray Hill, NJ.

Johns, M. V. (1988). Importance sampling for bootstrap confidence intervals. *JASA* **83**, 709-714.

Kendall, M., and Stuart, A. (1958). *The Advanced Theory of Statistics.* London: Charles W. Griffin.

Lachenbruch, P., and Mickey, N, (1968). Estimation of error rates in discriminant analysis. *Technometrics* **10**, 1-11.

Mardia, K. V., Kent, J. T., and Bibby, J. M. (1979). *Multivariate Analysis.* New York: Academic Press.

Miller, R. G. (1986). *Beyond ANOVA*. New York: John Wiley.

Ogbonmwan, S. M., and Wynn, H. P. (1986). Resampling generated likelihoods. *Proc. Fourth Purdue Symp. Decision Theory* (Vol. 1), eds. S. Gupta and J. Berger, New York: Springer-Verlag, pp. 137-147.

Stone, M. (1974). Cross-validatory choice and assessment of statistical predictions. *J. Royal Stat. Soc., Ser. B* **36**, 111-147.

Therneau, T. (1983). Variance reduction techniques for the bootstrap. Unpublished Ph.D. dissertation, Stanford University, Dept. of Statistics.

Welch, B. L., and Peers, H. W. (1963). On formulae for confidence points based on integrals of weighted likelihoods. *J. Royal Stat. Soc., Ser. B* **25**, 318-329.

EFFICIENT BOOTSTRAP SIMULATION

Peter Hall[1]
Australian National University

Abstract. We survey a variety of methods for efficient bootstrap simulation, including linear approximation, the centring method, bootstrap resampling, antithetic resampling and importance resampling.

1. Introduction

In many problems of practical interest, the nonparametric bootstrap is employed to estimate an expected value. For example, if $\hat{\theta}$ is an estimate of an unknown quantity θ then we might wish to estimate bias, $E(\hat{\theta} - \theta)$, or the distribution function of $\hat{\theta}$, $E\{I(\hat{\theta} \leq x)\}$. Generally, suppose we wish to estimate $E(U)$, where U is a random variable which will often be a functional of both the population distribution function F_0 and the empirical distribution function F_1 of the sample $\mathcal{X}$: $U = f(F_0, F_1)$. Let F_2 denote the empirical distribution function of a resample $\mathcal{X}^*$ drawn randomly, with replacement, from $\mathcal{X}$, and put $U^* = f(F_1, F_2)$. Then, $\hat{u} = E\{f(F_1, F_2) \,|\, F_1\} = E(U^* \,|\, \mathcal{X})$ is "the bootstrap estimate" of $u = E(U)$.

In the bias example considered above, we would have $U = \hat{\theta} - \theta$ and $U^* = \hat{\theta}^* - \hat{\theta}$, where $\hat{\theta}^*$ is the version of $\hat{\theta}$ computed for $\mathcal{X}^*$ rather than $\mathcal{X}$. In the case of a distribution function, $U = I(\hat{\theta} \leq x)$ and $U^* = I(\hat{\theta}^* \leq x)$.

Our aim in this paper is to describe some of the available methods for approximating $\hat{u}$ by Monte Carlo simulation, and to provide a little theory for each. The methods which we treat are uniform resampling, linear approximation, the centring method, balanced resampling, antithetic resampling and importance resampling, and are discussed in Sections 2 to 7 respectively. This account is not exhaustive; for example, we do not treat Richardson extrapolation (Bickel and Yahav 1988), computation by saddlepoint methods (Davison and Hinkley 1988, Reid 1988) or balanced importance resampling (Hall 1990b). Section 8 will briefly describe the problem of quantile estimation, which does not quite fit into the format of approximating $\hat{u} = E(U^* \,|\, \mathcal{X})$.

[1] Department of Statistics, Faculty of Economics and Commerce, Australian National University, G.P.O. Box 4, Canberra, ACT 2601, Australia.

Exploring the Limits of Bootstrap. Edited by Raoul LePage and Lynne Billard.
 ISBN: 0471-53631-8.

2. Uniform resampling

Since $\hat{u}$ is defined in terms of uniform resampling — that is, random resampling with replacement, in which each sample value is drawn with the same probability n^{-1} — then uniform resampling is the most obvious approach to simulation. Conditional on $\mathcal{X}$, draw B independent resamples $\mathcal{X}_1^*, \dots, \mathcal{X}_B^*$ by resampling uniformly, and let $\hat{\theta}_b^*$ denote the version of $\hat{\theta}$ computed for $\mathcal{X}_b^*$ rather than $\mathcal{X}$. Then

$$\hat{u}_B^* = B^{-1} \sum_{b=1}^{B} U_b^*$$

is a Monte Carlo approximation to $\hat{u}$. With probability one conditional on $\mathcal{X}$, $\hat{u}_B^*$ converges to $\hat{u}_\infty^* = \hat{u}$ as $B \to \infty$. We refer to $\hat{u}_B^*$ as an approximation rather than an estimate. The bootstrap estimate $\hat{u}$ is based on $B = \infty$, and only its approximate form is concerned with finite B's. Thus, we draw a distinction between the statistical problem of estimating u, and the numerical problem of approximating $\hat{u}$.

The expected value of $\hat{u}_B^*$, conditional on $\mathcal{X}$, equals $\hat{u}$. Therefore $\hat{u}_B^*$ is an unbiased approximation to $\hat{u}$, and the performance of $\hat{u}_B^*$ may be reasonably described in terms of conditional variance:

$$\operatorname{var}(\hat{u}_B^* \mid \mathcal{X}) = B^{-1} \operatorname{var}(U^* \mid \mathcal{X}) = B^{-1} \{E(U^{*2} \mid \mathcal{X}) - \hat{u}^2\}.$$

In many problems of practical importance, $\operatorname{var}(U^* \mid \mathcal{X})$ is asymptotic to either a constant or a constant multiple of n^{-1}, as $n \to \infty$. For example, if we are estimating a distribution function then U^* is an indicator function, and so $U^{*2} = U^*$, whence

$$\operatorname{var}(U^* \mid \mathcal{X}) = \hat{u}(1 - \hat{u}) \to u_0(1 - u_0),$$

where u_0 (typically a value of a Normal distribution function) equals the limit as $n \to \infty$ of $\hat{u}$. If we are estimating bias, and $\hat{\theta} = g(\bar{\mathbf{X}})$ is a smooth function of a d-variate sample mean, then $U^* = \hat{\theta}^* - \hat{\theta} = g(\bar{\mathbf{X}}^*) - g(\bar{\mathbf{X}})$, where $\bar{\mathbf{X}}^*$ is the resample mean. By Taylor expansion,

$$U^* = g(\bar{\mathbf{X}}^*) - g(\bar{\mathbf{X}}) \approx \sum_{j=1}^{d} (\bar{\mathbf{X}}^* - \bar{\mathbf{X}})^{(j)} g_j(\bar{\mathbf{X}}), \tag{2.1}$$

where $x^{(j)} = (\mathbf{x})^{(j)}$ denotes the j'th element of a d-vector $\mathbf{x}$ and $g_j(\mathbf{x}) = (\partial/\partial x^{(j)})\, g(\mathbf{x})$. Thus,

$$\begin{aligned}\operatorname{var}(U^* \mid \mathcal{X}) &= \operatorname{var}\Big\{ \sum_{j=1}^{d} (\bar{\mathbf{X}}^* - \bar{\mathbf{X}})^{(j)} g_j(\bar{\mathbf{X}}) \Big| \mathcal{X} \Big\} + O(n^{-2}) \\ &= n^{-1} \hat{\sigma}^2 + O(n^{-2}),\end{aligned}$$

where

$$\hat{\sigma}^2 = n^{-1} \sum_{i=1}^{n} \Big\{ \sum_{j=1}^{d} (\mathbf{X}_i - \bar{\mathbf{X}})^{(j)} \, g_j(\bar{\mathbf{X}}) \Big\}^2$$

$$\rightarrow \sigma^2 = E \Big\{ \sum_{j=1}^{d} (\mathbf{X} - \boldsymbol{\mu})^{(j)} \, g_j(\boldsymbol{\mu}) \Big\}^2$$

(as $n \to \infty$) and $\boldsymbol{\mu} = E(\mathbf{X})$ is the population mean.

These two examples describe situations which are typical of a great many that arise in statistical problems. When the target $\hat{u}$ is a distribution function or a quantile, the conditional variance of the uniform bootstrap approximant is roughly equal to CB^{-1} for large n and large B; and when the target is the expected value of a smooth function of a mean, the variance is approximately $CB^{-1}\,n^{-1}$. In both cases, C is a constant not depending on B or n. Efficient approaches to Monte Carlo approximation can reduce the value of C in the case of estimating a distribution function, or increase the power of n^{-1} (say, from n^{-1} to n^{-2}) in the case of estimating the expected value of a smooth function of a mean. Most importantly, they usually do not increase the power of B^{-1}. Therefore, generally speaking, even the most efficient of Monte Carlo methods has mean squared error which decreases like B^{-1} as B increases, for a given sample.

3. Linear approximation

We motivate linear approximation by considering the bias estimation example of the previous section. Suppose our aim is to approximate $\hat{u} = E(U^* \,|\, \mathcal{X})$, where $U^* = \hat{\theta}^* - \hat{\theta} = g(\bar{\mathbf{X}}^*) - g(\bar{\mathbf{X}})$ and g is a smooth function of d variables. Let us extend the Taylor expansion (2.1) to another term:

$$U^* = \sum_{j=1}^{d} (\bar{\mathbf{X}}^* - \bar{\mathbf{X}})^{(j)} \, g_j(\bar{\mathbf{X}}) + \tfrac{1}{2} \sum_{j=1}^{d} \sum_{k=1}^{d} (\bar{\mathbf{X}}^* - \bar{\mathbf{X}})^{(j)} \, (\bar{\mathbf{X}}^* - \bar{\mathbf{X}})^{(k)} \, g_{jk}(\bar{\mathbf{X}}) + \ldots \,, \tag{3.1}$$

where $g_{j_1 \ldots j_r}(\mathbf{x}) = (\partial^r / \partial x^{(j_1)} \ldots \partial x^{(j_r)})\, g(\mathbf{x})$. As we noted in Section 2, the conditional variance of U^* is determined asymptotically by the variance of the first term on the right-hand side of (3.1), which is the linear component in the Taylor expansion. Now, our aim is to approximate $E(U^* \,|\, \mathcal{X})$, and the linear component does not contribute anything to that expectation:

$$E \Big\{ \sum_{j=1}^{d} (\bar{\mathbf{X}}^* - \bar{\mathbf{X}})^{(j)} \, g_j(\bar{\mathbf{X}}) \,\Big|\, \mathcal{X} \Big\} = \sum_{j=1}^{d} E\{(\bar{\mathbf{X}}^* - \bar{\mathbf{X}})^{(j)} \,|\, \mathcal{X}\} \, g_j(\bar{\mathbf{X}}) = 0\,. \tag{3.2}$$

Therefore it makes sense to remove the linear component. That is, base the uniform resampling approximation on

$$V^* = U^* - \sum_{j=1}^{d} (\bar{\mathbf{X}}^* - \bar{\mathbf{X}})^{(j)} \, g_j(\bar{\mathbf{X}}),$$

instead of on U^*. Conditional on $\mathcal{X}$, draw B independent resamples $\mathcal{X}_1^*, \ldots, \mathcal{X}_B^*$ by resampling uniformly (exactly as in Section 2), and put

$$\begin{aligned} V_b^* &= U_b^* - \sum_{j=1}^{d} (\bar{\mathbf{X}}_b^* - \bar{\mathbf{X}})^{(j)} \, g_j(\bar{\mathbf{X}}) \\ &= g(\bar{\mathbf{X}}_b^*) - g(\bar{\mathbf{X}}) - \sum_{j=1}^{d} (\bar{\mathbf{X}}_b^* - \bar{\mathbf{X}})^{(j)} \, g_j(\bar{\mathbf{X}}), \end{aligned}$$

$1 \le b \le B$, where $\bar{\mathbf{X}}_b^*$ denotes the mean of the resample $\mathcal{X}_b^*$. Define

$$\hat{v}_B^* = B^{-1} \sum_{b=1}^{B} V_b^* .$$

Then $\hat{v}_B^*$ is the linear approximation method approximation of $\hat{u}$. With probability one conditional on $\mathcal{X}$, $\hat{v}_B^* \to \hat{u}$ as $B \to \infty$.

In view of (3.2), $E(V^* \mid \mathcal{X}) = E(U^* \mid \mathcal{X}) = \hat{u}$, and so $\hat{v}_B^*$ is an unbiased approximant of $\hat{u}$. The conditional variance of $\hat{v}_B^*$ is dominated by that of the first remaining term in the Taylor expansion, just as in the case of $\hat{u}_B^*$:

$$\operatorname{var}(\hat{v}_B^* \mid \mathcal{X}) = B^{-1} \operatorname{var}(V^* \mid \mathcal{X}),$$

and

$$\begin{aligned} \operatorname{var}(V^* \mid \mathcal{X}) &= \operatorname{var}\Big\{ \tfrac{1}{2} \sum_{j=1}^{d} \sum_{k=1}^{d} (\bar{\mathbf{X}}^* - \bar{\mathbf{X}})^{(j)} (\bar{\mathbf{X}}^* - \bar{\mathbf{X}})^{(k)} \, g_{jk}(\bar{\mathbf{X}}) \,\Big|\, \mathcal{X} \Big\} + O(n^{-3}) \\ &= n^{-2} \hat{\beta} + O(n^{-3}), \end{aligned} \tag{3.3}$$

with probability one as $n \to \infty$, where[2]

$$\hat{\beta} = \tfrac{1}{2} \sum_{j_1=1}^{d} \sum_{k_1=1}^{d} \sum_{j_2=1}^{d} \sum_{k_2=1}^{d} g_{j_1 k_1}(\bar{\mathbf{X}}) \, g_{j_2 k_2}(\bar{\mathbf{X}}) \, \hat{\sigma}^{j_1 j_2} \, \hat{\sigma}^{k_1 k_2}, \tag{3.4}$$

$$\hat{\sigma}^{jk} = n^{-1} \sum_{i=1}^{n} (\mathbf{X}_i - \bar{\mathbf{X}})^{(j)} (\mathbf{X}_i - \bar{\mathbf{X}})^{(k)} . \tag{3.5}$$

[2] In Hall (1989a), the factor $\frac{1}{2}$ was inadvertently omitted from the right-hand side of (3.4).

Therefore, the order of magnitude of the variance has been reduced from $B^{-1}n^{-1}$ (in the case of $\hat{u}_B^*$) to $B^{-1}n^{-2}$ (in the case of $\hat{v}_B^*$). The numerical value of the reduction, for a given problem and sample, will depend on values of the first and second derivatives of g, as well as on the higher-order terms which our asymptotic argument has ignored.

More generally, we may approximate U^* by an arbitrary number of terms in the Taylor expansion (3.1), and thereby compute a general "polynomial approximation" to $\hat{u}$. For example, if $m \geq 1$ is an integer then we may define

$$W_b^* = U_b^* - \sum_{r=1}^{m} \frac{1}{r!} \sum_{j_1=1}^{d} \cdots \sum_{j_r=1}^{d} (\bar{\mathbf{X}}_b^* - \bar{\mathbf{X}})^{(j_1)} \cdots (\bar{\mathbf{X}}_b^* - \bar{\mathbf{X}})^{(j_r)} \, g_{j_1 \ldots j_r}(\bar{\mathbf{X}})$$

(a generalization of V_b^*),

$$\tilde{w} = \sum_{r=1}^{m} \frac{1}{r!} \sum_{j_1=1}^{d} \cdots \sum_{j_r=1}^{d} E\{(\bar{\mathbf{X}}^* - \bar{\mathbf{X}})^{(j_1)} \cdots (\bar{\mathbf{X}}^* - \bar{\mathbf{X}})^{(j_r)} \mid \mathcal{X}\} \, g_{j_1 \ldots j_r}(\bar{\mathbf{X}}),$$

and

$$\hat{w}_B^* = B^{-1} \sum_{b=1}^{B} W_b^* + \tilde{w}.$$

Then $\hat{w}_B^*$ is an unbiased approximation of $\hat{u}$. Of course, the approximation is only practicable if we can compute $\tilde{w}$, which is a linear form in the first m central sample moments, with coefficients equal to the derivatives of g. In the special case $m = 1$, $\hat{w}_B^*$ reduces to our linear approximation $\hat{v}_B^*$, and there $\tilde{w} = 0$. The conditional variance of $\hat{w}_B^*$ is of order $B^{-1}n^{-(m+1)}$ as $n \to \infty$. Of course, $\hat{w}_B^* \to \hat{u}$ as $B \to \infty$, for fixed $\mathcal{X}$. If the function g is analytic then for fixed $\mathcal{X}$ and B, $\hat{w}_B^* \to \hat{u}$ as $m \to \infty$. Each of these limits is attained with probability one, conditional on $\mathcal{X}$.

Monte Carlo methods based on functional approximation have been discussed by Oldford (1985) and Davison, Hinkley and Schechtman (1986).

4. Centring method

To motivate the centring method approximation, recall that the linear approximation method produces the approximant

$$\hat{v}_B^* = B^{-1} \sum_{b=1}^{B} V_b^* = B^{-1} \sum_{b=1}^{B} \left\{ g(\bar{\mathbf{X}}_b^*) - g(\bar{\mathbf{X}}) - \sum_{j=1}^{d} (\bar{\mathbf{X}}_b^* - \bar{\mathbf{X}})^{(j)} \, g_j(\bar{\mathbf{X}}) \right\}$$

$$= \hat{u}_B^* - \sum_{j=1}^{d} (\bar{\mathbf{X}}_\cdot^* - \bar{\mathbf{X}})^{(j)} \, g_j(\bar{\mathbf{X}}),$$

where $\hat{u}_B^*$ is the uniform resampling approximation of $\hat{u}$ and

$$\bar{\mathbf{X}}_{\cdot}^{*} = B^{-1} \sum_{b=1}^{B} \bar{\mathbf{X}}_b^*$$

is the grand mean of all the resamples. Now,

$$\sum_{j=1}^{d} (\bar{\mathbf{X}}_{\cdot}^{*} - \bar{\mathbf{X}})^{(j)}\, g_j(\bar{\mathbf{X}}) \approx g(\bar{\mathbf{X}}_{\cdot}^{*}) - g(\bar{\mathbf{X}}),$$

by Taylor expansion, and so

$$\hat{v}_B \approx \hat{u}_B^* - \{g(\bar{\mathbf{X}}_{\cdot}^{*}) - g(\bar{\mathbf{X}})\} = \hat{x}_B^*,$$

where

$$\hat{x}_B^* = B^{-1} \sum_{b=1}^{B} g(\bar{\mathbf{X}}_b^*) - g(\bar{\mathbf{X}}_{\cdot}^{*}).$$

We call $\hat{x}_B^*$ the centring method approximation to $\hat{u}$. It differs from $\hat{u}_B^*$ in that the mean of the $g(\bar{\mathbf{X}}_b^*)$'s is now centred at $g(\bar{\mathbf{X}}_{\cdot}^{*})$ rather than $g(\bar{\mathbf{X}})$; recall that

$$\hat{u}_B^* = B^{-1} \sum_{b=1}^{B} g(\bar{\mathbf{X}}_b^*) - g(\bar{\mathbf{X}}).$$

The centring method approximation was suggested by Efron (1990).

As expected, $\hat{x}_B^* \to \hat{u}$ as $B \to \infty$, conditional on $\mathcal{X}$. The approximant $\hat{x}_B^*$ is not unbiased for $\hat{u}$, although the bias is generally very low, of smaller order than the error about the mean. Indeed, it may be proved that

$$E(\hat{x}_B^* \mid \mathcal{X}) - \hat{u} = -(Bn)^{-1}\,\hat{\alpha} + O\{(Bn)^{-2}\}, \tag{4.1}$$

$$\operatorname{var}(\hat{x}_B^* \mid \mathcal{X}) = (Bn^2)^{-1}\,\hat{\beta} + O\{(Bn)^{-2} + (Bn^3)^{-1}\}, \tag{4.2}$$

with probability one as $n \to \infty$, where

$$\hat{\alpha} = \tfrac{1}{2} \sum_{j=1}^{d} \sum_{k=1}^{d} g_{jk}(\bar{\mathbf{X}})\, \hat{\sigma}^{jk}, \tag{4.3}$$

and $\hat{\beta}$, $\hat{\sigma}^{jk}$ are given by (3.4), (3.5) respectively. See Hall (1989a). Note particularly that by (3.3) and (4.2), the conditional asymptotic variances of $\hat{v}_B^*$ and $\hat{x}_B^*$ are identical. Since $\hat{v}_B^*$ is an unbiased approximant, and the bias of $\hat{x}_B^*$ is negligible relative to the error about the mean (order $B^{-1} n^{-1}$ relative to $B^{-\frac{1}{2}} n^{-1}$), then the approximations $\hat{v}_B^*$ and $\hat{x}_B^*$ have asymptotically equivalent mean squared error.

5. Balanced resampling

If we could ensure that the grand mean of the bootstrap resamples was identical to the sample mean, i.e.

$$\bar{\mathbf{X}}_{\cdot}^* = \bar{\mathbf{X}}\,, \tag{5.1}$$

then the uniform approximation $\hat{u}_B^*$, the linear approximation $\hat{v}_B^*$ and the centring approximation $\hat{x}_B^*$ would all be identical. The only practical way of guaranteeing (5.1) is to resample in such a way that each data point $\mathbf{X}_i$ occurs the same number of times in the union of the resamples $\mathcal{X}_b^*$. To achieve this end, write down each of the sample values $\mathbf{X}_1, \ldots, \mathbf{X}_n$ B times, in a string of length Bn; then randomly permute the elements of this string; and finally, divide the permuted string into B chunks of length n, putting all the sample values lying between positions $(b-1)n+1$ and bn of the permuted string into the b'th resample $\mathcal{X}_b^\dagger$, for $1 \le b \le B$. This is balanced resampling, and amounts to random resampling subject to the constraint that $\mathbf{X}_i$ appears just B times in $\cup_b \mathcal{X}_b^\dagger$. Balanced resampling was introduced by Davison, Hinkley and Schechtman (1986), and high-order balance has been discussed by Graham, Hinkley, John and Shi (1990). See also Ogbonmwan and Wynn (1986, 1988). An algorithm for performing balanced resampling has been described by Gleason (1988). The method of Latin hypercube sampling (McKay, Beckman and Conover 1979; Stein 1987), used for Monte Carlo simulation in a non-bootstrap setting, is closely related to balanced resampling.

The balanced resampling approximation of $\hat{u} = E\{g(\bar{\mathbf{X}}^*) \mid \mathcal{X}\}$ is

$$\hat{u}_B^\dagger = B^{-1} \sum_{b=1}^{B} g(\bar{\mathbf{X}}_b^\dagger)\,,$$

where $\bar{\mathbf{X}}_b^\dagger$ denotes the mean of $\mathcal{X}_b^\dagger$. Once again, $\hat{u}_B^\dagger \to \hat{u}$ as $B \to \infty$, with probability one conditional on $\mathcal{X}$. The balanced resampling approximation shares the asymptotic bias and variance formulae of the centring approximation $\hat{x}_B^*$, introduced in Section 4:

$$E(\hat{u}_B^\dagger \mid \mathcal{X}) - \hat{u} = -(Bn)^{-1}\,\hat{\alpha} + O(B^{-1} n^{-2})\,, \tag{5.2}$$

$$\mathrm{var}(\hat{u}_B^\dagger \mid \mathcal{X}) = (Bn^2)^{-1}\,\hat{\beta} + O\{(Bn)^{-2} + (Bn^3)^{-1}\}\,, \tag{5.3}$$

with probability one as $n \to \infty$, where $\hat{\alpha}$, $\hat{\beta}$ are given by (4.3), (3.4) respectively; compare (4.1) and (4.2). In particular, in the context of bias estimation for a smooth function of a mean, balanced resampling reduces the orders of magnitude of variance and mean squared error by a factor of n^{-1}. Formulae (5.2) and (5.3) are not entirely trivial to prove, and the asymptotic equivalence of bias and variance formulae for the centring method and balanced resampling is not quite obvious; see Hall (1989a).

In Sections 3 and 4, and so far in the present section, we have treated only the case of approximating a smooth function of a mean. The methods of linear approximation and centring do not admit a wide range of other applications. For example, linear approximation relies on Taylor expansion, and that demands a certain level of smoothness of the statistic U. However, balanced resampling is not constrained in this way, and in principle applies to a much wider range of problems, including distribution function and quantile estimation. In those cases the extent of improvement of variance and mean squared error is generally by a constant factor, not by a factor n^{-1}.

Suppose that U is an indicator function of the form $U = I(S \le x)$ or $U = I(T \le x)$, where $S = n^{\frac{1}{2}}(\hat{\theta} - \theta)/\sigma$ and $T = n^{\frac{1}{2}}(\hat{\theta} - \theta)/\hat{\sigma}$ are statistics which are asymptotically Normal $N(0,1)$. The bootstrap versions are $U^* = I(S^* \le x)$ and $U^* = I(T^* \le x)$ respectively, where $S^* = n^{\frac{1}{2}}(\hat{\theta}^* - \hat{\theta})/\hat{\sigma}$ and $T^* = n^{\frac{1}{2}}(\hat{\theta}^* - \hat{\theta})/\hat{\sigma}^*$. (The pros and cons of pivoting are not relevant to the present discussion.) To construct a balanced resampling approximation to $\hat{u} = E(U^* \mid \mathcal{X})$, first draw B balanced resamples $\mathcal{X}_b^\dagger$, $1 \le b \le B$, as described two paragraphs earlier. Let $\hat{\theta}_b^\dagger$, $\hat{\sigma}_b^\dagger$ denote the versions of $\hat{\theta}$, $\hat{\sigma}$ respectively computed from $\mathcal{X}_b^\dagger$ instead of $\mathcal{X}$, and put

$$S_b^\dagger = n^{\frac{1}{2}}(\hat{\theta}_b^\dagger - \hat{\theta})/\hat{\sigma}, \qquad T_b^\dagger = n^{\frac{1}{2}}(\hat{\theta}_b^\dagger - \hat{\theta})/\hat{\sigma}_b^\dagger,$$

and $U_b^\dagger = I(S_b^\dagger \le x)$ or $I(T_b^\dagger \le x)$, depending on whether $U = I(S \le b)$ or $I(T \le x)$. (We define $T_b^\dagger = c$, for an arbitrary but fixed constant c, if $\sigma_b^\dagger = 0$.) The balanced resampling approximation to $\hat{u}$ is

$$\hat{u}_B^\dagger = B^{-1} \sum_{b=1}^{B} U_b^\dagger .$$

Recall from Section 2 that the uniform resampling approximation $\hat{u}_B^*$ is unbiased for $\hat{u}$, in the sense that $E(\hat{u}_B^* \mid \mathcal{X}) = \hat{u}$, and has variance

$$\mathrm{var}(\hat{u}_B^* \mid \mathcal{X}) = B^{-1}\,\hat{u}(1-\hat{u}) \sim B^{-1}\,\Phi(x)\,\{1-\Phi(x)\}$$

as $n \to \infty$. The balanced resampling approximant is slightly biased, although the bias is low relative to the error about the mean:

$$E(\hat{u}_B^\dagger \mid \mathcal{X}) - \hat{u} = O(B^{-1})$$

with probability one. The asymptotic variance of $\hat{u}_B^\dagger$ is less than that of $\hat{u}_B^*$ by a constant factor $\rho(x)^{-1} < 1$, since

$$\mathrm{var}(\hat{u}_B^\dagger \mid \mathcal{X}) \sim B^{-1}\left[\Phi(x)\{1 - \Phi(x)\} - \phi(x)^2\right];$$

see Hall (1990a). Thus, the asymptotic efficiency of balanced resampling relative to uniform resampling, in this context, is

$$\rho(x) = \Phi(x)\{1 - \Phi(x)\}[\Phi(x)\{1 - \Phi(x)\} - \phi(x)^2]^{-1},$$

which reaches a maximum at $x = 0$ and decreases to 1 as $|x| \uparrow \infty$. See Hall (1990a).

The same asymptotic efficiencies apply to the case of quantile estimation, which we discuss in more detail in Section 8.

6. Antithetic resampling

The method of antithetic resampling dates back at least to Hammersley and Morton (1956) and Hammersley and Mauldon (1956). See Snijders (1984) for a recent account in connection with Monte Carlo estimation of probabilities. Antithetic resampling may be described as follows. Suppose we have two estimates $\hat{\theta}_1$ and $\hat{\theta}_2$ of the same parameter θ, with identical means and variances but negative covariance. Assume that the costs of computing $\hat{\theta}_1$ and $\hat{\theta}_2$ are identical. Define

$$\hat{\theta}_3 = \tfrac{1}{2}(\hat{\theta}_1 + \hat{\theta}_2).$$

Then $\hat{\theta}_3$ has the same mean as either $\hat{\theta}_1$ or $\hat{\theta}_2$, but less than half the variance, since

$$\begin{aligned}\mathrm{var}(\hat{\theta}_3) &= \tfrac{1}{4}\{\mathrm{var}\,\hat{\theta}_1 + \mathrm{var}\,\hat{\theta}_2 + 2\mathrm{cov}(\hat{\theta}_1, \hat{\theta}_2)\}\\ &= \tfrac{1}{2}\{\mathrm{var}(\hat{\theta}_1) + \mathrm{cov}(\hat{\theta}_1, \hat{\theta}_2)\} < \tfrac{1}{2}\mathrm{var}(\hat{\theta}_1).\end{aligned}$$

Since the cost of computing $\hat{\theta}_3$ is scarcely more than twice the cost of computing either $\hat{\theta}_1$ or $\hat{\theta}_2$, but the variance is more than halved, then there is an advantage from the viewpoint of cost-effectiveness in using $\hat{\theta}_3$, rather than either $\hat{\theta}_1$ or $\hat{\theta}_2$, to estimate θ. Obviously, the advantage increases with increasing negativity of the covariance, all other things being equal.

To appreciate how this idea may be applied to the case of resampling, let U^* denote the version of a statistic U computed from a (uniform) resample $\mathcal{X}^* = \{X_1^*, \ldots, X_n^*\}$ rather than the original sample $\mathcal{X} = \{X_1, \ldots, X_n\}$. Let π be an arbitrary but fixed permutation of the integers $1, \ldots, n$, and let $j_1, \ldots, j_n$ be the random integers such that $X_i^* = X_{j_i}$ for $1 \le i \le n$. Define $X_i^{**} = X_{\pi(j_i)}$, $1 \le i \le n$, and put $\mathcal{X}^{**} = \{X_1^{**}, \ldots, X_n^{**}\}$. That is, $\mathcal{X}^{**}$ is the (uniform) resample obtained by replacing each appearance of X_k in $\mathcal{X}^*$ by $X_{\pi(k)}$. If U^{**} denotes the version of U computed from $\mathcal{X}^{**}$ instead of $\mathcal{X}$, then U^* and U^{**} have the same distributions, conditional on $\mathcal{X}$. In particular, they have the same conditional mean and variance. If we choose the permutation π in such a way that the conditional covariance of U^* and U^{**} is negative, we may apply the antithetic argument to the pair (U^*, U^{**}). That is, the approximant

$$U^{@} = \tfrac{1}{2}(U^* + U^{**})$$

will have the same conditional mean as U^* but less than half the conditional variance of U^*.

If the X_i's are scalars then in many cases of practical interest, the "asymptotically optimal" permutation π (which asymptotically minimizes the covariance of U^* and U^{**}), is that which takes the largest X_i into the smallest X_i, the second largest X_i into the second smallest X_i, and so on. That is, if we index the X_i's such that $X_1 \le \ldots \le X_n$, then $\pi(i) = n - i + 1$ for $1 \le i \le n$. For example, this is true when $U = g(\bar{X}) - g(\mu)$, where g is a smooth function, and also when

$$U = I[n^{\frac{1}{2}}\{g(\bar{X}) - g(\mu)\} \le x]. \tag{6.1}$$

The asymptotically optimal permutation π in the case of a d-dimensional sample, where $U = g(\bar{\mathbf{X}}) - g(\boldsymbol{\mu})$ or

$$U = I[n^{\frac{1}{2}}\{g(\bar{\mathbf{X}}) - g(\boldsymbol{\mu})\} \le x], \tag{6.2}$$

is that one which reverses the order of the quantities

$$Y_i = \sum_{j=1}^{d} (\mathbf{X}_i - \bar{\mathbf{X}})^{(j)}\, g^{(j)}(\bar{\mathbf{X}}), \qquad 1 \le i \le n.$$

That is, if we index the $\mathbf{X}_i$'s such that $Y_1 \le \ldots \le Y_n$, then $\pi(i) = n - i + 1$. We shall call π the antithetic permutation. These results remain true if we Studentize the arguments of the indicator functions at (6.1) and (6.2); the pros and cons of pivoting do not have a role to play here. The reader is referred to Hall (1989b) for details.

The method of antithetic resampling may be used in the following way to approximate the bootstrap estimate $\hat{u} = E(U^* \mid \mathcal{X})$. Draw B independent, uniform resamples $\mathcal{X}_b^*$, $1 \le b \le B$; by applying the antithetic permutation, convert $\mathcal{X}_b^*$ into the corresponding antithetic resample $\mathcal{X}_b^{**}$; compute the versions of U_b^* and U_b^{**} for $\mathcal{X}_b^*$ and $\mathcal{X}_b^{**}$, respectively; and define

$$\hat{u}_B^{@} = \tfrac{1}{2} B^{-1} \sum_{b=1}^{B} (U_b^* + U_b^{**}).$$

Note that $\hat{u}_B^{@}$ is an unbiased approximant of $\hat{u}$, in the sense that $E(\hat{u}_B^{@} \mid \mathcal{X}) = \hat{u}$. The conditional variance of $\hat{u}_B^{@}$ is given by

$$\operatorname{var}(\hat{u}_B^{@} \mid \mathcal{X}) = (2B)^{-1}\{\operatorname{var}(U^* \mid \mathcal{X}) + \operatorname{cov}(U^*, U^{**} \mid \mathcal{X})\},$$

where (U^*, U^{**}) denotes a generic pair (U_b^*, U_b^{**}). In general, as $n \to \infty$,

$$\operatorname{var}(\hat{u}_B^{@} \mid \mathcal{X}) = q \operatorname{var}(\hat{u}_B^* \mid \mathcal{X}) + o\{\operatorname{var}(\hat{u}_B^* \mid \mathcal{X})\},$$

where $0 \le q < 1$. The exact value of q depends on the situation, it being greater for less symmetric distributions. For example, in the case of distribution function estimation where $\hat{u} = P(S^* \le x \mid \mathcal{X})$ or $P(T^* \le x \mid \mathcal{X})$, the value of

$$q = q(x) = \lim_{n\to\infty} \lim_{B\to\infty} \{\operatorname{var}(\hat{u}_B^{@} \mid \mathcal{X}) / \operatorname{var}(\hat{u}_B^* \mid \mathcal{X})\}$$

is an increasing function of both $|x|$ and the skewness of the sampling distribution. The minimum value $q = 0$ can only arise in this context when both $x = 0$ and the sampling distribution is symmetric.

Thus, generally speaking, antithetic resampling reduces variance by a constant factor. In this respect, balanced resampling is superior in the case of approximating the conditional mean of a smooth function, since it reduces variance by a factor which converges to zero as sample size increases. Again, the reader is referred to Hall (1989b) for details.

7. Importance resampling

The method of importance resampling is a standard technique for improving the efficiency of Monte Carlo approximations. See Hammersley and Handscomb (1964, p.60ff). It was first suggested in the context of bootstrap resampling by Johns (1988) and Davison (1988).

To introduce importance resampling, let $\mathcal{X} = \{X_1, \ldots, X_n\}$ denote the sample from which a resample will be drawn. (This notation is only for the sake of convenience, and in no way precludes a multivariate sample.) Under importance resampling, each X_i is assigned a probability p_i of being selected on any given draw, where $\Sigma\, p_i = 1$. Sampling is conducted with replacement, so that the chance of drawing a resample of size n in which X_i appears just m_i times $(1 \le i \le n)$, is given by a multinomial formula,

$$\frac{n!}{m_1! \dots m_n!} \prod_{i=1}^{n} p_i^{m_i} .$$

Of course, $\Sigma\, m_i = n$. Taking $p_i = n^{-1}$ for each i, we obtain the uniform resampling method of Section 2.

The name "importance" derives from the fact that resampling is designed to take place in a manner which ascribes more importance to some sample values than others. The aim is to select the p_i's so that the value assumed by a bootstrap statistic is relatively likely to be close to the quantity whose value we wish to approximate.

There are two parts to the method of importance resampling: firstly, a technique for passing from a sequence of importance resamples to an approximation of a quantity which would normally be defined in terms of a uniform resample; and secondly, a method for computing the appropriate values of p_i for the importance resampling algorithm. One would usually endeavour to choose the p_i's so as to minimize the error, or variability, of the approximation.

In most problems, if the parent distribution is continuous then there are $N = \binom{2n-1}{n}$ different possible resamples. Let these be $\mathcal{X}_1, \dots, \mathcal{X}_N$, indexed in any order, and let m_{ji} denote the number of times X_i appears in $\mathcal{X}_j$. The probability of obtaining $\mathcal{X}_j$ after n resampling operations, under uniform resampling or importance resampling, equals

$$\pi_j = \frac{n!}{m_{j1}! \dots m_{jn}!} n^{-n}$$

or

$$\pi_j' = \frac{n!}{m_{j1}! \dots m_{jn}!} \prod_{i=1}^{n} p_i^{m_{ji}} = \pi_j \prod_{i=1}^{n} (np_i)^{m_{ji}} , \tag{7.1}$$

respectively. Let U be the statistic of interest, a function of the original sample. We wish to construct a Monte Carlo approximation to the bootstrap estimate $\hat{u}$ of the mean of U, $u = E(U)$.

Let $\mathcal{X}^*$ denote a resample drawn by uniform resampling, and write U^* for the value of U computed for $\mathcal{X}^*$. Of course, $\mathcal{X}^*$ will be one of the $\mathcal{X}_j$'s. Write u_j for the value of U^* when $\mathcal{X}^* = \mathcal{X}_j$. In this notation,

$$\hat{u} = E(U^* \,|\, \mathcal{X}) = \sum_{j=1}^{N} u_j \, \pi_j = \sum_{j=1}^{N} u_j \, \pi_j' \prod_{i=1}^{n} (np_i)^{-m_{ji}} , \tag{7.2}$$

the last identity following from (7.1).

Let $\mathcal{X}^\dagger$ denote a resample drawn by importance resampling, write $U^\dagger$ for the value of U computed from $\mathcal{X}^\dagger$, and let $M_i^\dagger$ be the number of times X_i appears in $\mathcal{X}^\dagger$. Then by (7.2),

$$\hat{u} = E\Big\{U^\dagger \prod_{i=1}^{n} (np_i)^{-M_i^\dagger} \,\Big|\, \mathcal{X}\Big\}.$$

Therefore it is possible to approximate $\hat{u}$ by importance resampling. In particular, if $\mathcal{X}_b^\dagger$, $1 \le b \le B$, denote independent resamples drawn by importance resampling, if $U_b^\dagger$ equals the value of U computed for $\mathcal{X}_b^\dagger$, and if $M_{bi}^\dagger$ denotes the number of times X_i appears in $\mathcal{X}_b^\dagger$, then the importance resampling approximant of $\hat{u}$ is given by

$$\hat{u}_B^\dagger = B^{-1} \sum_{b=1}^{B} U_b^\dagger \prod_{i=1}^{n} (np_i)^{-M_{bi}^\dagger}.$$

This approximation is unbiased, in the sense that $E(\hat{u}_B^\dagger \,|\, \mathcal{X}) = \hat{u}$. Note too that conditional on $\mathcal{X}$, $\hat{u}_B^\dagger \to \hat{u}$ with probability one as $B \to \infty$.

If we take each $p_i = n^{-1}$ then $\hat{u}_B^\dagger$ is just the usual uniform resampling approximant $\hat{u}_B^*$. We wish to choose $p_1, \ldots, p_n$ to optimize the performance of $\hat{u}_B^\dagger$. Since $\hat{u}_B^\dagger$ is unbiased then the performance of $\hat{u}_B^\dagger$ may be described in terms of variance:

$$\begin{aligned} \operatorname{var}(\hat{u}_B^\dagger \,|\, \mathcal{X}) &= B^{-1} \operatorname{var}\Big\{U_b^\dagger \prod_{i=1}^{n} (np_i)^{-M_{bi}^\dagger} \,\Big|\, \mathcal{X}\Big\} \\ &= B^{-1}(\hat{v} - \hat{u}^2), \end{aligned} \tag{7.3}$$

where

$$\begin{aligned} \hat{v} = \hat{v}(p_1, \ldots, p_n) &= E\Big[\Big\{U_B^\dagger \prod_{i=1}^{n} (np_i)^{-M_{bi}^\dagger}\Big\}^2 \,\Big|\, \mathcal{X}\Big] \\ &= \sum_{j=1}^{N} \pi'_j u_j^2 \prod_{i=1}^{n} (np_i)^{-2m_{ji}} \\ &= \sum_{j=1}^{N} \pi_j u_j^2 \prod_{i=1}^{n} (np_i)^{-m_{ji}} \\ &= E\Big\{U^{*2} \prod_{i=1}^{n} (np_i)^{-M_i^*} \,\Big|\, \mathcal{X}\Big\}. \end{aligned} \tag{7.4}$$

On the last line, M_i^* denotes the number of times X_i appears in the uniform resample $\mathcal{X}^*$. Ideally we would like to choose $p_1, \ldots, p_n$ so as to minimize $\hat{v}(p_1, \ldots, p_n)$, subject to $\Sigma\, p_i = 1$.

In the case of estimating a distribution function there can be a significant advantage in choosing non-identical p_i's, the amount of improvement depending on the argument of the distribution function. To appreciate the extent of improvement, consider the case of estimating the distribution function of a Studentized statistic, $T = n^{\frac{1}{2}}(\hat{\theta} - \theta_0)/\hat{\sigma}$, assuming the "smooth function model" introduced in Section 2. Other cases, such as that where $S = n^{\frac{1}{2}}(\hat{\theta} - \theta_0)/\sigma$ is the subject of interest, are similar; the issue of Studentizing does not play a role here.

Take $U^* = I(T^* \le x)$, where $T^* = n^{\frac{1}{2}}(\hat{\theta}^* - \hat{\theta})/\hat{\sigma}^*$,

$$\hat{\sigma}^{*2} = n^{-1} \sum_{i=1}^{n} \Big\{ \sum_{j=1}^{d} (\mathbf{X}_i^* - \bar{\mathbf{X}}^*)^{(j)}\, g_j(\bar{\mathbf{X}}^*) \Big\}^2 .$$

(We agree to define $T^* = c$, for an arbitrary but fixed constant c, in the event that $\hat{\sigma}^* = 0$.) Now, $\hat{u} = P(T^* \le x \,|\, \mathcal{X})$, and for this particular x we wish to choose $p_1, \ldots, p_n$ to minimize

$$\hat{v} = \hat{v}(p_1, \ldots, p_n) = E\Big\{ I(T^* \le x) \prod_{i=1}^{n} (np_i)^{-M_i^*} \,\Big|\, \mathcal{X} \Big\} .$$

A straightforward asymptotic analysis, given for example by Do and Hall (1990), shows that the asymptotic minimum of $\hat{v}$ is attained with $p_i = n^{-1} e^{-\delta_i}$, where $\delta_i = A\,\epsilon_i + C$, $\epsilon_i = n^{-\frac{1}{2}}\, \hat{\sigma}^{-1}\, \Sigma_j (X_i - \bar{X})^{(j)}\, g_j(\bar{X})$, $A = A(x) > 0$ is chosen to minimize $\Phi(x - A)\, e^{A^2}$, and $C = (A)$ is chosen to ensure that $\Sigma\, p_i = 1$. Values of $A = A(x)$, for selected Standard Normal quantiles, are given in the table. For this choice of the p_i's,

$$r(x) = \frac{\text{variance of } \hat{u}_B^* \text{ (under uniform resampling)}}{\text{variance of } \hat{u}_B^{\dagger} \text{ (under importance resampling)}} \sim \frac{\Phi(x)\{1 - \Phi(x)\}}{\Phi(x - A)\, e^{A^2} - \Phi(x)^2} .$$

The function r is strictly decreasing, with $r(-\infty) = \infty$, $r(0) = 1.7$ and $r(+\infty) = 1$. Therefore, importance resampling can be considerably more efficaceous for negative x than positive x. If we wish to approximate $\hat{G}(x) = P(T^* \le x \,|\, \mathcal{X})$ for a value of $x > 0$, it is advisable to work throughout with $-T^*$ rather than T^* and use importance resampling to calculate

$$P(-T^* \le -x \,|\, \mathcal{X}) = 1 - \hat{G}(x) .$$

$\Phi(x)$	$A(x)$	$r(x)$
0.005	2.6561	69.0
0.01	2.5704	38.0
0.025	2.1787	17.6
0.05	1.8940	10.0
0.1	1.5751	5.8
0.5	0.6120	1.7
0.9	0.1150	1.1
0.95	0.0602	1.1
0.975	0.0320	1.0
0.99	0.0139	1.0
0.995	0.0074	1.0

TABLE: Optimal values of $A = A(x)$, and values of asymptotic efficiency $r(x)$, for selected x (taken to be quantiles of the Standard Normal distribution).

8. Quantile estimation

Here we consider the problem of estimating the α'th quantile, ξ_α, of the distribution of a random variable R such as $S = n^{\frac{1}{2}}(\hat{\theta} - \theta_0)/\sigma$ or $T = n^{\frac{1}{2}}(\hat{\theta}-\theta_0)/\hat{\sigma}$. We define ξ_α to be the solution of the equation $P(R \leq \xi_\alpha) = \alpha$. Now, the bootstrap estimate of ξ_α is the solution $\hat{\xi}_\alpha$ of

$$P(R^* \leq \hat{\xi}_\alpha \,|\, \mathcal{X}) = \alpha\,, \tag{8.1}$$

where $R^* = S^* = n^{\frac{1}{2}}(\hat{\theta}^* - \hat{\theta})/\hat{\sigma}$ (if $R^* = S$) or $R^* = T^* = n^{\frac{1}{2}}(\hat{\theta}^* - \hat{\theta})/\hat{\sigma}^*$ (if $R = T$). The equation (8.1) usually cannot be solved exactly, although the error will be an exponentially small function of n. For the sake of definiteness we shall define

$$\hat{\xi}_\alpha = \hat{H}^{-1}(\alpha) = \inf\{x : \hat{H}(x) \geq \alpha\}\,, \tag{8.2}$$

where $\hat{H}(x) = P(R^* \leq x \,|\, \mathcal{X})$.

Approximation of the function $\hat{H}(x)$ falls neatly into the format described in earlier sections, where we discussed Monte Carlo methods for approximating a conditional expectation of the form $\hat{u} = E(U^* \,|\, \mathcal{X})$. Take $U^* = I(R^* \leq x)$, and apply any one of the methods of uniform resampling, balanced resampling, antithetic resampling or importance resampling to obtain an approximation $\hat{H}_B(x)$ to $\hat{H}(x)$. Then, following the prescription suggested by (8.2), define

$$\hat{\xi}_{\alpha,B} = \hat{H}_B^{-1}(\alpha) = \inf\{x : \hat{H}_B(x) \geq \alpha\}. \tag{8.3}$$

For each of these methods, the asymptotic performance of $\hat{\xi}_{\alpha,B}$ as an approximation to $\hat{\xi}_\alpha$ may be represented in terms of the performance of $\hat{H}_B(\hat{\xi}_\alpha)$ as an approximation to $\hat{H}(\hat{\xi}_\alpha) = \alpha$, by using the method of Bahadur representation:

$$\hat{\xi}_{\alpha,B} - \hat{\xi}_\alpha = \{\alpha - \hat{H}_B(\hat{\xi}_\alpha)\}\, \phi(z_\alpha)^{-1}\, \{1 + o_p(1)\},$$

where ϕ is the Standard Normal density and z_α the α-level Standard Normal quantile. See Hall (1990a, 1990c). (Note from Section 7 that if $\alpha > 0.5$ and we are using importance resampling, it is efficaceous to approximate instead the $(1-\alpha)$'th quantile of $-T^*$, and then change sign.)

This makes it clear that the asymptotic gain in performance obtained by using an efficient resampling algorithm to approximate $\hat{H}^{-1}(\alpha)$ is equivalent to the asymptotic improvement when approximating $\hat{H}(x)$ at the point $x = \hat{\xi}_\alpha$. In the case of importance resampling, where optimal choice of the resampling probabilities p_i depends very much on the value of $\hat{\xi}_\alpha$, one implication of this result is that we may select the p_i's as though we were approximating $\hat{H}(z_\alpha)$ (bearing in mind that $\hat{\xi}_\alpha \to z_\alpha$ as $n \to \infty$). Alternatively, we might construct an approximation $\tilde{\xi}_\alpha$ of $\hat{\xi}_\alpha$ via a pilot program of uniform resampling, and then select the p_i's as though we were approximating $\hat{H}(\tilde{\xi}_\alpha)$.

References

BICKEL, P.J. and YAHAV, J.A. (1988). Richardson extrapolation and the bootstrap. *J. Amer. Statist. Assoc.* **83**, 387–393.

DAVISON, A.C. (1988). Discussion of papers by D.V. Hinkley and by T.J. DiCiccio and J.P. Romano. *J. Roy. Statist. Soc. Ser. B* **50**, 356–357.

DAVISON, A.C. and HINKLEY, D.V. (1988). Saddlepoint approximations in resampling methods. *Biometrika* **75**, 417–431.

DAVISON, A.C., HINKLEY, D.V. and SCHECHTMAN, E. (1986). Efficient bootstrap simulation. *Biometrika* **73**, 555–566.

DO, K.-A. and HALL, P. (1990). On importance resampling for the bootstrap.

EFRON, B. (1990). More efficient bootstrap computations. *J. Amer. Statist. Assoc.*.

GLEASON, J.R. (1988). Algorithms for balanced bootstrap simulations. *Amer. Statistician* **42**, 263–266.

GRAHAM, R.L., HINKLEY, D.V., JOHN, P.W.M. and SHI, S. (1990). Balanced design of bootstrap simulations. *J. Roy. Statist. Soc. Ser. B* **52**, 185–202.

HALL, P. (1989a). On efficient bootstrap simulation. *Biometrika* **76**, 613–617.

HALL, P. (1989b). Antithetic resampling for the bootstrap. *Biometrika* **76**, 713–724.

HALL, P. (1990a). Performance of bootstrap balanced resampling in distribution function and quantile problems. *Probab. Th. Rel. Fields*.

HALL, P. (1990b). Balanced importance resampling for the bootstrap.

HALL, P. (1990c). Bahadur representations for uniform resampling and importance resampling, with applications to asymptotic relative efficiency. *Ann. Statist.*

HAMMERSLEY, J.M. and HANDSCOMB, D.C. (1964). *Monte Carlo Methods*. London: Methuen.

HAMMERSLEY, J.M. and MORTON, K.W. (1956). A new Monte Carlo technique: antithetic variates. *Proc. Camb. Phil. Soc.* **52**, 449–475.

HAMMERSLEY, J.M. and MOULDON, J.G. (1956). General principles of antithetic variates. *Proc. Camb. Phil. Soc.* **52**, 476–481.

JOHNS, M.V. Jr (1988). Importance sampling for bootstrap confidence intervals. *J. Amer. Statist. Assoc.* **83**, 709-714.

OLDFORD, R.W. (1985). Bootstrapping by Monte Carlo versus approximating the estimator and bootstrapping exactly: Cost and performance. *Comm. Statist. Ser. B* **14**, 395–424.

SNIJDERS, T.A.B. (1984). Antithetic variates for Monte Carlo estimation of probabilities. *Statist. Neerland.* **38**, 55-73.

BOOTSTRAPPING U-QUANTILES

R. HELMERS [1]
Centre for Mathematics and Computer Science
Amsterdam, The Netherlands

P. JANSSEN and N. VERAVERBEKE[2]
Limburgs Universitair Centrum
Diepenbeek, Belgium

ABSTRACT

The asymptotic consistency of the bootstrap approximation for generalized quantiles of U-statistic structure (U-quantiles for short) is established. The same method of proof also yields the asymptotic accuracy of the bootstrap approximation in this case. Applications to location and spread estimators, such as the classical sample quantile, the Hodges-Lehmann estimator of location and a spread estimator proposed by Bickel and Lehmann are given.

1. INTRODUCTION

Let $X_1, X_2, \dots$ be independent random variables defined on a common probability space $(\Omega, \mathcal{A}, P)$, having common unknown distribution function (df) F. Let $h(x_1, \dots, x_m)$ be a kernel of degree m (i.e. a real-valued measurable function symmetric in its m arguments) and let

$$H_F(y) = P(h(X_1, \dots, X_m) \le y), \qquad y \in \mathbb{R} \tag{1}$$

denote the df of the random variable $h(X_1, \dots, X_m)$. Define, for each $n \ge m$ and real y,

$$H_n(y) = \binom{n}{m}^{-1} \sum_{1 \le i_1 < \dots < i_m \le n} \cdots \sum I(h(X_{i_1}, \dots, X_{i_m}) \le y) \tag{2}$$

the empirical df of U-statistic structure.

Let, for $0 < p < 1$, $\xi_p = H_F^{-1}(p)$, denote the p-th quantile corresponding to H_F, and let $\hat{\xi}_{pn} = H_n^{-1}(p)$ denote its empirical counterpart. Generalized

[1] R. Helmers, Centre for Math. and Comp. Science, Kruislaan 413, 1098 SJ Amsterdam (The Netherlands)
[2] P. Janssen, N. Veraverbeke, Limburgs Universitair Centrum, Universitaire Campus, B-3590 Diepenbeek (Belgium)

Exploring the Limits of Bootstrap. Edited by Raoul LePage and Lynne Billard.

quantiles of the form $\hat{\xi}_{pn} = H_n^{-1}(p)$ are called U-quantiles. Choudhury and Serfling (1988) note that $\hat{\xi}_{pn} \to \xi_p$, a.s. $[P]$, as $n \to \infty$, and, in addition, that, as $n \to \infty$,

$$n^{\frac{1}{2}}(\hat{\xi}_{pn} - \xi_p) \xrightarrow{d} N(0, \sigma^2) \tag{3}$$

where

$$\sigma^2 = m^2 \zeta_p h_F^{-2}(\xi_p) \tag{4}$$

with

$$\zeta_p = Var(g_p(X_1)) > 0 \tag{5}$$

and

$$g_p(X_1) = E(I(h(X_1, \ldots, X_m) \leq \xi_p) \mid X_1) - p \tag{6}$$

provided H_F has density h_F positive at ξ_p.

In applications one often wishes to establish a confidence interval for $\xi_p = H_F^{-1}(p)$ and a studentized version of (3) is required. A strongly consistent estimator of the asymptotic variance σ^2 is proposed by Choudhury and Serfling (1988). It requires $0(n^{2m-1})$ computational steps. They also propose a strongly consistent but less efficient estimator requiring only $0(n)$ computational steps.

The aim of this paper is to employ bootstrap methods for the construction of a confidence interval for $\xi_p = H_F^{-1}(p)$. In Section 2 we establish a bootstrap analog of (3), under a slightly more stringent smoothness condition on H_F and in Section 3 we establish the asymptotic accuracy of this bootstrap approximation. Applications to certain estimators of location and spread, such as the classical sample quantile, the Hodges-Lehmann estimator of location and a spread estimator proposed in Bickel and Lehmann (1979) are discussed in Section 4.

2. CONSISTENCY OF THE BOOTSTRAP FOR U-QUANTILES

Let $\hat{F}_n$ denote the empirical df based on $X_1, \ldots, X_n$. Define $\hat{\xi}_{pn}^* = H_n^{*-1}(p), 0 < p < 1$, where H_n^* denotes the empirical df of U-statistic structure based on the bootstrap sample $X_1^*, \ldots, X_n^*$. Here and elsewhere $X_1^*, \ldots, X_n^*$

denotes a random sample of size n from $\hat{F}_n$, conditionally given $X_1, \ldots, X_n$.

Our first main result is as follows :

Theorem 2.1. *Suppose that H_F is continuously differentiable (with density h_F) on a neighborhood of ξ_p with $h_F(\xi_p) > 0$. Then, for almost every sample sequence $X_1, X_2, \ldots$*

$$n^{\frac{1}{2}}(\hat{\xi}^*_{pn} - \hat{\xi}_{pn}) \xrightarrow{d} N(0, \sigma^2) \tag{7}$$

with σ^2 as in (4).

Proof. With $\hat{P}_n$ the probability measure corresponding to $\hat{F}_n$, we have

$$\begin{aligned} &\hat{P}_n(n^{\frac{1}{2}}(\hat{\xi}^*_{pn} - \hat{\xi}_{pn}) \leq x) \\ &\quad = \hat{P}_n(n^{\frac{1}{2}}(H_n^{*-1}(p) - H_n^{-1}(p)) \leq x) \\ &\quad = \hat{P}_n(H_n^{*-1}(p) \leq H_n^{-1}(p) + xn^{-\frac{1}{2}}) \\ &\quad = \hat{P}_n(H_n^*(H_n^{-1}(p) + xn^{-\frac{1}{2}}) \geq p) \\ &\quad = \hat{P}_n(W_n^* \geq -\bar{D}_n) \end{aligned} \tag{8}$$

where

$$W_n^* = n^{\frac{1}{2}}\{H_n^*(H_n^{-1}(p) + xn^{-\frac{1}{2}}) - \bar{H}_n(H_n^{-1}(p) + xn^{-\frac{1}{2}})\} \tag{9}$$

with, for each $n \geq m$ and real y,

$$\bar{H}_n(y) = n^{-m} \sum_{i_1=1}^{n} \cdots \sum_{i_m=1}^{n} I(h(X_{i_1}, \ldots, X_{i_m}) \leq y) \tag{10}$$

the empirical df of von Mises structure, and

$$\bar{D}_n = n^{\frac{1}{2}}\{\bar{H}_n(H_n^{-1}(p) + xn^{-\frac{1}{2}}) - p\}. \tag{11}$$

We first consider $\bar{D}_n$. Note that

$$\bar{D}_n = \sum_{i=1}^{3} \bar{D}_{in} \tag{12}$$

where

$$\begin{aligned} \bar{D}_{1n} = n^{\frac{1}{2}}\{&\bar{H}_n(H_n^{-1}(p) + xn^{-\frac{1}{2}}) - H_F(H_n^{-1}(p) + xn^{-\frac{1}{2}})\} \\ &- n^{\frac{1}{2}}\{\bar{H}_n(H_n^{-1}(p)) - H_F(H_n^{-1}(p))\} \end{aligned} \tag{13}$$

$$\bar{D}_{2n} = n^{\frac{1}{2}}\{H_F(H_n^{-1}(p) + xn^{-\frac{1}{2}}) - H_F(H_n^{-1}(p))\} \tag{14}$$

and

$$\bar{D}_{3n} = n^{\frac{1}{2}}\{\bar{H}_n(H_n^{-1}(p)) - p\}. \tag{15}$$

To treat $\bar{D}_{1n}$ note first that $\bar{D}_{1n} = D_{1n} + O(n^{-\frac{1}{2}})$ a.s. $[P]$, as $n \to \infty$, where D_{1n} is obtained from $\bar{D}_{1n}$ by replacing $\bar{H}_n$ by H_n, with H_n as in (2). Suppose without loss of generality that $x > 0$. Clearly, for n sufficiently large,

$$|D_{1n}| \leq \sup_{\substack{|t-s| \leq xn^{-\frac{1}{2}} \\ s,t \in J}} |U_n(t) - U_n(s)| \quad \text{a.s. } [P] \tag{16}$$

where J is the neighborhood of ξ_p on which H_F is continuously differentiable, and

$$U_n(t) = n^{\frac{1}{2}}(H_n(t) - H_F(t)), \quad t \in I\!R \tag{17}$$

denotes the empirical process of U-statistic structure.

Similarly as in Silverman (1983) it is easy to see that

$$|D_{1n}| \leq (n!)^{-1} \sum_{\alpha} \sup_{\substack{|t-s| \leq xn^{-\frac{1}{2}} \\ t,s \in J}} |U^{\alpha}_{[\frac{n}{m}]}(t) - U^{\alpha}_{[\frac{n}{m}]}(s)| \tag{18}$$

where, for any given permutation α of $\{1, 2, \ldots, n\}$ $U^{\alpha}_{[\frac{n}{m}]}(t)$ denotes the empirical process based on the $[\frac{n}{m}]$ independent random variables $h(X_{\alpha(mj+1)}, \ldots, X_{\alpha(mj+m)}), j = 0, 1, \ldots, [\frac{n}{m}] - 1$, all with common df H_F. With impunity we may replace at stage $n+1$ any of the $n.n!$ permutations α of $\{1, \ldots, n+1\}$ which do not extend those of $\{1, \ldots, n\}$ by one of the $n!$ permutations which do extend those of $\{1, \ldots, n\}$. Application of relation (2.13) of Stute (1982) to each of the resulting $n!$ terms appearing on the r.h.s. of (18) directly yields that $D_{1n} = O(n^{-\frac{1}{4}}(\ln n)^{\frac{1}{2}})$ a.s. $[P]$, as $n \to \infty$, hence,

$$\bar{D}_{1n} = O(n^{-\frac{1}{4}}(\ln n)^{\frac{1}{2}}) \quad \text{a.s. } [P], \quad \text{as } n \to \infty. \tag{19}$$

Here we have used the smoothness of H_F as well as the inequality (18).

Next we consider $\bar{D}_{2n}$. Using again the smoothness assumption on H_F and employing the a.s. [P] convergence of $\hat{\xi}_{pn}$ to ξ_p, as $n \to \infty$, we easily obtain from the mean value theorem that

$$\bar{D}_{2n} \to x h_F(\xi_p) \quad \text{a.s. } [P], \quad \text{as } n \to \infty. \tag{20}$$

Finally note that $\bar{D}_{3n} = O(n^{-\frac{1}{2}})$. We can conclude that

$$\bar{D}_n \to x h_F(\xi_p) \text{a.s. } [P], \quad \text{as } n \to \infty. \tag{21}$$

Next we consider the limit behaviour of W_n^*, $n = m, m+1, \ldots$ (cf (9)), conditionally given $\hat{F}_n$. Obviously, given $\hat{F}_n$, W_n^* is a normalized U-statistic of degree m, with bounded kernel, depending on n, of the form

$$h_n(x_1, \ldots, x_m) = I(h(x_1, \ldots, x_m) \leq \hat{\xi}_{pn} + xn^{-\frac{1}{2}}) - \bar{H}_n(\hat{\xi}_{pn} + xn^{-\frac{1}{2}}). \tag{22}$$

Of course $E_{\hat{F}_n} W_n^* = E_{\hat{F}_n} h_n(X_1^*, \ldots, X_m^*) = 0$, a.s. [P], whereas it is easily checked that

$$Var_{\hat{F}_n}(W_n^*) \sim m^2 E_{\hat{F}_n} g_n^2(X_1^*) \quad \text{a.s. } [P], \text{ as } n \to \infty, \tag{23}$$

where

$$g_n(X_1^*) = E_{\hat{F}_n}(h_n(X_1^*, \ldots, X_m^*)|X_1^*). \tag{24}$$

A simple argument involving the strong law for U-statistics with estimated parameters (Theorem 2.9 of Iverson and Randles (1989)) directly yields that

$$E_{\hat{F}_n} g_n^2(X_1^*) \to \zeta_p \quad \text{a.s. } [P], \text{ as } n \to \infty, \tag{25}$$

with ζ_p as in (5).

At this point we apply the Berry-Esseen bound for U-statistics of degree m of van Zwet (1984) to find that

$$\sup_y |\hat{P}_n(W_n^* \leq y) - \Phi(ym^{-1}\zeta_p^{-\frac{1}{2}})| = O\{(\frac{E_{\hat{F}_n}|g_n(X_1^*)|^3}{(E_{\hat{F}_n} g_n^2(X_1^*))^{\frac{3}{2}}} + \frac{E_{\hat{F}_n} h_n^2(X_1^*, \ldots, X_m^*)}{E_{\hat{F}_n} g_n^2(X_1^*)})n^{-\frac{1}{2}}\}. \tag{26}$$

Note that, in contrast to Corollary 4.1 of van Zwet (1984), the asymptotic variance instead of the exact variance of W_n^* is employed. It is easy to see that this does not affect the bound (26). The different standardization will give rise to an additional term of type

$$\frac{E_{\hat{F}_n} h_n^2(X_1^*, \ldots, X_m^*)}{E_{\hat{F}_n} g_n^2(X_1^*)} n^{-\frac{1}{2}} \tag{27}$$

which is already present in van Zwet's bound. Because h_n is bounded by 1, for all n, and combining (25) with the fact that $\zeta_p > 0$, we easily see that the moments appearing on the r.h.s. of (26) are $O(1)$ a.s. $[P]$, as $n \to \infty$. Hence the r.h.s. of (26) is $O(n^{-\frac{1}{2}})$ a.s. $[P]$, as $n \to \infty$.

From (8), (21) and (26) we obtain

$$\begin{aligned} \hat{P}_n(n^{\frac{1}{2}}(\hat{\xi}_{pn}^* - \hat{\xi}_{pn}) \le x) \\ &= 1 - \Phi(-\bar{D}_n m^{-1} \zeta_p^{-\frac{1}{2}}) + O(n^{-\frac{1}{2}}) \\ &= \Phi(x\sigma^{-1}) + o(1) \end{aligned} \tag{28}$$

a.s. $[P]$, as $n \to \infty$. This completes the proof of theorem. □

For the special case $m = 1, h(x) = x, p = \frac{1}{2}$, the classical sample median, our result reduces to Proposition 5.1 of Bickel and Freedman (1981). An insightful proof of their proposition is given in an unpublished note by Sheehy and Wellner (1988). Our proof is in part inspired by their argument.

3. ACCURACY OF THE BOOTSTRAP FOR U-QUANTILES

From (3) and (7) we know that the bootstrap approximation for a normalized U-quantile is asymptotically consistent. In this section we investigate the a.s. rate at which the difference between the bootstrap approximation and the exact distribution of a normalized U-quantile tends to zero, as the sample size gets large.

Theorem 3.1. *Suppose that the assumptions of Theorem 2.1 are satisfied. Suppose, in addition, that h_F satisfies a Lipschitz condition of order $\ge \frac{1}{2}$ on a neighborhood of ξ_p. Then*

$$\sup_x |\hat{P}_n(n^{\frac{1}{2}}(\hat{\xi}_{pn}^* - \hat{\xi}_{pn}) \le x) - P(n^{\frac{1}{2}}(\hat{\xi}_{pn} - \xi_p) \le x)| = O(n^{-\frac{1}{4}}(\ln n)^{\frac{3}{4}}) \tag{29}$$

a.s.$[P]$, *as* $n \to \infty$.

For the special case $m = 1, h(x) = x$, the classical p-th sample quantile, Singh (1981) obtained a slightly better a.s. rate : the factor $(\ln n)^{\frac{3}{4}}$ in (29) is replaced by $(ln ln\, n)^{\frac{1}{2}}$ in this case. Whenever the same improvement holds true for U-quantiles appears to be an interesting open problem.

Proof. First note that

$$\sup_x |\hat{P}_n(n^{\frac{1}{2}}(\hat{\xi}^*_{pn} - \hat{\xi}_{pn}) \le x) - P(n^{\frac{1}{2}}(\hat{\xi}_{pn} - \xi_p) \le x)| \le \sum_{i=1}^{3} I_{in} \tag{30}$$

where, for some constant $K > 0$,

$$I_{1n} = \sup_{|x| \le K(ln\, n)^{\frac{1}{2}}} |\hat{P}_n(n^{\frac{1}{2}}(\hat{\xi}^*_{pn} - \hat{\xi}_{pn}) \le x) - \Phi(x\sigma^{-1})| \tag{31}$$

and

$$I_{2n} = \sup_{|x| > K(ln\, n)^{\frac{1}{2}}} |\hat{P}_n(n^{\frac{1}{2}}(\hat{\xi}^*_{pn} - \hat{\xi}_{pn}) \le x) - \Phi(x\sigma^{-1})| \tag{32}$$

and

$$I_{3n} = \sup_x |P(n^{\frac{1}{2}}(\hat{\xi}_{pn} - \xi_p) \le x) - \Phi(x\sigma^{-1})|.$$

We first consider I_{1n}. Going through the proof of Theorem 2.1 we easily verify that

$$\sup_{|x| \le K(ln\, n)^{\frac{1}{2}}} |\bar{D}_n - xh_F(\xi_p)| = 0(n^{-\frac{1}{4}}(ln\, n)^{\frac{3}{4}}) \quad \text{a.s. } [P], \text{ as } n \to \infty. \tag{33}$$

Here we have used (see (18)) that

$$\begin{aligned}\sup_{|x| \le K(ln\, n)^{\frac{1}{2}}} |D_{1n}| &\le (n!)^{-1} \sum_{\alpha} \sup_{\substack{|t-s| \le Kn^{-\frac{1}{2}}(\ln n)^{\frac{1}{2}} \\ t,s \in J}} |U^{\alpha}_{[\frac{n}{m}]}(t) - U^{\alpha}_{[\frac{n}{m}]}(s)| \\ &= 0(n^{-\frac{1}{4}}(ln\, n)^{\frac{3}{4}}) \quad \text{a.s. } [P],\ n \to \infty\end{aligned}$$

by application of relation (2.13) in Stute (1982). Also (20) is replaced by the stronger assertion that

$$\sup_{|x| \le K(ln\, n)^{\frac{1}{2}}} |\bar{D}_{2n} - xh_F(\xi_p)| = 0(n^{-\frac{1}{4}}(ln\, n)^{\frac{3}{4}}) \quad \text{a.s. } [P], \text{ as } n \to \infty.$$

For this we used Lemma 3.1 of Choudhury and Serfling (1988) and the Lipschitz condition on h_F. Combining (33) with (28) directly yields

$$I_{1n} = 0(n^{-\frac{1}{4}}(ln\, n)^{\frac{3}{4}}) \quad \text{a.s. } [P], \text{ as } n \to \infty. \tag{34}$$

For the quantity I_{2n} we have

$$\begin{aligned} I_{2n} \leq \hat{P}_n(\hat{\xi}^*_{pn} - \hat{\xi}_{pn} > Kn^{-\frac{1}{2}}(ln\, n)^{\frac{1}{2}}) \\ +\hat{P}_n(\hat{\xi}^*_{pn} - \hat{\xi}_{pn} < -Kn^{-\frac{1}{2}}(ln\, n)^{\frac{1}{2}}) \\ +2(1 - \Phi(K(ln\, n)^{\frac{1}{2}}\sigma^{-1}). \end{aligned} \tag{35}$$

The third term is $0(n^{-\frac{1}{2}})$ by taking K large enough. It remains to estimate the two other terms. Since the argument is the same for both, we only deal with the first term of the r.h.s. of (35). Similarly as in (8) we write

$$\begin{aligned} &\hat{P}_n(\hat{\xi}^*_{pn} - \hat{\xi}_{pn} > Kn^{-\frac{1}{2}}(ln\, n)^{\frac{1}{2}}) \\ &= \hat{P}_n(H^*_n(\hat{\xi}_{pn} + Kn^{-\frac{1}{2}}(ln\, n)^{\frac{1}{2}}) - \bar{H}_n(\hat{\xi}_{pn} + Kn^{-\frac{1}{2}}(ln\, n)^{\frac{1}{2}}) \\ &\qquad < p - \bar{H}_n(\hat{\xi}_{pn} + Kn^{-\frac{1}{2}}(ln\, n)^{\frac{1}{2}})) \end{aligned} \tag{36}$$

Application of Lemma 3.1 of Choudhury and Serfling (1988) directly yields that for all n sufficiently large,

$$p - \bar{H}_n(\hat{\xi}_{pn} + Kn^{-\frac{1}{2}}(lnn)^{\frac{1}{2}}) \leq p - \bar{H}_n(\xi_p + \frac{K}{2}n^{-\frac{1}{2}}(ln\, n)^{\frac{1}{2}}) \tag{37}$$

a.s. $[P]$, provided we take K large enough. A simple argument involving Corollary 2.1 of Helmers, Janssen and Serfling (1988) and the a.s. closeness of $\bar{H}_n$ and H_n gives us (with C_m as in the corollary)

$$\begin{aligned} &p - \bar{H}_n(\xi_p + \frac{K}{2}n^{-\frac{1}{2}}(lnn)^{\frac{1}{2}}) \\ &\leq p - H_F(\xi_p + \frac{K}{2}n^{-\frac{1}{2}}(ln\, n)^{\frac{1}{2}}) + C_m n^{-\frac{1}{2}}(ln\, n)^{\frac{1}{2}} + 0(n^{-1}) \end{aligned} \tag{38}$$

a.s. $[P]$. The smoothness assumption of the theorem directly implies that

$$\begin{aligned} &p - H_F(\xi_p + \frac{K}{2}n^{-\frac{1}{2}}(ln\, n)^{\frac{1}{2}}) \\ &= -\frac{K}{2}h_F(\xi_p)n^{-\frac{1}{2}}(ln\, n)^{\frac{1}{2}}(1 + o(1)) \quad \text{a.s. } [P], \text{ as } n \to \infty. \end{aligned} \tag{39}$$

Together (37), (38) and (39) yield that $p - \bar{H}_n(\hat{\xi}_{pn} + Kn^{-\frac{1}{2}}(ln\, n)^{\frac{1}{2}}) < 0$, for all n sufficiently large, a.s. $[P]$, provided we take K large enough.

We can now apply an exponential bound for U-statistics with bounded kernels of Hoeffding (1963) (see also Serfling (1980), p. 201) to find that

$$\begin{aligned}
&\hat{P}_n(H_n^*(\hat{\xi}_{pn} + Kn^{-\frac{1}{2}}(ln\, n)^{\frac{1}{2}}) - \bar{H}_n(\hat{\xi}_{pn} + Kn^{-\frac{1}{2}}(ln\, n)^{\frac{1}{2}}) \\
&\qquad\qquad < p - \bar{H}_n(\hat{\xi}_{pn} + Kn^{-\frac{1}{2}}(ln\, n)^{\frac{1}{2}})) \\
&\leq exp\{-\tfrac{1}{8}[\tfrac{n}{m}]n^{-1} ln\, n K^2 h_F^2(\xi_p)\} \\
&= 0(n^{-\frac{1}{2}}) \quad \text{a.s. } [P], \quad \text{as } n \to \infty,
\end{aligned} \tag{40}$$

provided K is taken sufficiently large. This together with (35) and (36) implies that

$$I_{2n} = 0(n^{-\frac{1}{2}}) \quad \text{a.s. } [P], \quad \text{as } n \to \infty. \tag{41}$$

Hence I_{2n} is of negligible order for our purposes. It remains to consider I_{3n}. Clearly, as $n \to \infty$,

$$I_{3n} = \sup_x |P(n^{\frac{1}{2}}(\hat{\xi}_{pn} - \xi_p) \leq x) - \Phi(x\sigma^{-1})| = O(n^{-\frac{1}{2}}), \tag{42}$$

i.e. the Berry-Esseen bound for U-quantiles is valid. To check (42) is an easy matter in view of the classical proof of a Berry-Esseen bound for ordinary sample quantiles (see, e.g. Serfling (1980), p.81-84). We have to apply instead of the Lemma on p.75 of Serfling (1980), the exponential bound of Hoeffding (1963) for U-statistics with bounded kernels. Also a Berry-Esseen bound for U-statistics is needed.

Combining (34), (41) and (42) with (30), we find that the theorem is proved. □

4. APPLICATIONS

In this section we indicate briefly applications of our results to the problem of obtaining confidence intervals for $\xi_p = H_F^{-1}(p)$. Let $u_{\frac{\alpha}{2}} = \Phi^{-1}(1 - \frac{\alpha}{2})$. The normal approximation (3) yields an approximate two-sided confidence interval

$$(\hat{\xi}_{pn} - n^{-\frac{1}{2}}\hat{\sigma}_n u_{\frac{\alpha}{2}} \ , \ \hat{\xi}_{pn} + n^{-\frac{1}{2}}\hat{\sigma}_n u_{\frac{\alpha}{2}}) \tag{43}$$

for ξ_p. Here $\hat{\sigma}_n^2$ denotes a consistent estimator (e.g., the one proposed by Choudhury and Serfling (1988)) of the asymptotic variance σ^2. Clearly, the error rates corresponding to the upper and lower confidence limits in (43) will depend on the rate at which $\hat{\sigma}_n^2$ approaches σ^2.

A bootstrap based confidence interval for ξ_p is given by

$$(\hat{\xi}_{pn} - n^{-\frac{1}{2}} c^*_{n,1-\frac{\alpha}{2}} \ , \ \hat{\xi}_{pn} - n^{-\frac{1}{2}} c^*_{n,\frac{\alpha}{2}}) \tag{44}$$

where $c^*_{n,\frac{\alpha}{2}}$ and $c^*_{n,1-\frac{\alpha}{2}}$ denote the $\frac{\alpha}{2}$-th and $(1-\frac{\alpha}{2})$-th percentile of the (simulated) bootstrap approximation. It is easily verified that the upper and lower confidence limits in (44) have error rates equal to $\frac{\alpha}{2} + 0(n^{-\frac{1}{4}}(\ln n)^{\frac{3}{4}})$.

We discuss a few specific examples of U-quantiles. In the first of these we take $m = 1$, $h(x) = x$ and obtain the classical p-th sample quantile $\hat{\xi}_{pn} = \hat{F}_n^{-1}(p)$, $0 < p < 1$. Our second example is obtained by taking $p = \frac{1}{2}$, $m = 2$, $h(x_1, x_2) = (x_1 + x_2)/2$. In this case $\hat{\xi}_{\frac{1}{2}n} = H_n^{-1}(\frac{1}{2})$ becomes the well-known Hodges-Lehmann location estimator. In the third and final example we take $p = \frac{1}{2}$, $m = 2$, $h(x_1, x_2) = |x_1 - x_2|$. In this case $\hat{\xi}_{\frac{1}{2}n} = H_n^{-1}(\frac{1}{2})$ reduces to an estimator of spread proposed by Bickel and Lehmann (1979).

A further investigation into the relative merits of the normal and bootstrap based confidence intervals (43) and (44) for U-quantiles appears to be worthwhile. The authors hope to report on these matters elsewhere.

REFERENCES

Bickel, P.J. and Freedman, D. (1981). Some asymptotic theory for the bootstrap. Ann. Statist. 9, 1196-1217.

Bickel, P.J. and Lehmann, E.L. (1979). Descriptive statistics for nonparametric models. IV. Spread. Contributions to Statistics (J. Hájek Memorial Volume), 33-40. (ed. J. Jurečková). Academia, Prague.

Choudhury, J. and Serfling, R. (1988). Generalized order statistics, Bahadur representations, and sequential nonparametric fixed-width confidence intervals. J. Statist. Planning Inf. 19, 269-282.

Helmers, R., Janssen, P. and Serfling, R. (1988). Glivenko-Cantelli properties of some generalized empirical df's and strong convergence of generalized L-statistics. Probab. Th. Rel. Fields 79, 75-93.

Hoeffding, W. (1963). Probability inequalities for sums of bounded random variables. J. Amer. Statist. Assoc. 58, 13-30.

Iverson, H. and Randles, R. (1989). The effect on convergence of substituting parameter estimates into U-statistics and other families of statistics. Probab. Th. Rel. Fields 81, 453-471.

Serfling, R.J. (1980). Approximation theorems of mathematical statistics. J. Wiley, New York.

Sheehy, A. and Wellner, J. (1988). Almost sure convergence of the bootstrap for the median. Unpublished manuscript.

Silverman, B.W. (1983). Convergence of a class of empirical distribution functions of dependent random variables. Ann. Probability 11, 745-751.

Singh, K. (1981). On the asymptotic accuracy of Efron's bootstrap. Ann. Statist. 9, 1187-1195.

Stute, W. (1982). The oscillation behavior of empirical processes. Ann. Probability 10, 86-107.

van Zwet, W.R. (1984). A Berry-Esseen bound for symmetric statistics. Z. Wahrsch. verw. Gebiete 66, 425-440.

AN INVARIANCE PRINCIPLE APPLICABLE TO THE BOOTSTRAP

John G. Kinateder[1]

Michigan State University

1 Introduction

Suppose $X_1, X_2, \ldots$ are independent random variables distributed according to a distribution function F with location parameter θ. In order to make inferences about θ, we may consider the distribution of the sample mean about θ :

$$\bar{X}_n - \theta.$$

For example, if $EX_1^2 < \infty$, the well-known Lindeberg-Levy Central Limit Theorem [Bil86] tells us that

$$n^{1/2}(\bar{X}_n - EX_1) \to_d N(0, \sigma^2)$$

where σ^2 represents the variance of X_1. In the finite variance case, we can use this to make inferences about $\theta = EX_1$.

Of course $EX_1^2 < \infty$ is not necessary for convergence in distribution of the sample mean.

Definition 1 *F is said to be in the domain of attraction of a distribution μ (not concentrated at one point) if there exist constants $a_n > 0, b_n$, and a random variable Y with distribution μ, such that*

$$S_n = a_n^{-1} \sum_{j=1}^{n} (X_j - b_n) \to_d Y, \tag{1}$$

Necessarily $a_n \sim cn^{1/\alpha}$ for some $\alpha \in (0, 2]$ and $c > 0$; Y and μ are said to be α-stable. (See [Fel71].)

In what follows, we assume that F is in the domain of attraction of an α-stable distribution and $X_1, X_2, \ldots$ is a sequence of i.i.d. F random variables. If $0 < \alpha < 2$, then we fix a sequence $a_n > 0$ such that for each $y > 0$, $n(1 - F(a_n y)) \to y^{-\alpha}$ as $n \to \infty$. For such a_n, (1) holds with $b_n = 0$ for

[1]Research partially supported by the Office of Naval Research under grant N000014-85-K-0150

Exploring the Limits of Bootstrap. Edited by Raoul LePage and Lynne Billard.
 ISBN: 0471-53631-8.

$0 < \alpha < 1$, $b_n = EX_1$ if $1 < \alpha < 2$, and $b_n = E\sin(X/a_n)$ if $\alpha = 1$. (For existence of such a sequence, see Feller [Fel71].) If $\alpha = 2$, then we choose a_n such that

$$a_n^{-1}\sum_{j=1}^{n}(X_j - EX_1) \rightarrow_d N(0,1).$$

In either case, let μ denote the limit distribution.

Since the distribution F is generally unknown, so is the distribution of S_n. Thus, if we are to use S_n to estimate θ, then we need to have some idea of the variability of S_n. As was suggested by Efron [Efr79], in a wide variety of situations, we can use resampling of the data $X_1, \ldots, X_n$ to estimate the distribution of an estimator. This is the essence of bootstrap.

Let F_n be the empirical distribution function of $X_1, \ldots, X_n$:

$$F_n(x) = \frac{1}{n}\sum_{k=1}^{n} I(X_k \leq x).$$

For each observation of the data $X_1, \ldots, X_n$, we consider the distribution of

$$S_m^* = a_m^{-1}\sum_{j=1}^{m}(X_j^* - \bar{X}_n)$$

where $X_1^*, \ldots, X_m^*$ are independent and distributed according to F_n. This is equivalent to simple random sampling from the original sample, $X_1, \ldots, X_n$, with replacement, and applying the same statistic to the resampled data as we would to the original data. The resampled data is often called the *bootstrap sample*, and the conditional distribution of the statistic applied to the bootstrap sample (given the data $(X_1, \ldots, X_n)$) is referred to as the *bootstrap distribution.*

Bickel and Freedman [BF81] showed that in the finite variance case, the bootstrap distribution of S_m^*, given $(X_1, \ldots, X_n)$, converges weakly to $N(0,1)$ as long as $m, n \rightarrow \infty$. (Recall that the variance is removed here by the choice of a_n.) Singh [Sin81]) showed that under the added assumptions that $E|X_1|^3 < \infty$ and F is non-lattice, then asymptotically the bootstrap of the pivoted sample mean is actually a better approximation to the true distribution than the normal, based on the Edgeworth expansion.

Hall [Hal88] showed that when X_1 has finite variance but $E|X_1|^3 = \infty$, the general situation is that the normal approximation and bootstrap approximation are asymptotically equivalent. Therefore, in this case, it is better to use the normal approximation in lieu of the computational cost of the bootstrap.

For the case of $0 < \alpha < 2$, when the bootstrap sample size m is taken to be the same as the original sample size n, it was shown by Athreya ([Ath84] and [Ath87]) that the bootstrap distribution of S_n^* does not converge weakly to a constant distribution along almost all sample sequences. He showed that it converges in distribution (with respect to the weak topology on the space of bounded measures) to a random distribution. He gave a representation for this random limit distribution in terms of Poisson random measures.

Notice that if $(M_{n1}^*, \ldots, M_{nn}^*)$ has the multinomial distribution with parameters $(n, (\frac{1}{n}, \ldots, \frac{1}{n}))$ and is independent of the sample sequence, then

$$\mathcal{L}(S_n^* \mid X_1, \ldots, X_n) = \mathcal{L}(\sum_{k=1}^{n} X_k(M_{nk}^* - 1) \mid X_1, \ldots, X_n). \tag{2}$$

The M_{nj}^* can be thought of as *counts*; X_j is chosen M_{nj}^* times in the bootstrap sample.

Following Athreya's work, Knight [Kni89] gave a different representation for this limit law. Using the distributional relationship (2) above, and the sample sequence representation provided by LePage, Woodroofe, and Zinn [LWZ81], he gave the following explicit representation of the limit law: As in [LWZ81], define p by

$$p = \lim_{y \to \infty} \frac{1 - F(y)}{1 - F(y) + F(-y-)}.$$

Let $\epsilon_1, \epsilon_2, \ldots$ be i.i.d., $P(\epsilon_1 = 1) = p = 1 - P(\epsilon_1 = -1)$. Let $\Gamma = \langle \Gamma_1, \Gamma_2, \ldots \rangle$ represent the arrival times of a Poisson process with unit rate; $\Gamma_k = \sum_{j=1}^{k} \xi_i$ where $P(\xi_i \geq x) = e^{-x}$ for all i ($\xi_1, \xi_2, \ldots$ are independent). Let $M_1^*, M_2^*, \ldots$ be independent Poisson mean 1 random variables. Finally, assume that $\{\epsilon_j\}$, $\{\Gamma_j\}$, $\{M_j^*\}$ are mutually independent. Then

$$\mathcal{L}(S_n^* \mid X_1, \ldots, X_n) \to_d \mathcal{L}(\sum_{k=1}^{\infty} \epsilon_k \Gamma_k^{-1/\alpha}(M_k^* - 1) \mid \epsilon_k \Gamma_k^{-1/\alpha}, k \geq 1). \tag{3}$$

It is important to note that the above convergence is *in distribution*. In fact, Giné and Zinn [GZ89] show that in the infinite variance case, this cannot be strengthened to almost sure convergence (which does occur in the finite variance case).

In the infinite variance case, the bootstrap distribution of S_n^* does not converge in distribution to the limit distribution μ obtained in the limit of the original S_n sequence. That is, the bootstrap distribution is not a consistent estimate of μ. Because of this phenomenon, it has been said that the bootstrap does not *work* in this case. (Although it is understood why this claim was made, Section 5 suggests that the method may actually be viable for applications in the $\alpha < 2$ case.)

What happens when the bootstrap sample size m is allowed to differ from the sample size n? Athreya [Ath85] showed that in the $0 < \alpha < 2$ case, the bootstrap can still be made to *work* if the bootstrap sample size is chosen small enough in relation to n. More precisely, if the bootstrap sample size $m_n \to \infty$

such that $m_n/n \to 0$, then the bootstrap distribution of $S^*_{m_n}$ converges weakly to μ *in probability.*

Arcones and Giné [AG89] added to this answer by showing that if

$$m_n \log\log m_n/n \to 0,$$

then the bootstrap central limit theorem holds *almost surely.* That is, the conditional distribution of $S^*_{m_n}$ converges with respect to the weak topology to μ almost surely. But if

$$\liminf_n m_n \log\log m_n/n > 0,$$

then there is no almost sure convergence — not even to a random measure!

In this paper, we examine the relationship between the distribution of the partial sums and the resampling criteria. We give a decomposition of the bootstrap of the sample mean in the form of a stochastic integral. Then we state and prove an invariance principle explaining the behavior of the processes involved in the decomposition. When these processes are replaced by their limits in the stochastic integral, the integral obtained turns out to have the limiting distribution of the bootstrap *for all* α.[2] This affords a general representation of the bootstrap limit law encompassing both the normal and nonnormal domains of attraction in one expression.

We give the stochastic integral decomposition in Section 2. In Section 3 we introduce the invariance principle. The theorem is proved for the finite variance case as well as the $\alpha < 2$ symmetric case. Section 4 gives a decomposition of the limiting distributions of the same form as that given in Section 2. We give some simulation results in Section 5 which suggest that in the $\alpha < 2$ symmetric case the bootstrap of the sample mean actually performs very well. Section 6 contains some concluding remarks suggesting some of the value of the research as well as future directions. The Appendix contains the statements of some technical results which are referenced in the proof of the invariance principle. Some of the proofs have been omitted here but all have been included in the author's Ph.D. thesis [Kin90].

2 The Stochastic Integral Representation

Here we give a new decomposition of the bootstrap of the sample mean.

Definition 2 *For each pair of positive integers m, n, we define the following:* (*i*) $\widetilde{X}_n = (X_{(1)}, \ldots, X_{(n)})$ *are the absolutely ordered observations*

$$|X_{(1)}| \geq \ldots \geq |X_{(n)}|,$$

[2] It should be mentioned here that in the infinite variance case the limit result is restricted to distributions symmetric about the location parameter.

so that $X_{(i)}$ *is the* i^{th} *largest in absolute value.*
(ii) W^n *is the scaled partial sum process associated with* $(X_1, \ldots, X_n)$*:*

$$W^n(t) = a_n^{-1} \sum_{k=1}^{\lfloor nt \rfloor} X_k, \qquad t \in [0,1].$$

(iii) $(M^*_{m1}, \ldots, M^*_{mn})$ *is a multinomial* $(m, (\frac{1}{n}, \ldots, \frac{1}{n}))$ *vector independent of the observations* $(X_1, \ldots, X_n)$*.*
(iv) $H^{(m,n)}$ *is the empirical distribution function of the centered multinomial vector:*

$$H^{(m,n)}(x) = \frac{1}{n} \sum_{k=1}^{n} I(M^*_{mk} - \frac{m}{n} \le x).$$

Theorem 1 *Let n be the number of observations, and let m be the bootstrap sample size. Then*

$$\mathcal{L}\left(a_n^{-1} \sum_{k=1}^{m} (X^*_k - \bar{X}_n) \,|\, \widetilde{X}_n\right) = \mathcal{L}\left(\int t\, W^n \circ\ H^{(m,n)}(dt) \,|\, \widetilde{X}_n\right).$$

We will refer to $W^n \circ H^{(m,n)}$ as the *bootstrap partial sum process.*

It should be pointed out here that in reference to the bootstrap, since resampling with replacement has no dependence on the order of the data,

$$\mathcal{L}(\sum_{k=1}^{m} (X^*_k - \bar{X}_n) \,|\, \widetilde{X}_n) = \mathcal{L}(\sum_{k=1}^{m} (X^*_k - \bar{X}_n) \,|\, X_1, \ldots, X_n).$$

But this alternative conditioning is not valid for the stochastic integral representation;

$$\mathcal{L}(\int r\, W^n \circ\ H^{(m,n)}(dr) \,|\, \widetilde{X}_n) \ne \mathcal{L}(\int r\, W^n \circ\ H^{(m,n)}(dr) \,|\, X_1, \ldots, X_n).$$

3 The Invariance Principle

Since $\widetilde{X}_n$ is a function of the partial sum process W^n, and conditionally on $\widetilde{X}_n$, S^*_m has distribution dependent on the bootstrap partial sum process $W^n \circ H^{(m,n)}$, it is clear that the behavior of S^*_m jointly with $\widetilde{X}^n$ is dependent on the behavior of $W^n \circ H^{(m,n)}$ jointly with W^n. The following invariance principle helps explain this behavior for large n.

3.1 The Main Theorem

We define the Skorohod metric as it is defined in Billingsley [Bil68].

Definition 3 *For each pair of functions x and y in $D[0,1]$, define the distance $d_s(x,y)$ as the infimum of all those values of δ for which there exists a strictly increasing and onto transformation $\lambda : [0,1] \longrightarrow [0,1]$ such that*

$$\|\lambda - \imath\| \le \delta \quad and \quad \|x - y(\lambda)\| \le \delta,$$

where $\imath$ denotes the identity function on $[0,1]$.

Theorem 2 *Suppose H^n is a sequence of stochastic processes on R independent of W^n converging uniformly to H almost surely. Let W be the homogeneous independent increments symmetric α-stable process with scale determined by $S_n \to_d W(1)$.*

(A) *If X_1 has a symmetric distribution in the domain of attraction of an α-stable distribution ($\alpha < 2$) and H is the distribution function of a discrete random variable which takes values in a finite set or a set which can be written in the form $\{d_1, d_2, \ldots\}$ such that $d_1 < d_2 < \cdots$, then*

$$(W^n \circ H^n, W^n) \to_d (W \circ H, W)$$

in the product space $(D(R), \mathcal{U}) \times (D[0,1], \mathcal{S})$ where $\mathcal{U}$ denotes the uniform topology on $D(R)$ and $\mathcal{S}$ denotes the Skorohod topology on $D[0,1]$.

(B) *If $EX_1^2 < \infty$ then*

$$(W^n \circ H^n, W^n) \to_d (W \circ H, W)$$

in the product space $(D(R), \mathcal{U}_1) \times (D[0,1], \mathcal{U}_2)$ where $\mathcal{U}_1$ denotes the uniform topology on $D(R)$ and $\mathcal{U}_2$ denotes the uniform topology on $D[0,1]$.

Notice that in part (B), W is a Brownian motion.

3.2 Proof for the $\alpha < 2$ Case

Throughout this section we assume the hypotheses of Theorem 2(A).

Definition 4 *Let $\epsilon_1, \epsilon_2, \ldots$ be i.i.d.,*

$$P(\epsilon_1 = 1) = P(\epsilon_1 = -1) = 1/2.$$

Let Γ represent the arrival times of a unit rate Poisson process as described in the introduction.
Let $T_1, T_2, \ldots$ be i.i.d., uniformly distributed on $(0,1)$.
Define W by

$$W(t) = \sum_{k=1}^{\infty} \epsilon_k \Gamma_k^{-1/\alpha} I(T_k \le t), \qquad t \in [0,1].$$

LePage [LeP80] showed that W is a homogeneous independent increments symmetric α-stable process.

We start by exploring the way that a particular *LePage*-like representation $\widetilde{W}^n$ with the same distribution as W^n converges to W. Then we will use this convergence along with the almost sure uniform convergence of H^n to H to finish the proof.

Let $1-G$ be the distribution function of $|X_1|$. Let G^{-1} be the usual inverse: For real $y \in (0,1)$, $G^{-1}(y) = \inf\{x : G(x) \leq y\}$. Define for each n and $k = 1, 2, \ldots, n$,

$$Y_{nk} = a_n^{-1} G^{-1}(\Gamma_k / \Gamma_{n+1}).$$

Notice that

$$(\epsilon_1 Y_{n1}, \ldots, \epsilon_n Y_{nn}) =_d a_n^{-1}(X_{(1)}, \ldots, X_{(n)}). \tag{4}$$

As introduced by LePage [LeP80], we define random variables L_j^n in such a way that the processes $I(L_j^n \leq t)$, $j = 1, \ldots, n$ facilitate *scrambling* of the ordered random variables, $\epsilon_1 Y_{n1}, \ldots, \epsilon_n Y_{nn}$,

$$\begin{aligned} L_1^n(t) &= \min\{t : T_1 \leq \frac{[nt]}{n}\}; \\ L_j^n(t) &= \min\{t : T_j \leq \frac{[nt] - \sum_{i<j} I(L_i^n \leq t)}{n+1-j}\} \quad j = 2, \ldots, n. \end{aligned}$$

Lastly, define $\widetilde{W}^n$ to be the *scrambled* partial sum process associated with $(\epsilon_1 Y_{n1}, \ldots, \epsilon_n Y_{nn})$:

$$\widetilde{W}^n(t) = \sum_{k=1}^{n} \epsilon_k Y_{nk} I(L_k^n \leq t) \quad t \in [0,1]. \tag{5}$$

$\widetilde{W}^n(t)$ is constant on each of the intervals $[\frac{j}{n}, \frac{j+1}{n})$, $j = 1, \ldots, n$, adding a random selection of one of the $\epsilon_k Y_{nk}$'s at each of the times $t = \frac{1}{n}, \frac{2}{n}, \ldots, 1$. Thus $\widetilde{W}^n$ has the same distribution as W^n in $D[0,1]$.

We first examine the behavior of $\widetilde{W}^n$ truncated to its $N-1$ largest jumps. Let

$$V_N^n(t) = \sum_{j=1}^{N-1} \epsilon_j Y_{nj} I(L_j^n \leq t); \qquad \text{and} \qquad S_N = \sum_{j=1}^{N-1} \epsilon_j \Gamma_j^{-1/\alpha} I(T_j \leq t).$$

Proposition 1 *For each N, $V_N^n \to S_N$ a.s. in the Skorohod topology.*

Proof. By definition of a_n, for each j, with probability one, $Y_{nj} \to \Gamma_j^{-1/\alpha}$. Therefore the vector of ordered sizes of the $N-1$ jumps of V_N^n is approaching the vector of ordered sizes of the $N-1$ jumps of S_N.

But the vector of locations of the $N-1$ jumps of V_N^n is also converging to the corresponding vector for S_N. To see this note that

$$\min\{t : T_j \le \frac{[nt]}{n+1-(N-1)}\} \le L_j^n \le \min\{t : T_j \le \frac{[nt]-(N-1)}{n+1}\}.$$

The two outer terms converge to T_j almost surely as $n \to \infty$.

Define λ_n to be the piecewise linear function with endpoints given by $\lambda_n(0) = 0, \lambda_n(1) = 1$, and $\lambda_n(L_i^n) = T_i$, for $i = 1, \dots, N-1$. Eventually λ_n is strictly increasing in which case

$$d_s(V_N^n, S_N) \le \max\{\sum_{j=1}^{N-1} |Y_{nj} - \Gamma_j^{-1/\alpha}|, |L_j^n - T_j| \; : \; 1 \le j \le N-1\}.$$

The right side converges to zero almost surely by the previous analysis. □

We will proceed to show that the tail sums $\widetilde{W}^n - V_N^n$ and $W - T_N$ converge to zero in the uniform metric as $N \to \infty$ rapidly enough so that $d_s(\widetilde{W}^n, W) \to_p 0$.

To this end, we will use a weak version of a theorem from Pollard [Pol84]. But first we need to define conditional variance as Pollard does.

Definition 5 *The conditional variance process associated with the L^2-martingale ξ is the unique, increasing, predictable process V with $V(0) = 0$, such that $\xi^2 - V$ is a martingale. It is denoted by $\langle \xi \rangle_t$.*

Theorem 3 *Let (ξ_n) be a sequence of L^2-martingales with conditional variance processes $\langle \xi_n \rangle$. The following are sufficient conditions for convergence in distribution of (ξ_n) as random elements of $D[0,1]$ under the uniform topology to Brownian motion on $[0,1]$:*
As $n \to \infty$,
(i) $\xi_n(0) \to_p 0$;
(ii) $\langle \xi_n \rangle_t \to_p t$, for each fixed t;
(iii) $EJ(\xi_n)^2 \to 0$, where J denotes the functional giving the maximum jump of the path operated on.

Specifically, Pollard states the theorem for processes on $[0, \infty)$ and the convergence is with respect to the *uniform on compacta* topology. Thus he proves a generalized version of this statement for processes restricted to intervals of the form $[0, T]$.

Let

$$T_n(t) = \sum_{j=n}^{\infty} \epsilon_j \Gamma_j^{-1/\alpha} I(T_j \le t); \qquad s_n^2 = \sum_{k=n}^{\infty} \Gamma_k^{-2/\alpha}.$$

We will show that T_n/s_n converges in distribution to standard Brownian motion on $[0,1]$ in the uniform topology on $D[0,1]$. As a consequence, we shall get

$$\sup_t |T_n(t)| \to_p 0. \tag{6}$$

Theorem 3 seems almost tailored to our application; the paths of the processes we are dealing with are constructed in such a way that the maximum jump (and all others) can be read right off.

Notation. In what follows, $\|\cdot\|$ will denote the supremum norm for any space on which it makes sense; it will be subscripted to remove any ambiguity. For σ-fields $\mathcal{F}$, we will let $E^{\mathcal{F}}$ denote conditional expectation given the σ-field $\mathcal{F}$; E^{Γ} will denote conditional expectation given the σ-field generated by the process $\{\Gamma_j\}$.

Proposition 2 *T_n/s_n converges in distribution to Brownian motion with respect to the uniform topology on $D[0,1]$.*

Proof. For s in $[0,1]$, define

$$\mathcal{G}_s = \sigma\{\Gamma_j, \epsilon_j I\{T_j \le u\}, \, u \le s; \, j \ge 1\},$$

and let $\mathcal{F}_s$ be the P-completion of $\mathcal{G}_s$.

Lemma 1 *$\{(T_n(t)/s_n, \mathcal{F}_t) : t \in [0,1]\}$ is an L^2-martingale.*

This is proved in [Kin90] using the completeness of $\mathcal{F}_t$, the continuity of the operator $E^{\mathcal{F}_s}$ on $L^2(\mathcal{F}_t)$, and results from LePage [LeP80].

Lemma 2 *With probability one, the conditional variance process of T_n/s_n is given by*

$$\langle T_n/s_n\rangle_t = -s_n^{-2}\sum_{k=n}^{\infty}\Gamma_k^{-2/\alpha} ln(1 - T_k \wedge t) \text{ (for all } t\text{)}. \tag{7}$$

Sketch of Proof. Use the method provided by Aalen in [Aal78], and the monotone convergence theorem, and the Minkowski inequality for conditional expectation to verify the conditions given in definition 5. □

Now we check that the conditions of Theorem 3 hold for T_n/s_n.
Relation (i) is clear.

For (ii), by (2) and the MCT for conditional expectation,

$$\begin{aligned} E\langle T_n/s_n\rangle_t &= E\Big(E^{\Gamma}(-s_n^{-2}\sum_{k=n}^{\infty}\Gamma_k^{-2/\alpha}ln(1-T_k\wedge t))\Big) \\ &= E\Big(s_n^{-2}\sum_{k=n}^{\infty}\Gamma_k^{-2/\alpha}E^{\Gamma}(-ln(1-T_k\wedge t))\Big). \end{aligned}$$

But the processes $\{\Gamma_j\}$ and $\{T_j\}$ are independent and $E(-ln(1-T_k\wedge t)) = \int_0^1 ln(1-u\wedge t)\,du = t$. Hence

$$E\langle T_n/s_n\rangle_t = tE\sum_{k=n}^{\infty}(\Gamma_k^{-2/\alpha}/s_n^2) = t.$$

Lemma 7 shows that $\sigma^2(\langle T_n/s_n\rangle_t) \to 0$ as $n\to\infty$. Thus, Chebyshev's inequality gives $\langle T_n/s_n\rangle_t \to_p t$, for each t.

Finally, Lemma 6 in the Appendix shows that $\Gamma_n^{-2/\alpha}/s_n^2 \to 0$ with probability one. By the bounded convergence theorem (iii) follows. □

Corollary 1 $\sup|T_n(t)| \to_p 0.$

Proof. Noting that $s_n^2 \to 0$ almost surely, this follows by Proposition 2. □

Notice that this corollary is equivalent to the statement: In the uniform metric,

$$\sum_{j=1}^{n}\epsilon_j\Gamma_j^{-1/\alpha}I(T_j\le\cdot)\to_p W. \tag{8}$$

Now we will work on finding an appropriate bound for the tail sums of the processes $\widetilde{W}^n$ described earlier. Define

$$U_N^n(t) = \sum_{j=N}^{n}\epsilon_jY_{nj}I(L_j^n\le t); \qquad s_{Nn}^2 = \sum_{j=N}^{n}Y_{nj}^2.$$

We will apply a method similar to that which was used in Proposition 2. Notice that here we must control the starting and ending indeces, whereas before we only needed to keep track of the former.

Proposition 3 *Indexed by t, U_N^n/s_{Nn} is an L^2-martingale, with conditional variance*

$$\left\langle\frac{U_N^n}{s_{Nn}}\right\rangle_t = s_{Nn}^{-2}\sum_{j=N}^{n}Y_{nj}^2\sum_{k=0}^{[nt]-1}\frac{I(L_j^n>\frac{k}{n})}{n-k}.$$

Proof. For each n, t let

$$\mathcal{F}_{nt} = \sigma\{\Gamma_j, \epsilon_j I(L_j^n \leq u), u \leq t;\ 1 \leq j \leq n\}.$$

Notice that the process $(U_N^n / s_{Nn}(t))$ is adapted to $(\mathcal{F}_{nt})$, and that for each n, j, and t, s_{Nn} and Y_{nj} are measurable $\mathcal{F}_{nt}$. Use Lemma 5 to show U_N^n / s_{Nn} is an L^2-martingale.

As is pointed out by Pollard [Pol84, p.176], if an L^2-martingale Z changes only by jumps $\xi_1, \xi_2, \ldots$ occurring at fixed times $t_1 < t_2 < \ldots$, and if $\mathcal{F}_t = \mathcal{F}_{t_k}$ for $t_k \leq t < t_{k+1}$, then the conditional variance process is just a partial sum process of conditional variances: For $t_k \leq t < t_{k+1}$.

$$\langle Z \rangle_t = E^{\mathcal{F}_0} \xi_1^2 + E^{\mathcal{F}_{t_1}} \xi_2^2 + \ldots + E^{\mathcal{F}_{t_{k-1}}} \xi_k^2,$$

Our process and σ-fields satisfy these conditions, so for each N, n such that $N \leq n$,

$$\begin{aligned}
\left\langle \frac{U_N^n}{s_{Nn}} \right\rangle_t &= \sum_{k=0}^{[nt]-1} E^{\mathcal{F}_{k/n}} \left(s_{Nn}^{-2} \left(U_N^n\left(\frac{k+1}{n}\right) - U_N^n\left(\frac{k}{n}\right) \right)^2 \right) \\
&= \sum_{k=0}^{[nt]-1} s_{Nn}^{-2} E^{\mathcal{F}_{k/n}} \left(\sum_{j=N}^{n} Y_{nj} \epsilon_j I\left(L_j^n \in \left(\frac{k}{n}, \frac{k+1}{n}\right]\right) \right)^2 \\
&= \sum_{k=0}^{[nt]-1} s_{Nn}^{-2} \sum_{j=N}^{n} Y_{nj}^2 P\left(L_j^n = \frac{k+1}{n} \| \mathcal{F}_{k/n}\right) \\
&= \sum_{k=0}^{[nt]-1} s_{Nn}^{-2} \sum_{j=N}^{n} Y_{nj}^2 \frac{I(L_j^n > k/n)}{n-k} \\
&= s_{Nn}^{-2} \sum_{j=N}^{n} Y_{nj}^2 \sum_{k=0}^{[nt]-1} \frac{I(L_j^n > \frac{k}{n})}{n-k}. \qquad \square
\end{aligned}$$

Proposition 4 *For each t in $[0,1]$, the following relations hold:*

$$E\left\langle \frac{U_N^n}{s_{Nn}} \right\rangle_t = \frac{[nt]}{n} \tag{9}$$

$$Var\left\langle \frac{U_N^n}{s_{Nn}} \right\rangle_t \leq 2E\left(\frac{Y_{nN}^2}{s_{Nn}^2} \right) + \frac{[nt]^2}{n^2(n-1)}. \tag{10}$$

This is proved in [Kin90] using some tedious calculations with Proposition 3.

Proposition 5 *There exists a sequence* $(n_0(N))$ *such that*

$$\frac{Y_{n'(N)N}}{s_{Nn'(N)}} \to 0 \ a.s.$$

for every sequence $(n'(N))$ *such that* $n'(N) \geq n_0(N)$ *for each* N.

Proof. Since for each j, $Y_{nj} \to \Gamma_j^{-1/\alpha}$ a.s., we have for each N,

$$\frac{Y_{nN}^2}{\sum_{N+1}^{2N} Y_{nj}^2} \to \frac{\Gamma_N^{-2/\alpha}}{\sum_{N+1}^{2N} \Gamma_j^{-2/\alpha}} \text{ a.s..}$$

Hence we can choose $n_0(N) \geq 2N$ large enough that for $n \geq n_0(N)$,

$$P\left(|\frac{Y_{nN}^2}{\sum_{N+1}^{2N} Y_{nj}^2} - \frac{\Gamma_N^{-2/\alpha}}{\sum_{N+1}^{2N} \Gamma_j^{-2/\alpha}}| > 2^{-N} \right) < 2^{-N},$$

Fix a sequence $(n'(N))$ such that $n'(N) \geq n_0(N)$ for each N. By a simple application of a Borel-Cantelli lemma, as $N \to \infty$

$$\frac{Y_{n'(N)N}^2}{\sum_{N+1}^{2N} Y_{n'(N)j}^2} - \frac{\Gamma_N^{-2/\alpha}}{\sum_{N+1}^{2N} \Gamma_j^{-2/\alpha}} \to 0 \text{ a.s..} \tag{11}$$

By the strong law of large numbers,

$$\frac{\Gamma_N^{-2/\alpha}}{\sum_{N+1}^{2N} \Gamma_j^{-2/\alpha}} \leq \frac{\Gamma_N^{-2/\alpha}}{N\Gamma_{2N}^{-2/\alpha}} = \frac{(\Gamma_N/N)^{-2/\alpha}}{N(\Gamma_{2N}/N)^{-2/\alpha}} \to 0 \text{ a.s..} \tag{12}$$

Combining (11) and (12), we see that

$$\frac{Y_{n'(N)N}^2}{\sum_{N+1}^{2N} Y_{n'(N)j}^2} \to 0 \text{ a.s..}$$

But since $n'(N) \geq 2N$,

$$\frac{Y_{n'(N)N}^2}{s_{Nn'(N)}^2} \leq \frac{Y_{n'(N)N}^2}{\sum_{N+1}^{2N} Y_{n'(N)j}^2}. \qquad \square$$

Proposition 6 *For each* N, *as* $n \to \infty$,

$$\|V_N^n \circ H^n - S_N \circ H\| \to 0 \ a.s..$$

Proof. Fix N. Let B_N equal the set on which

$$\bigvee_{j=1}^{N} |L_j^n - T_j| \quad \to \quad 0,$$

$$\bigvee_{j=1}^{N} |Y_{nj} - \Gamma_j^{-1/\alpha}| \quad \to \quad 0,$$

$$T_i \neq H(t) \text{ for all } i \geq 1 \quad \text{and} \quad t \in R,$$

$$\|H^n - H\| \quad \to \quad 0.$$

By hypotheses, $P(B_N) = 1$. We claim that $\|V_N^n \circ H^n - S_N \circ H\| \to 0$ on B_N. To see this, fix $\omega \in B_N$ and let $\delta > 0$. Suppress the argument ω from the following relations. By the assumptions on H, there exists K such that for $k \geq K$,

$$H(d_k) \geq H(d_K) > \max\{T_i : i \leq N\}.$$

Thus, we can choose $\varepsilon > 0$ such that

$$\varepsilon < \bigwedge_{i \leq N, t \in R} |T_i - H(t)|. \tag{13}$$

Fix M' such that for $n \geq M'$

$$\bigvee_{j=1}^{N} |L_j^n - T_j| \quad < \varepsilon/3, \tag{14}$$

$$\bigvee_{j=1}^{N} |Y_{nj} - \Gamma_j^{-1/\alpha}| \quad < \delta/N,$$

$$\|H^n - H\| \quad < \varepsilon/3.$$

Now let $n \geq M$. We show that $|V_N^n \circ H^n(t) - S_N \circ H(t)| < \delta$, for all real t. Recall that S_N is constant on $[T_{(i)}, T_{(i+1)})$ for $i = 0, 1, \ldots, N$ (with $T_0 = 0$, and $T_{N+1} = 1$); and V_N^n is constant on $[L_{(i)}^n, L_{(i+1)}^n)$. We proceed to show that

$$L_i^n \leq H^n(t) \text{ if and only if } T_i \leq H(t). \tag{15}$$

Pick $t \in R$. Say $T_i \leq H(t)$. Then by (13), $T_i + \varepsilon < H(t)$. Since $|L_i^n - T_i| < \varepsilon/3$ and $|H^{(m_n)}(t) - H(t)| < \varepsilon/3$, we must have $L_i^n < H^n(t)$. The argument for the other direction of (15) is similar. Thus

$$|V_N^n(H^n(t)) - S_N(H(t))|$$

$$= |\sum_{i=1}^{N-1} \epsilon_i Y_{ni} I(L_i^n \leq H^n(t)) - \sum_{i=1}^{N-1} \epsilon_i \Gamma_i^{-1/\alpha} I(T_i \leq H(t))|$$

$$\leq \sum_{i=1}^{N-1} |Y_{ni} - \Gamma_i^{-1/\alpha}|$$

$$< \delta.$$

We chose $\delta > 0$ arbitrarily, so we are done. □

With existence guaranteed by Propositions 1 and 6, choose $n_1(N) \geq n_0(N)$ (n_0 is defined in Proposition 5) such that for $n \geq n_1(N)$,

$$P(\|V_N^n \circ H^n - S_N \circ H\| > N^{-1}) < N^{-1}, \tag{16}$$

and

$$P(d_s(V_N^n, S_N) > N^{-1}) < N^{-1}. \tag{17}$$

Define

$$N(n) = \max\{N \geq 1 : n_1(N) \leq n\} \vee 1.$$

Lemma 3 *$V_{N(n)}^n \to_p W$ in the Skorohod topology.*

Proof. Let (n_k) be a subsequence. Since $n_1(N) \geq n_0(N) \geq 2N \to \infty$, we can choose a subsequence (n_{k_j}) such that $N(n_{k_j}) < N(n_{k_{j+1}})$ for each j.

Let $\varepsilon > 0$. Now $n_{k_j} \geq n_1(N(n_{k_j}))$ by definition of $N(\cdot)$, and for j large enough, $N(n_{k_j}) > 1/varepsilon$, in which case

$$P(d_s(V_{N(n_{k_j})}^{n_{k_j}}, W) \geq 2\varepsilon)$$

$$\leq P(d_s(V_{N(n_{k_j})}^{n_{k_j}}, S_{N(n_{k_j})}) > \varepsilon) + P(d_s(S_{N(n_{k_j})}, W) > \varepsilon)$$

$$\leq \frac{1}{N(n_{k_j})} + P(d_s(S_{N(n_{k_j})}, W) > \varepsilon),$$

Since $N(n_{k_j}) \to \infty$, the claim is proved. □

Lemma 4 *$U_{N(n)}^n / s_{N(n)n}$ converges in distribution to Brownian motion with respect to the uniform topology on $D[0,1]$.*

Proof. As in the proof of Lemma 3, choose subsequences (n_k) and (n_{k_j}). Again we use Theorem 3. By Proposition 3, $U_{N(n)}^n / s_{N(n)n}$ is an L^2-martingale. Surely $U_{N(n)}^n / s_{N(n)n}(0) \equiv 0$. Also, since

$$n_{k_j} \geq n_1(N(n_{k_j})) \geq n_0(N(n_{k_j})),$$

Propositions 4 and 5 give us that the conditional variance condition is satisfied. Finally, Proposition 5 gives us that the expected squared maximum jump also converges to 0. □

Theorem 4 $\widetilde{W}^n \rightarrow_p W$ *in the Skorohod topology on* $D[0,1]$.

Proof. Notice that $\widetilde{W}^n = V^n_{N(n)} + U^n_{N(n)}$. By the corollary to Proposition 8, $s_{N(n)n} \rightarrow_p 0$. Therefore, by Lemma 4, $\|U^n_{N(n)}\| \rightarrow_p 0$. By definition of d_s it is easy to show that if $d_s(x_n, x) \rightarrow 0$, and $\|y_n\| \rightarrow 0$, then $d_s(x_n + y_n, x) \rightarrow 0$. Apply Lemmas 3 and 4 to complete the proof. □

Proposition 7 $\widetilde{W}^n \circ H^n \rightarrow_p W \circ H$ *in the uniform topology on* $D(R)$.

Proof. For any n,

$$\|\widetilde{W}^n \circ H^n - W \circ H\| \leq \underbrace{\|\widetilde{W}^n \circ H^n - V^n_{N(n)} \circ H^n\|}_{\Delta_1}$$

$$+ \underbrace{\|V^n_{N(n)} \circ H^n - S_{N(n)} \circ H\|}_{\Delta_2} + \underbrace{\|S_{N(n)} \circ H - W \circ H\|}_{\Delta_3}.$$

$\Delta_1 \leq \|\widetilde{W}^n - V^n_{N(n)}\| = \|U^n_{N(n)}\| \rightarrow_p 0$ as was shown in the proof of Theorem 4. $\Delta_3 \leq \|V_{N(n)} - W\| = \|T_{N(n)}\| \rightarrow_p 0$ by (2.2) because $N(n) \rightarrow \infty$. Also, since $n_1(N(n)) \leq n$, and $N(n) \rightarrow \infty$, $\Delta_2 \rightarrow_p 0$. □

Combining Theorem 4 and Proposition 7,

$$(\widetilde{W}^n \circ H^n, \widetilde{W}^n) \rightarrow_p (W \circ H, W)$$

in the prescribed space. Since these processes have the same joint distribution as $(W^n \circ H^n, W^n)$, Theorem 2(A) is proved. □

3.3 Proof in the Finite Variance Case

Assume the hypotheses of Theorem 2(B). Here $EX^2 < \infty$ so $a_n = n^{1/2}\sigma$ and Donsker's Theorem applies:

$$W^n \rightarrow_d W,$$

under the uniform topology on $D[0,1]$, where W is Brownian motion (see Billingsley [Bil68]).

Let $\widetilde{W}^n$ be a process with the same distribution as W^n such that $\widetilde{W}^n$ converges uniformly almost surely to a continuous Brownian motion $\widetilde{W}$.

$$\begin{aligned}
\|\widetilde{W}^n \circ H^n - \widetilde{W} \circ H\|_R &\leq \|\widetilde{W}^n \circ H^n - \widetilde{W} \circ H^n\|_R + \|\widetilde{W} \circ H^n - \widetilde{W} \circ H\|_R \\
&\leq \|\widetilde{W}^n - \widetilde{W}\|_{[0,1]} + \|\widetilde{W} \circ H^n - \widetilde{W} \circ H\|_R
\end{aligned}$$

By construction, the first term on the right converges to 0 almost surely. With probability one $\widetilde{W}$ is uniformly continuous on [0,1]. Since $H^n \to H$ uniformly almost surely, the last term converges to 0 almost surely.

With this representation, we have $(\widetilde{W}^n \circ H^n, \widetilde{W}^n) \to (\widetilde{W} \circ H, \widetilde{W})$ uniformly almost surely. Therefore,

$$(W^n \circ H^n, W^n) \to_d (W \circ H, W)$$

under the uniform topology in each coordinate. □

4 The Limit Laws

Under the usual resampling scheme, (i.i.d. resampling from the data), when we replace the processes involved in the stochastic integral decomposition by their limits obtained in the invariance principle, we get the limiting distribution of the bootstrap.

Throughout this section, let M^* be a Poisson (mean 1) random variable and let $H(x) = P(M^* - 1 \leq x)$.

4.1 Infinite Variance ($\alpha < 2$): Symmetric Case

In light of the results of Bickel and Freedman [BF81] and the result (3) by Knight [Kni89], and the invariance principle, the following theorem shows why the stochastic integral decomposition may be the natural way to view the bootstrap.

Let ϵ, Γ, and W be as in Definition 4. Let $M_1^*, M_2^*, \ldots$ be i.i.d. $\sim M^*$ independent of ϵ, Γ.

Theorem 5

$$\mathcal{L}(\sum_{k=1}^{\infty} \epsilon_k \Gamma_k^{-1/\alpha}(M_k^* - 1) \,|\, \epsilon_k \Gamma_k^{-1/\alpha}, k \geq 1)$$

$$= \mathcal{L}\left(\int_{-\infty}^{\infty} t\, W \circ H(dt) \,|\, \epsilon_k \Gamma_k^{-1/\alpha}, k \geq 1\right).$$

4.2 Finite Variance Case

The following proposition shows that the above representation for the limit law carries over naturally to the finite variance case.

Theorem 6 *If W is a Brownian motion then $\int t\, W \circ H(dt)$ has the standard normal distribution.*

Proof. To see this, look at the characteristic functions again. The stochastic integral can be written as $\sum_{r=-1}^{\infty} r[W(H(r)) - W(H(r-))]$. By independence of the increments of Brownian motion, its characteristic function is

$$\begin{aligned}\lim_{n\to\infty} \prod_{r=-1}^{n} \exp\Big(-\frac{r^2t^2[H(r) - H(r-)]}{2}\Big) \\ &= \lim_{n\to\infty} \exp\Big(-\frac{t^2}{2} \sum_{r=-1}^{n} r^2[H(r) - H(r-)]\Big) \\ &= \exp\Big(-\frac{t^2}{2} \int r^2\, dH(r)\Big) \\ &= \exp(-t^2/2).\end{aligned}$$

This is the characteristic function of the standard normal distribution, so the assertion is proved. □

Assume W is a version of Brownian motion with continuous sample paths. Since conditioning on the ordered jumps of such a process provides us with no additional information,

$$\mathcal{L}\Big(\int t\, W \circ H(dt) \mid \textit{ ordered jumps of } W\Big) = \mathcal{L}\Big(\int t\, W \circ H(dt)\Big).$$

By applying Bickel and Freedman's results [BF81], using the proper scaling, $a_n = n^{1/2}\sigma$, we see that in the symmetric case the result concerning a distribution in the domain of attraction of an infinite variance stable random variable can be viewed as an extension of what was already known in the finite variance case.

5 Simulation Results

For random variables which have infinite variance, we found that in the symmetric case, the bootstrap of the sample mean does not perform so badly. In fact, in some ways, the method gives better results than it does in the finite variance case.

For various indices of stability α, and confidence levels γ, we simulated observations of random variables X_i, symmetric about θ, in the domain of attraction of an α-stable random variable and applied the bootstrap of the sample mean to create symmetric γ-confidence intervals for θ in the following manner:

For a given sample $\mathbf{X_n} = (X_1, \ldots, X_n)$, the confidence interval $C_\gamma(\mathbf{X_n})$ is given by

$$C_\gamma(\mathbf{X_n}) = [\bar{X}_n - T_\gamma(\mathbf{X_n}), \bar{X}_n + T_\gamma(\mathbf{X_n})],$$

where $T_\gamma(\mathbf{X_n})$ is estimated as a quantity which satisfies

$$P[|\bar{X}_n^* - \bar{X}_n| \leq T_\gamma(\mathbf{X_n}) \| \mathbf{X_n}] = \gamma.$$

A suprising observation was that the empirical coverage of this method was consistently higher than γ for $\alpha < 2$. Figure 1 shows the observed coverage on the bootstrap method applied 1000 times each for $\gamma = .95$ confidence with sample size $n = 50$, bootstrap resample size $m = 50$, and 500 bootstrap observations for various values of α. In our simulations, $X_1 \sim \epsilon U^{-1/\alpha}$ where $P(\epsilon = 1) = 1/2 = P(\epsilon = -1)$, U uniform on (0,1). Notice that for $\alpha > 2$, F has finite variance, and hence it is expected that the coverage should be approximately .95.

Consider the confidence radii obtained by the above method, scaled by $n^{1-1/\alpha}$ because

$$S_n^* = na_n^{-1}(\bar{X}_n^* - \bar{X}_n),$$

and $a_n \sim cn^{1/\alpha}$ (see Feller [Fel71]).

In the finite variance case, the bootstrap distribution of the scaled and centered bootstrapped sample mean converges weakly to a fixed (normal) distribution almost surely. Since the limit distribution is continuous, the confidence radii given by the above method and then scaled as indicated converge almost surely to a fixed number as the sample size n tends to infinity.

But by Athreya's early results [Ath84] one might not expect this phenomenon to occur in the infinite variance case. Our simulation results exemplify this. Figure 2 shows a frequency histogram of the observed confidence radii (logarithmically scaled) with $n = 50$, $\alpha = 1.0$. The vertical line represents the logarithm of the radius necessary for an unconditional confidence interval with confidence level equal to the coverage observed by applying this method.

Figure 3 shows more of the same phenomena for various values of α.

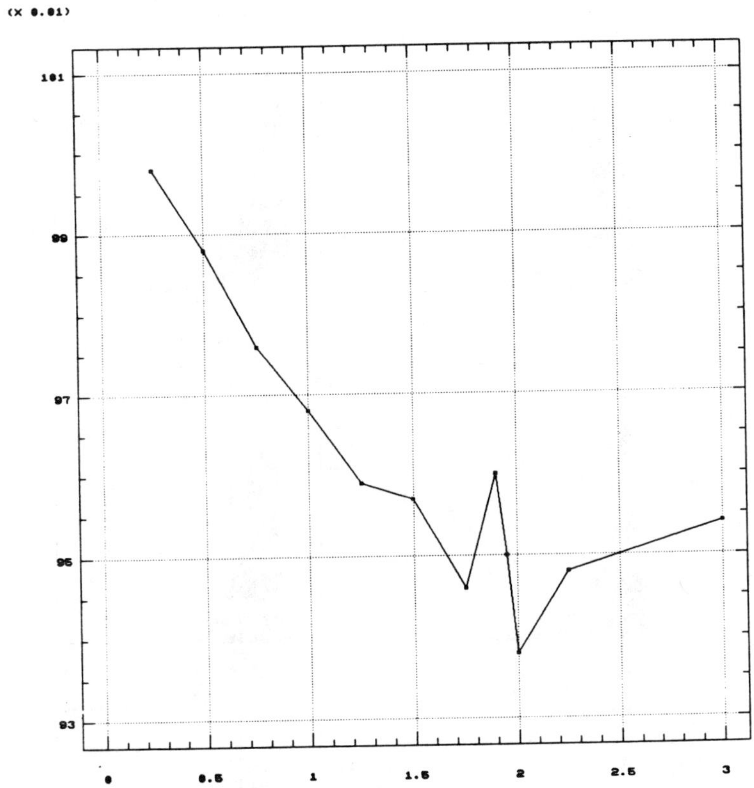

Figure 1: Coverage of bootstrap method for various α.

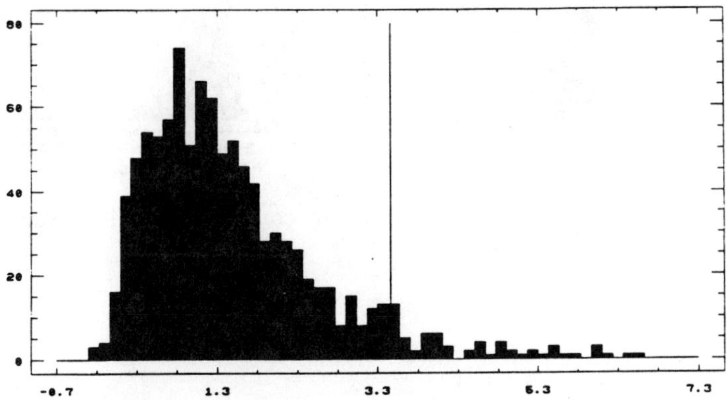

Figure 2: Distribution of bootstrap confidence radii for $\alpha = 1$, $n = 50$.

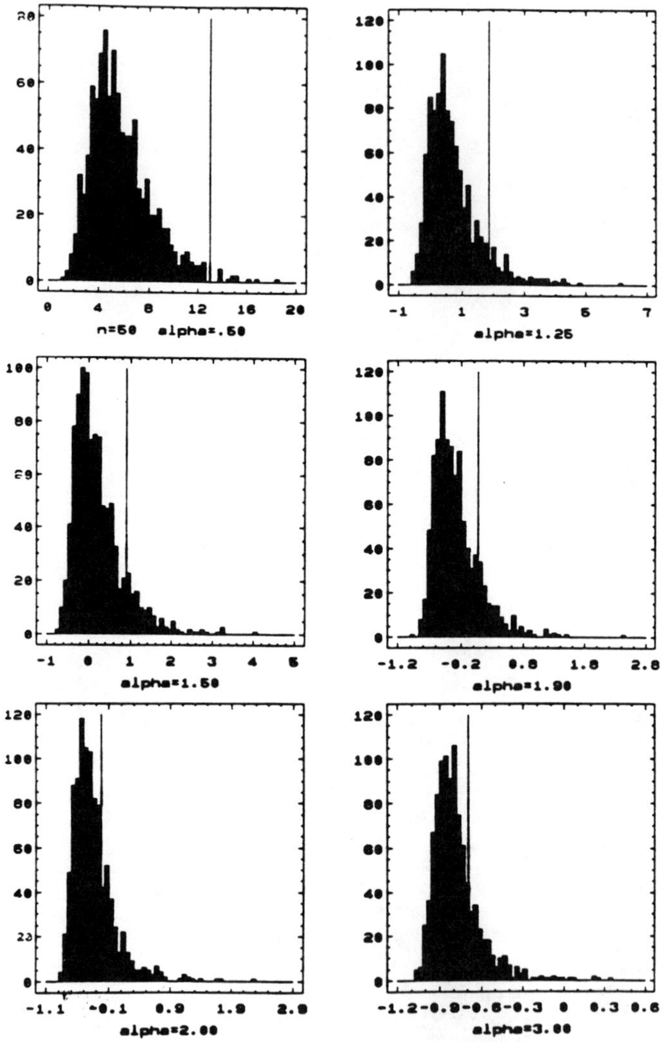

Figure 3: Distribution of bootstrap confidence radii for various α.

As expected, even as n gets large, the distribution of the scaled radii obtained by this method is dispersed apparently continuously over the positive real axis. Figure 4 shows what happens when $n \doteq 200$.

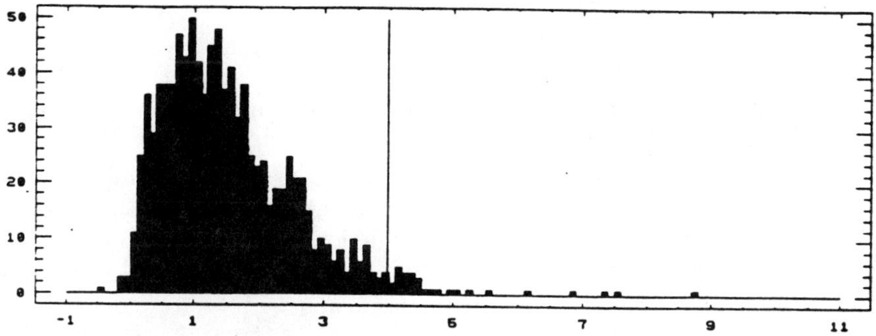

Figure 4: Distribution of bootstrap confidence radii for $\alpha = 1$, $n = 200$.

More is observed. Since the confidence widths generated by the method have such a wide distribution, we examined how well the procedure compares with the procedure which simply uses the unconditional distribution of the sample mean about θ.

That is, we estimated T_γ such that

$$P(|\bar{X}_n - \theta| \leq T_\gamma) = \gamma.$$

Since the $X_i's$ are in the domain of attraction of the stable random variable Y, we can do this not only by Monte Carlo simulation, but also by using the quantiles of Y, because $S_n \to_d Y$ (see (1)).

The results were startling. For a very small proportion of the applications, the confidence radii obtained by the method $T_\gamma(\mathbf{X_n})$, were larger than T_γ. But for the complement, the $T_\gamma(\mathbf{X_n})$ was extremely small compared to T_γ, suggesting that the bootstrap performs well.

Example. Here is an illustration of how the bootstrap confidence radii compare with confidence radii obtained using the unconditional distribution of the sample mean.

We ran a simulation with 1000 observations for $\alpha = 1$, $n = 50$, and 2000 bootstrap resamplings for each observation with $m = 50$. The observed coverage of bootstrap method was 0.968. The radius about sample mean necessary for unconditional confidence of .968 is $T_{.968} \doteq 32.13$.

The implications of these results could be very far reaching. 94% of the times the method was applied, the radius of the confidence interval for θ was

Table 1: Analysis of sizes of bootstrap confidence radii.

c	Observed proportion of times $T_{.95}(\mathbf{X_n}) < cT_{.968}$
1	.940
1/2	.881
1/4	.770
.107	.500

less than the radius necessary to give confidence equivalent to the empirical coverage obtained by the method.

Maybe more substantial is *how much* smaller the observed radii were. Half of the time the bootstrap confidence interval radius was less than about a tenth of the radius necessary for unconditional confidence.

More needs to be studied in this direction. The invariance principle proved in this paper will help to explain the phenomenon.

6 Remarks

6.1 Knight's Result Follows in the Symmetric Case

In [Kin90] it is shown that Knight's result follows in the symmetric case from the technique used in proving the invariance principle. This is a different approach than that used by Knight — bounding the multipliers as opposed to bounding the scaled sample. It reveals in another way what is important about the resampling plan.

6.2 Other Resampling Plans

One of the problems with the usual bootstrap in the infinite variance case is its inconsistency.

The stochastic integral representation given in Section 2 together with the invariance principle proved in Section 3 provide a new way of studying the resampling problem.

6.3 Only Off by a Scale.

Consider the symmetric case.

Recall the result given by Knight [Kni89]. The representation he gives for the random limit law is the random conditional distribution of

$$\sum_{k=1}^{\infty} \epsilon_k \Gamma_k^{-1/\alpha}(M_k^* - 1)$$

given the sequence $\epsilon_1\Gamma_1^{-1/\alpha}, \epsilon_2\Gamma_2^{-1/\alpha}, \ldots$.

It is shown by LePage in [LWZ81] that for our choice of a_n and b_n,

$$S_n \to_d \sum_{k=1}^{\infty} \epsilon_k \Gamma_k^{-1/\alpha}.$$

By Lemma 5.1 in [LeP80], the unconditional distribution of the sum given by Knight is only a scale away from the correct limit law μ :

$$\sum_{k=1}^{\infty} \epsilon_k \Gamma_k^{-1/\alpha}(M_k^* - 1) =_d (E|M_1^* - 1|^\alpha)^{1/\alpha} \sum_{k=1}^{\infty} \epsilon_k \Gamma_k^{-1/\alpha}.$$

A Appendix

Below are stated some of the technical results referenced during the proof of the invariance principle. These are all proved in the author's Ph.D. thesis.

Lemma 5 *Let T be an arbitrary random variable taking on values in $(0,1)$ and ϵ be a random variable independent of T with $P(\epsilon = 1) = \frac{1}{2} = P(\epsilon = -1)$. For s in $(0,1)$ let $\mathcal{F}_s = \sigma\{\epsilon I(T \le u), u \le s\}$. Then for $B \subseteq (s, 1]$, with probability one,*

$$P[T \in B \| \mathcal{F}_s] = \frac{P(T \in B)}{P(T > s)} I(T > s),$$

and

$$E^{\mathcal{F}_s} \epsilon I(T \in B) = 0.$$

Lemma 6 *With probability one, $\Gamma_n^{-2/\alpha}/s_n^2 \to 0$.*

Lemma 7 *For each t, $\sigma^2(V_n(t)) \to 0$.*

Lemma 8 *For $0 < s < t \le 1$*

$$E^{\mathcal{F}_s} \ln(1 - T_k \wedge t) = \ln(1 - T_k \wedge s) - \frac{t-s}{1-s}[T_k > s].$$

Proposition 8

$$\sum_{j=1}^{n} Y_{nj}^2 \to_p \sum_{k=1}^{\infty} \Gamma_k^{-2/\alpha}.$$

Acknowledgements. The author would like to thank Professor Raoul LePage for the suggestion of the problem and all of the helpful direction in solving it. He would also like to thank Professors Schlomo Levental and Anil Jain for their time and interest. Professor Hira Koul was particularly helpful with his meticulous reading of the original manuscript and many helpful corrections and suggestions. Finally, the author would like to thank the Office of Naval Research for supporting him for the last year and one half of his doctoral study.

References

[Aal78] Odd Aalen. Nonparametric inference for a family of counting processes. *Annals of Statistics*, 6:701–726, 1978.

[AG89] Miquel A. Arcones and Evarist Giné. The bootstrap of the mean with arbitrary sample size. *Annals of Inst. of H. Poincare*, 1989.

[Ath84] K. B. Athreya. Bootstrap of the mean in the infinite variance case. Technical report, Iowa State University, 1984.

[Ath85] K. B. Athreya. Bootstrap of the mean in the infinite variance case-ii. Technical report, Iowa State University, 1985.

[Ath87] K.B. Athreya. Bootstrap of the mean in the infinite variance case. *Annals of Statistics*, 15:724–731, 1987.

[BF81] Peter J. Bickel and David A. Freedman. Some asymptotic theory for the bootstrap. *Annals of Statistics*, 9:1196–1217, 1981.

[Bil68] Patrick Billingsley. *Convergence of Probability Measures.* John Wiley, New York, 1968.

[Bil86] Patrick Billingsley. *Probability and Measure.* John Wiley, New York, 1986.

[Efr79] Bradley Efron. Bootstrap methods: Another look at the jackknife. *Annals of Statistics*, 7:1–26, 1979.

[Fel71] William Feller. *An Introduction to Probability Theory and Applications.* John Wiley, New York, 1971.

[GZ89] Evarist Giné and Joel Zinn. Necessary conditions for the bootstrap of the mean. *Annals of Statistics*, 17:684–691, 1989.

[Hal88] Peter Hall. Rate of convergence in bootstrap approximations. *Annals of Probability*, 16:1665–1684, 1988.

[Kin90] John G. Kinateder. *An Invariance Principle Applicable to the Bootstrap.* PhD thesis, Michigan State University, 1990.

[Kni89] Keith Knight. On the bootstrap of the sample mean in the infinite variance case. *Annals of Statistics*, 17:1168–1175, 1989.

[LeP80] Raoul LePage. Multidimensional infinitely divisible variables and processes. part i: Stable case. Technical report, Stanford University, 1980.

[LWZ81] Raoul LePage, Michael Woodroofe, and Joel Zinn. Convergence to a stable distribution via order statistics. *Annals of Probability*, 9:624–632, 1981.

[Pol84] Henry Pollard. *Convergence of Stochastic Processes.* Springer-Verlag, 1984.

[Sin81] Kesar Singh. On the asymptotic accuracy of Efron's bootstrap. *Annals of Statistics*, 9:1187–1195, 1981.

Edgeworth Correction by 'Moving Block' Bootstrap
for Stationary and Nonstationary Data

S.N. Lahiri
Iowa State University

ABSTRACT

This paper considers second order properties of the 'moving block' bootstrap procedure, proposed by Künsch (1989: Annals of Statistics, 17, 1217-). In the case of sample mean of weakly dependent stationary data, the exact rate of approximation by Künsch's bootstrapped statistic is determined and observed to be worse than the rate of normal approximation. However, a suitable modification in the definition of the bootstrapped statistic removes this deficiency and provides a second order correct approximation. Similar optimality is proved for the studentized sample mean of m-dependent data. Furthermore, the modified version is shown to be "robust" in the cases where the observations are not necessarily stationary.

1. Introduction: It has been more than a decade since Efron (1979) introduced the 'bootstrap' procedure for estimating unknown sampling distributions of statistics based on independent and identically distributed (iid) observations. It is well known that, in the iid set up, bootstrap provides more accurate approximations to the distributions of many regular statistics than the classical large sample approximations (e.g. Singh (1981), Babu (1986)). However, the situation may be totally different when the observations are not necessarily independent. Remark 2.1 of Singh (1981)) shows that the iid resampling scheme of usual bootstrap method fails to capture the

Exploring the Limits of Bootstrap. Edited by Raoul LePage and Lynne Billard.
 ISBN: 0471-53631-8.

dependence structure even for the sample mean of m-dependent data. Following this observation, there have been several attempts in the literature to modify and extend Efron's idea to dependent models. In most cases, the modifications cleverly exploit certain special properties of the specific models under consideration. For ARMA models, the bootstrap procedure is extended by using iid innovations, e.g. Freedman (1984), Efron and Tibshirani (1986). In the case of Markov chains with finite state space, Basawa et al. (1990) defines a version of the bootstrap method by using a well known representation of the Markov chain in terms of independent random variables. Also, see Kulperger and Prakasa Rao (1989), Datta and McCormick (1990). The extension to the countable Markov chains has been done by Athreya and Fuh (1989) using iid cycles.

Unlike the above works, a general bootstrap procedure for weak dependent stationary data is formulated by Künsch (1989). There it is shown that the proposed method provides a valid approximation to the unknown distribution of normalized univariate sample mean. It should be pointed out that the same modification is also put forward in an independent work by Liu and Singh (1988) on bootstrapping m-dependent data. Some other versions of bootstrap are proposed by Shi and Shao (1988) (for m-dependent data) and by Moore and Rais (1990) (for observations satisfying the uniform mixing condition). In this paper we consider the second order properties of the 'moving block' bootstrap procedure of Künsch (1989) and show that, under appropriate conditions, this procedure is second order correct even for the sample mean of nonstationary data.

We begin with a brief description of Künsch's bootstrap. Let $X_1, X_2, \ldots$ be a sequence of stationary random vectors in $\mathbb{R}^p$. For $n \geq 1$, let ρ_n denote the empirical distribution (e.d.) of $X_1, \ldots, X_n$. Suppose that the random vector of interest $T_n = T(\rho_n)$ derives from a functional T, defined on a (rich) set of probability measures on $\mathbb{R}^p$. Given the observations $X_1, \ldots, X_n$, and an integer $\ell = \ell_n$, $1 \leq \ell \leq n$, form the blocks $\xi_t = (X_t, \ldots, X_{t+\ell-1})$ for $t = 1, 2, \ldots, b$, where $b = n - \ell + 1$. Set $k = [n/\ell]$. Here, $[x]$ denotes the integer part of a real number x. For the stationary bootstrap procedure proposed by Künsch (1989), one selects a random, with replacement sample $\xi_1^*, \ldots, \xi_k^*$ of size k from the observed blocks $\xi_1, \xi_2, \ldots, \xi_b$ and defines the bootstrapped statistic as $T_n^* = T(\rho_n^*)$ where ρ_n^* denotes the empirical distribution of the $(\ell \cdot k)$ components of $\xi_1^*, \ldots, \xi_k^*$. The distribution of T_n is then approximated by the bootstrap distribution of T_n^* letting ℓ increase to infinity with n at a suitable rate.

Künsch (1989) proves the validity of his procedure for the statistic $\sqrt{n}(\bar{X}_n - \mu)$ in the univariate case. Here, $n\bar{X}_n = \sum_{i=1}^n X_i$ and $\mu = EX_1$. In the next three sections we shall show that even in the multivariate set up, this procedure is second order correct for a wide class of estimators based on sample means. Write $M_n = \mathrm{Cov}(n^{-1/2}(X_1 + \ldots + X_n))$ and $T_n = \sqrt{n}A_n(\bar{X}_n - \mu)$ where $A_n = M_n^{-1/2}$ (assuming M_n to be nonsingular). In the case of normalized sample mean T_n, an equivalent formulation of Künsch's bootstrap can be given as follows. Let $\xi_i^* = (\xi_{i1}^*, \ldots, \xi_{i\ell}^*)$ and $U_i^* = \ell^{-1} \sum_{j=1}^{\ell} \xi_{ij}^*$, $1 \leq i \leq k$. Note that U_i^*'s are iid with $P(U_1^* = U_t) = b^{-1}$ for $1 \leq t \leq b$ where $U_t = (X_t + \ldots + X_{t+\ell-1})/\ell$. Define $\hat{\mu}_n = E_* U_1^*$ and $\hat{M}_n = E_*(U_1^* - \hat{\mu}_n)(U_1^* - \hat{\mu}_n)'$ where

E_* denotes the expectation under the bootstrap probability. Set $\bar{X}_n^*$ $= k^{-1} \sum_{i=1}^k U_i^*$ and $\hat{A}_n = \hat{M}_n^{-1/2}$. Then, according to Künsch (1989)'s definition, the bootstrapped statistic T_n^* is given by $T_n^* = \sqrt{k} \cdot \hat{A}_n(\bar{X}_n^* - \bar{X}_n)$. However, this statistic suffers from a problem which is not very serious for the first order results but becomes predominant in the second order analysis. The bootstrapped statistic T_n^* has a random bias B_n which indeed determines the rate of bootstrap approximation. Under the assumptions of Künsch (1989), $\|B_n\|$ tends to zero almost surely but not at the desired rate $o(n^{-1/2})$. Theorem 2.1 of Section 2 shows that $\sqrt{n}\, B_n/\sqrt{\ell}$ actually has a limiting normal distribution, so that the rate of approximation for T_n^* is much worse than the usual normal approximation. However, this problem can be resolved by redefining the bootstrapped statistic as $T_{1n}^* = \sqrt{k}\, \hat{A}_n(\bar{X}_n^* - \hat{\mu}_n)$. As shown in Lahiri (1990), the bootstrap distribution of T_{1n}^* approximates the distribution of T_n at the rate $o(n^{-1/2})$ almost surely. Similar results continue to hold for more general statistics that are smooth functions of the sample mean. In particular, it is shown that the rate of bootstrap approximation to the distribution of studentized $\bar{X}_n$ is also $o(n^{-1/2})$ almost surely when the data are m-dependent. In this case as well, one has to use a modification of Künsch's original formulation to get hold of the second order correct version of the bootstrapped statistic.

Next we study the performance of Künsch's bootstrap for non-identically distributed, weak-dependent data. For independent observations, Liu (1988) showed that Efron's bootstrap retains its second order optimality in certain non-iid models. In the dependent set up, similar optimality holds for Künsch's bootstrap as well. In fact, it is shown that the bootstrap approximation to the normalized sample mean is second order correct for observations which may not be stationary, but satisfy some uniformity conditions only. See Section 4 for details.

The layout of the paper is as follows. Section 2 contains the results on the normalized sample mean and Section 3 on studentized mean. The nonstationary case is treated in Section 4. Proofs of all the results are given in Section 5.

2. Results on $\bar{X}_n$: As in the iid case, the key step in proving the second order correctness of stationary bootstrap procedure is to derive Edgeworth expansions for the original statistic and the corresponding bootstrapped statistic. In the dependent situation, derivations of Edgeworth expansions are usually very much involved. Depending on the dependency structure, they require regularity conditions which are often difficult to verify. The most general results in this direction are obtained by Götze and Hipp (1983) (hereafter referred to as [GH]). We formulate our results under regularity conditions similar to those of [GH]. Suppose that $X_1, X_2 \ldots$ are defined on a common probability space $(\Omega, \mathcal{A}, P)$ and a sequence $\mathcal{D}_0, \mathcal{D}_{\pm 1}, \mathcal{D}_{\pm 2}, \ldots$ of sub σ-fields of $\mathcal{A}$ are given. We shall assume that the conditions (C.1)-(C.5) below hold throughout the paper.

(C.1) $E\|X_1\|^4 < \infty$ and $M = \lim_{n\to\infty} M_n$ is nonsingular.

(C.2) There exists a positive constant d such that for n,m=1,2,... with $m>d^{-1}$, there exists a $\mathcal{D}_{n-m}^{n+m}$ measurable p-variate random vector $Y_{n,m}$ for which (i) $E\|X_n-Y_{n,m}\| \le d^{-1}\exp(-dm)$ and (ii) with $m=a_n$, $E\|Y_{n,m}\|^{12}I(\|Y_{n,m}\| < n^{1/4}) < d^{-1}$. Here $\mathcal{D}_a^b$ is the σ-field generated by $\{\mathcal{D}_j, a\le j\le b\}$, $\{a_n\}$ is a sequence of real numbers satisfying $\log n = o(a_n)$ and $a_n = O((\log n)^{1+1/d})$ as $n \to \infty$, and for any set A, I(A) denotes the indicator of A.

(C.3) There exists d>0 such that for all m,n=1,2,... $A\in\mathcal{D}_{-\infty}^{n}, B\in\mathcal{D}_{n+m}^{\infty}$,

$$|P(A\cap B)-P(A)P(B)| \le d^{-1}e^{-dm}.$$

(C.4) There exists d>0 such that for all m,n=1,2..., $d^{-1}<m<n$, and all $t\in\mathbb{R}^p$ with $\|t\| \ge d$,

$$E|E(e^{it'(X_{n-m}+X_{n-m+1}+\ldots+X_{n+m})}|\mathcal{D}_j:j\ne n)|\le e^{-d}.$$

(C.5) There exists d>0 such that for all m,n,q=1,2,... and $A\in\mathcal{D}_{m-q}^{m+q}$

$$E|P(A|\mathcal{D}_j:j\ne n)-P(A|\mathcal{D}_j:0<|n-j|\le m+q)| \le d^{-1}e^{-dm}.$$

<u>Remark 2.1</u> Except for (C.2) (ii), all other conditions have been used by [GH] for obtaining the Edgeworth expansion for the unbootstrapped statistic T_n. Here, the moment condition is used for proving the almost sure convergence of the third order bootstrap moments of T_n^*.

In comparison, even in the univariate case, Künsch (1989) requires $EX_1^8 < \infty$ for almost sure convergence of the second moment. Condition (C.2) (ii) somewhat relaxes the moment requirement on the original observations X_i's. A sufficient condition for (C.2) (ii) is that $E\|X_1\|^{12} < \infty$.

As for the other assumptions, (C.4) is a conditional Cramer's condition and (C.3) is the strong mixing condition stated in terms of the σ-fields D_j's. To verify conditions (C.2)-(C.5), one usually chooses these σ-fields appropriately. As an example, if $X_j = \Sigma_{i=1}^{\infty} C_i Y_{i+j}$, $j \geq 1$ for some sequences of iid random variables $\{Y_n\}$ and constants $\{C_n\}$, one may define D_j's in terms of Y_j's, rather than in terms of X_j's. See [GH] for a number of interesting examples where conditions (C.2) (i), (C.3)-(C.5) are verified.

For stating the theorems, we need to develop some notation. Let $\mathbf{Z}^+$ denote the set of all nonnegative integers. For functions $f: \mathbb{R}^p \to \mathbb{R}$, and $r \geq 0$, write $M_r(f) = \sup\{(1+\|x\|)^{-r}|f(x)|: x \in \mathbb{R}^p\}$ and $\bar{w}(f,r,\mu) = \int w(f,r,\cdot)d\mu$ where μ is a finite signed measure and for $x \in \mathbb{R}^p$, $w(f,r,x) = \sup\{|f(x)-f(y)|: y \in \mathbb{R}^p, \|x-y\| \leq r\}$. For any set $B \in \mathbb{R}^p$ and $\eta > 0$, define ∂B = boundary of B and $(B)^\eta = \{x \in \mathbb{R}^p: \|x-y\| < \eta$ for some $y \in B\}$. Denote the distribution function and the density of a standard normal random vector on the relevant Euclidean space by Φ and ϕ respectively. Write $N_p(\mu,A)$ for the normal distribution on $\mathbb{R}^p$ with mean μ and dispersion matrix A. Unless otherwise specified, all the integrals will range over the entire Euclidean space in question and the limits in the order symbols are taken by letting $n \to \infty$. Finally, let P_* stand for the bootstrap probability and E_* for the expectation under P_*.

Our first result gives the exact rate of approximation for the bootstrap statistic $T_n^* = \sqrt{k}\,\hat{A}_n(\bar{X}_n^* - \bar{X}_n)$ considered by Künsch (1989). As pointed out in the introduction, in this case the convergence rate is entirely determined by the bias $B_n = \sqrt{k}\,\hat{A}_n(\bar{X}_n - \hat{\mu}_n)$ of T_n^*.

More precisely, one has

Theorem 2.1. Suppose that there exist constants $C_1>0$, $C_2>0$, $0<\alpha<\beta<1/4$ such that $C_1 n^{\alpha}<\ell<C_2 n^{\beta}$. Let $\mathcal{B}$ be a class of Borel sets in $\mathbb{R}^p$ such that for some a > 0,

$$\sup_{B\in\mathcal{B}} \Phi((\partial B)^{\eta}) = O(\eta^{a}) \text{ as } \eta \downarrow 0. \tag{2.1}$$

Then, for almost all sample points $\omega \in \Omega$,

(a) $\operatorname{Sup}_{B\in\mathcal{B}} |P_*(T_n^* \in B) - P(T_n \in B)| = O(\|B_n\|)$, where $B_n = \sqrt{k}\,\hat{A}_n(\bar{X}_n-\hat{\mu}_n)$,

(b) there exist constants $C_3 > 0$, $C_4 > 0$ such that

$$C_1 \|B_n\| < \sup_{x\in \mathbb{R}^p} |P_*(T_n^*\le x) - P(T_n\le x)| < C_4 \cdot\|B_n\|.$$

(c) $(6n/\ell)^{1/2} B_n$ converges in distribution to the standard normal distribution on $\mathbb{R}^p$.

Under the assumed conditions, the bound $\|B_n\|$ tends to zero almost surely (See Lemma 5.2 (ii)-(iv)), but not even at the rate $O(n^{-1/2}\ell^{1/4})$ a.s.(P). Consequently, the bootstrap approximation provided by the statistic T_n^* is worse than the usual normal approximation, which has an error $O(n^{-1/2})$ only. However, a significantly better convergence rate can be obtained if one centers $\bar{X}_n^*$ at $\hat{\mu}_n$ and considers the statistic $T_{1n}^* = \sqrt{k}\,\hat{A}_n(\bar{X}_n^* - \hat{\mu}_n)$. The following result from Lahiri (1990) shows that the bootstrap distribution of T_{1n}^* captures the second term in the Edgeworth expansion for T_n and thus outperforms the normal approximation.

Theorem 2.2. Suppose there exist constants $C_1>0$, $C_2>0$, $\alpha>0$, $\alpha<\beta<1/4$ such that $C_1 n^{\alpha} < \ell < C_2 n^{\beta}$. Then,

(a) Then, for any Borel measurable $f: \mathbb{R}^p \to \mathbb{R}$ with $M_2(f)<\infty$, and any positive real number t,

$$|E_* f(T^*_{1n}) - Ef(T_n)| \le C[M_2(f)o(n^{-1/2}) + \bar{w}(f, n^{-t};\Phi)] \quad \text{a.s.}(P)$$

(b) Let $\mathcal{B}$ be a class of Borel sets in $\mathbb{R}^p$ satisfying (2.1). Then,

$$\sup_{B\in\mathcal{B}} |P_*(T^*_{1n}\in B) - P(T_n\in B)| = o(n^{-1/2}) \quad \text{a.s.}(P).$$

<u>Remark 2.2</u> Theorem 2.2 shows that the stationary bootstrap procedure is second order correct if the block size ℓ is not too small or not too big compared to the sample size n. The lower bound on the order of ℓ is required for ensuring the error bound $o(n^{-1/2})$ in Theorem 2.2(b) over the class $\mathcal{B}$. Without it, the error of approximation is of the order $O(\ell^{-k})$ for any positive number k. The upper bound on the order of ℓ is imposed to guarantee the almost sure convergence of the third order sample moments. One can allow ℓ to increase at a faster rate if X_1 has finite higher order moments.

<u>Remark 2.3</u> In the iid set up, works of Singh (1981) and Babu and Singh (1984) imply that a strong nonlatticeness condition for the distribution of X_1 is sufficient for second order correctness of the iid bootstrap procedure. In comparison Theorem 2.2 requires a stronger (viz. Cramer-type) condition for the second order optimality of stationary bootstrap. Using the arguments in [GH], one can derive a two-term Edgeworth expansion for the unbootstrapped statistic T_n under a conditional strong-nonlatticeness condition. However it is not clear if this is enough to guarantee the conclusions of Theorem 2.2.

<u>**3. Smooth functions of mean:**</u> In this section we consider the asymptotic accuracy of Künsch's bootstrap procedure for more general statistics based on the means of transformed observations. Let $f: \mathbb{R}^p \to \mathbb{R}^q$ be a Borel measureable function. Let $Y_i = f(X_i)$, $\alpha = EY_1$ and $\bar{Y}_n = n^{-1} \sum_{i=1}^n Y_i$. Suppose that $H: \mathbb{R}^q \to \mathbb{R}^m$ is twice continuously differentiable in a neighborhood of α. Here we shall consider the normalized version of the statistic $\sqrt{n}(H(\bar{Y}_n) - H(\alpha))$. Let D denote the matrix of its first order partial derivatives at α. Set $\Sigma_n = nD(\text{Disp } \bar{Y}_n)D'$. If Σ_n is positive definite, define the normalized statistic $T_{2n} = \sqrt{n}\, \Sigma_n^{-1/2}(H(\bar{Y}_n) - H(\alpha))$. Examples of such statistics include normalized sample correlation coefficient and normalized sample coefficient of variation. The bootstrapped statistic corresponding to T_{2n} can be defined in the same way as before. Given the bootstrap sample $\{\xi_{ij}^*: 1 \le i \le k, 1 \le j \le \ell\}$, write

$$\underset{\sim}{U}_i = \ell^{-1} \sum_{j=1}^{\ell} f(\xi_{ij}^*),\ \underset{\sim}{Y}_n^* = k^{-1} \sum_{i=1}^{k} \underset{\sim}{U}_i^*,\ \underset{\sim}{\alpha}_n = E_* \underset{\sim}{U}_1^* \text{ and } \hat{\Sigma}_n = D(\text{Disp } \underset{\sim}{U}_1^*)D'.$$

Then the bootstrapped version of T_{2n} is given by $T_{2n}^* = \sqrt{k}\, \hat{\Sigma}_n^{-1/2}(H(\underset{\sim}{Y}_n^*) - H(\underset{\sim}{\alpha}_n))$. The following result of Lahiri (1990) asserts the second order correctness of Künsch's bootstrap for T_{2n}.

<u>**Theorem 3.1.**</u> Suppose that conditions (C.1) - (C.5) hold with X_i's replaced by Y_i's. If the block size ℓ satisfies the requirements of Theorem 2.2 and D is of full row rank, then

$$\sup_{B \in \mathcal{B}} |P(T_{2n} \in B) - P_*(T_{2n} \in B)| = o(n^{-1/2}) \quad \text{a.s. } (P)$$

where $\mathcal{B}$ is a class of Borel subsets of $\mathbb{R}^m$ satisfying (2.1).

<u>**Remark 3.1.**</u> In some applications, the transformation $Y_i = f(X_i)$ may

force the limiting dispersion matrix of $\sqrt{n}\ \bar{Y}_n$ to have rank $r < m$. If $r \geq 1$, one may still use a suitable linear transformation on the Y_i's and guarantee conditions (C.1) for the transformed variables.

Theorem 3.1 covers most of the commonly used normalized statistics based on sample means. However, unlike the iid case, one cannot directly apply this result to show the second order correctness of the bootstrap procedure for studentized sample mean. Here, the main problem arise from the form of the asymptotic covariance matrix M of $\sqrt{n}\ \bar{X}_n$. Since any sequence of estimators of M must cover an increasing number of lagged covarianes, the standard technique of transforming X_i's by a fixed q-dimensional function f does not apply. In the special case of m-dependent observations, however, it is possible to use a similar technique with suitable manipulations to prove the second order correctness of the bootstrap. Accordingly, for the rest of this section, we shall assume that $X_1, X_2, \ldots$ are m-dependent random variables with values in $\mathbb{R}^1$. Then by the CLT for m-dependent random r.v.'s, the asymptotic variance τ^2 of $\sqrt{n}(\bar{X}_n - \mu)$ is given by

$$\tau^2 = EX_1^2 + 2\sum_{i=1}^m EX_1X_{1+i} - (2m+1)\mu^2.$$

Hence, a natural estimator of M is

$$s_n^2 = n^{-1}\sum_{i=1}^n X_i^2 + 2n^{-1}\sum_{i=1}^m \sum_{j=1}^{n-i} X_iX_{j+i} - (2M+1)\ \bar{X}_n^2. \qquad (3.1)$$

We shall use s_n to studentize the sample mean $\bar{X}_n$ and write $T_{3n} = \sqrt{n}(\bar{X}_n - \mu)/s_n$, $n \geq 1$. Since $s_n \to \tau$ a.s. (P), T_{3n} is well defined for large values of n. To define the corresponding bootstrapped statistic, first note that s_n and hence T_{3n} are not simple averages of the observation $X_1,\ldots,X_n$. Rather they are defined in terms of averages of the products X_jX_{j+i}, $0 \le i \le m$, which involve (m+1) observations $(X_j,\ldots,X_{j+m})$ at a time. Following Künsch's original formulation, one may define the bootstrap versions of s_n and T_{3n} as follows. First redefine the blocks of size ℓ with $Z_i = (X_i,\ldots,X_{i+m})$, $1 \le i \le n - m = N$ (say) as observations and select $k_1 = [N/\ell]$ iid blocks $\zeta_i^* = (\zeta_{i1}^*,\ldots,\zeta_{i\ell}^*)$, $1 \le i \le k_1$ from the 'observed' blocks $\{\zeta_t = (Z_{t+1},\ldots,Z_{t+\ell}): 0 \le t \le N-\ell\}$. Note that each ζ_{ij}^* is a (m+1) dimensional vector. Let ζ_{ijr}^* denote the rth component of ζ_{ijr}^*, $1 \le r \le m+1$. Define $\underset{\sim}{U}_{ri}^* = \ell^{-1}\sum_{j=1}^{\ell} \zeta_{ij1}^*\,\zeta_{ijr-1}^*$, $2 \le r \le m+2$, $\underset{\sim}{U}_{1i}^* = \ell^{-1}\sum_{j=1}^{\ell}\zeta_{ij1}^*$, $\underset{\sim}{U}_i^* = (\underset{\sim}{U}_{1i}^*,\ldots,\underset{\sim}{U}_{m+2,i}^*)'$, $1 \le i \le k_1$. Let $\underset{\sim}{\alpha}_n = E_*(\underset{\sim}{U}_1^*)$, $\underset{\sim}{\alpha}_{in} = E_*(\underset{\sim}{U}_{i1}^*)$, $1 \le i \le m+2$. Write $\underset{\sim}{X}_n^* = k_1^{-1}(\underset{\sim}{U}_{11}^*+\ldots+\underset{\sim}{U}_{1k_1}^*)$ and $\underset{\sim}{\mu}_n = E_*\underset{\sim}{X}_n^* = \underset{\sim}{\alpha}_{1n}$. In accordance with the original statistic, one may define the bootstrap versions as

$$s_n^{*2} = \sum_{r=0}^{m} a_r\left(k_1^{-1}\sum_{i=1}^{k_1}\underset{\sim}{U}_{ri}^*\right) - (2m+1)\underset{\sim}{X}_n^{*2}$$

and $$\underset{\sim}{T}_{3n}^* = (k_1\ell)^{1/2}\,(\underset{\sim}{X}_n^* - \underset{\sim}{\mu}_n)/s_n^*$$

where $a_r = 1$ or 2 according as $r=0$ or $r \ge 1$. Unfortunately, the statistic $\underset{\sim}{T}_{3n}^*$ also has a demerit similar to T_n^*. In this case the problem is not with the bias, rather with the normalizing variable itself. For second order correctness, one has to use an additional multiplicative factor which depends on certain bootstrap moments. Let $\underset{\sim}{\tau}_n^2 = k_1\ell\operatorname{Var}((\underset{\sim}{X}_n^*)$ and $\hat{\tau}_n^2 = \sum_{r=0}^{m} a_r\underset{\sim}{\alpha}_{r+2,n} - (2m+1)\underset{\sim}{\alpha}_{1n}^2$. Then, the corrected bootstrapped statistic corresponding to T_{3n} is given by

and $$T_{3n}^* = \sqrt{k_1\ell}\;\hat{\tau}_n(\underset{\sim}{X}_n^* - \underset{\sim}{\mu}_n)/\underset{\sim}{\tau}_n s_n^*. \tag{3.2}$$

Note that $\underset{\sim}{\tau}_n^2/k_1\ell$ is the actual variance of the bootstrap sample mean $\underset{\sim}{X}_n^*$. On the other hand, writing $f_0(y) = \sum_{r=0}^{m} a_r y_{y+2} - (2m+1)y_1^2$, $y = (y_1,\ldots,y_{m+2}) \in \mathbb{R}^{m+2}$ and $\bar{U}_n^* = k_1^{-1}(\underset{\sim}{U}_1^* + \ldots + \underset{\sim}{U}_{k_1}^*)$, it follows that $s_n^{*2} = f_0(\bar{U}_n^*)$ and $\underset{\sim}{\tau}^2 = f_0(E_*\bar{U}_n^*)$. Under the assumption of Theorem 3.2 below, one can show that $\hat{\tau}_n - \underset{\sim}{\tau}_n = o(1)$ a.s.(P). However, the rate of convergence in this case is even worse than $\|B_n\|$. Since $\underset{\sim}{X}_n^*$ and s_n^* determine the approximate variance of $\underset{\sim}{T}_{3n}^*$ through $\underset{\sim}{\tau}_n$ and $\hat{\tau}_n$ respectively, one must remove the effect of the difference $|(\underset{\sim}{\tau}_n/\hat{\tau}_n)-1|$ for attaining faster convergence rates. The next result shows that the modified version of the bootstrap is indeed second order correct.

Theorem 3.2. Let $\lambda(X_1,\ldots,X_{m+1})$ denote a linearly independent subset (in $L^2(P)$) of $\{X_1, X_1^2, X_1X_2, \ldots, X_1X_{m+1}\}$. For $j \geq 1$, define $Y_j = \lambda(X_j,\ldots,X_{j+m})$. Assume that conditions (C.1), (C.2) and (C.4) hold with X_j's replaced by Y_j's and $\mathcal{D}_j = \sigma(Y_j)$. If there exists constants $C_1>0$, $C_2>0$, $0<\alpha<\beta\leq 1/4$ such that $C_1 n^{\alpha} < \ell < C_2 n^{\beta}$, then, for the bootstrapped statistic T_{3n}^* defined by (3.2),

$$\sup_{B\in\mathcal{B}} |P_*(T_{3n}^*\in B) - P(T_{3n}\in B)| = o(n^{-1/2}) \qquad \text{a.s. (P)}$$

Here $\mathcal{B}$ is a class of Borel subsets of $\mathbb{R}$ satisfying (2.1).

Remark 3.2: For studentized sample mean, it is possible to define the corresponding bootstrapped statistic in more than one ways, depending on the choice of s_n. See Göetz and Künsch (1990) for a different formulation in the strong mixing case.

Section 4 Nonstationary Data: In many applications, the stationarity assumption on the observations $\{X_1,X_2,\ldots\}$ does not hold. Time series data observed over a long period often tend to show lack of uniformity in their distributions. Consequently, it is important to investigate the validity of the bootstrap procedure for nonstationary data. It is well known that Efron's (1979) iid bootstrap works reasonably well even when the data are independent but not identically distributed (see Freedman (1981), Liu (1988), Lahiri (1989), Liu and Singh (1989)). In fact, the works of Liu (1988) and Lahiri (1989) establish the second order correctness of the bootstrap procedure under some mild conditions on the degree of heterogeneity of the data. In the dependent set up, Künsch's bootstrap also enjoys a similar 'robustness' property. It turns out that one can obtain 'good' approximations even when the observations are nonstationary, but satisfy some uniformity conditions.

Here only the case of sample mean is considered. In view of the nonstationarity of X_i's, some of the conditions of Section 2 need to be restated. However, the basic set up remains the same. As in Section 2, assume that X_i's are defined on the common probability space $(\Omega,\mathcal{A},P)$, but are not necessarily stationary. Replace condition (C.1) by condition (C.1)′ on the moments of X_i's:

(C.1)′ $\sup_{j\geq 1} E\|X_j\|^4 < \infty$; $M = \lim_{n\to\infty} M_n$ exists and is nonsingular.

In addition to this, some more conditions are needed on the joint behaviour of the moments of X_i's. For $j\geq 1$, write $\mu_j = EX_j$ and $\bar{\mu}_j = j^{-1}\sum_{i=1}^{j}\mu_i$. Let $m_t = EU_t = \ell^{-1}\sum_{j=1}^{\ell}\mu_{t+j-1}$ and $M_{nt} = \mathrm{Disp}(\sqrt{\ell}U_t)$ where $U_t = \ell^{-1}\sum_{j=1}^{\ell}X_{t+j-1}$, $1\leq t\leq b$. Denote the smallest eigen-value of M_{nt} by λ_{nt} and let $\lambda_n = \text{Min}\{\lambda_{nt}: 1 \le t \le b\}$. The next two conditions restrict the heterogeneity of m_t's and M_{nt}'s.

(C.6) $\max\{\ell^2 \|m_t - \bar{\mu}_n\| : 1 \le t \le b\} = o(1)$

(C.7) $\liminf_{n\to\infty} \lambda_n > 0.$

Condition (C.7) is equivalent to one of the conditions used by Liu (1988) in the independent case with block size $\ell=1$. In view of (C.1)′, (C.7) holds if $\text{Max}\{\|M_{nt} - M_n\| : 1 \le t \le b\} = o(1)$. Condition (C.6) is somewhat stronger than necessary. It can be replaced with a weaker condition which requires $b^{-1}\sum_{t=1}^{b} \ell^2 \|m_t - \bar{\mu}_n\|^3 = o(1)$ and the uniform boundedness of $|\sum_{t\in A(k,j)} (m_t - \bar{\mu}_n)|$ over the index sets $A(k,j) = \{i: 1 \le i \le b,\ j+k-\ell-1 \le i \le j\}$, $0 \le k \le \ell-1$, $1 \le j \le n$ for sufficiently large n. The following result proves the second order correctness of Künsch's bootstrap for the normalized sample mean $T_n = \sqrt{n} M_n^{-1/2} (\bar{X}_n - \bar{\mu}_n)$ of nonstationary data.

<u>**Theorem 4.1**</u> Assume that conditions (C.1)′, (C.2)-(C.7) hold and there exist constants $C_1 > 0$, $C_2 > 0$, $0 < \alpha \le \beta < 1/4$ such that $C_1 n^\alpha < \ell < C_2 n^\beta$. Then

$$\sup_{B\in\mathcal{B}} |P(T_n \in B) - P_*(T_{1n}^* \in B)| = o(n^{-1/2}) \quad \text{a.s. (P)}$$

where T_n, T_{1n}^* and $\mathcal{B}$ are as in the statement of Theorem 2.2.

<u>**Remark 4.1.**</u> Under some additional assumptions, one can prove similar results for the statistics of the form $\sqrt{n}(H(\bar{X}_n) - H(\bar{\mu}_n))$ of Section 3.

<u>**Remark 4.2.**</u> Results of this section can be readily applied to the least square estimators of multiple linear regression parameters.

<u>**5. Proofs:**</u> For proving the theorems of Section 2, we need the

following lemmas. All throughout this section, C, C(.) will denote generic constants depending only on their arguments, if any. Also, for $a,b \in \mathbb{R}$ and any set A, we write $a \wedge b = \min\{a,b\}$, $a \vee b = \max\{a,b\}$ and $|A|$ = size of A. The following result from Lahiri (1990) will be frequently used in the proofs.

Lemma 5.1: Let $\{W_1,W_2,\ldots,W_n\}$ be a collection of random variables on a probability space $(S,\mathcal{S},Q)$ such that for $i=1,\ldots,n$, $EW_i=0$ and $|W_i|\le B$ a.s. (Q) for some $B>0$. Then, for any positive integer r and $1\le m<C(r)\bullet n$,

$$E(\textstyle\sum_{i=1}^n W_i)^{2r} \le C(r)B^{2r}[n^r m^{2r}+n^{2r}\alpha(m)]$$

where $\alpha(m)=\sup\{|P(A_1\cap A_2)-P(A_1)P(A_2)|:A_1\in F_1^i, A_2\in F_j^n,\ j=i+m+1,\ 1\le i\le j\le n\}$ and F_s^t = the σ-field generated by W_i, $s\le i\le t$.

Proof of Lemma 5.1. Since some of the arguments in the proof of Lemma 5.1 were omitted in Lahiri (1990), we give a complete proof here. Fix $r\ge 1$ and let $k=2r$. Then

$$(\textstyle\sum_{i=1}^n W_i)^k = \sum_{j=1}^k \sum_1 C(\alpha_1\ldots\alpha_j) \sum_2 \prod_{t=1}^j W_{i_t}^{\alpha_t} \qquad (5.1)$$

where for each fixed $j\in\{1,2,\ldots,k\}$, $\sum_1$ extends over all j-jtuples of positive integers $(\alpha_1,\ldots,\alpha_j)$ such that $\alpha_1+\ldots+\alpha_j=k$ and $\sum_2$ extends over all ordered j-tuples $(i_1\ldots i_j)$ of integers such that $1\le i_1<\ldots<i_j\le n$.

By Lemma 6.1 of Bhattacharya and Rao (1986) (hereafter referred to as [BR]]), it follows that

$$|\textstyle\sum_{j=1}^r \sum_1 C(\alpha_1,\ldots,\alpha_j) \sum_2 E \prod_{t=1}^j W_{i_t}^{\alpha_t}| \le C(r)n^r EW_1^k \qquad (5.2)$$

For proving Lemma 3.1, it is now enough to show that for each fixed

$j>r$ and fixed j-tuple $(\alpha_1 \ldots \alpha_j)$ under Σ_1,

$$|E \sum_2 \Pi_{t=1}^{j} W_{i_t}^{\alpha_t}| \le C(r)[n^k B^k \alpha(m) + n^r m^k B^k]$$

Given j and $(\alpha_1,\ldots,\alpha_j)$ as above, define $s=j-r$, $A=\{t:\alpha_t=1\}$ and β_0 = size of A. Clearly, $1\le s\le r$. Next note that $k=\sum_{t=1}^{j} \alpha_t \ge \beta_0 + 2(j-\beta_0)$. Hence, it follows that $2s\le\beta_0\le k$. Given an integer m, $1\le m<n/3r$, define the set $B_m=\{(i_1,\ldots i_j):1\le i_1<\ldots<i_j\le n,\ |i_{t-1}-i_t|>m$ and $|i_{t+1} - i_t| > m$ for some $t\in A\}$. Next, rearrange the sum $\sum_2$ as $\sum_3+\sum_4$ where $\sum_3$ extends over all indices $(i_1\ldots i_j)\in B_m$.

By Lemma 27.2 of Billingsley (1986), it follows that

$$|\sum_3 E \Pi_{t=1}^{j} W_{i_t}^{\alpha_t}| \le C(r)n^k B^k \alpha(m).$$

Hence, it remains to find an upper bound on the sum $\sum_4$. Since $\beta_0 \ge 2s$, it is enough to find a bound on the size of $\Theta\equiv\{(i_1\ldots i_j): 1\le i_1<\ldots<i_j\le n, |i_{t+\epsilon_t} - i_t|\le m$ for all $t\in\Gamma\}$, for some $\Gamma\subseteq\{1,\ldots,j\}$,

$|\Gamma|=2s$ and $\epsilon_1,\ldots,\epsilon_{2s}\in\{-1,1\}$.

To that end note that the indices t, $t+\epsilon_t$, $t\in\Gamma$ are not necessarily distinct. Let $1\le\sigma_1<\ldots<\sigma_\nu\le j$ be the increasing arrangement of the distinct indices in $\{t, t+\epsilon_t:t\in\Gamma\}$. Then, $2s\le\nu\le(4s\wedge j)$. Let q be the number of unbroken integer sequences in the arrangement $\sigma_1,\ldots,\sigma_\nu$ and let $\beta_1,\ldots,\beta_q$ respectively denote their lengths. It is easy to see that $1\le q\le 2s$, $2\le\beta_i\le 2s$ and $\beta_1+\ldots+\beta_q=\nu$. Also, using a straightforward argument, one can show that $|\Theta|\le n^q m^{\nu-q} n^{j-\nu}$. To obtain a bound independent of q, ν and j, first note that $j+q-\nu \le r$ if and only if $s+q \le \nu$. If $q \le s$, then the bounds on ν imply that $s+q \le 2s \le \nu$. On the other hand, if $q > s$, then $s+q < 2q \le$

$\beta_1+\ldots+\beta_q=\nu$. Hence, it follows that $|\theta| \le m^k\, n^{q+j-\nu} \le n^r m^k$. This proves Lemma 5.1.

For Lemmas 5.2 and 5.3, we need to introduce a special truncation function used by [GH]. Let $g \in C^\infty[0,\infty)$ satisfy $g(x) = x$ if $x \le 1$, $g(x) = 2$ for $x \ge 2$ and g is increasing. Then, define the function $T(n;\theta,\cdot)$ by $T(n;\theta,x) = n^\theta x\, g(n^{-\theta}\|x\|)/\|x\|$, for $x \in \mathbb{R}^p$, $\theta>0$, $n\ge1$. For $\nu = (\nu_1,\ldots,\nu_p)' \in (\mathbb{Z}^+)^p$ and $t = (t_1,\ldots,t_p)' \in \mathbb{R}^p$, write $|\nu| = \nu_1+\ldots+\nu_p$, $\nu! = \nu_1!\ldots\nu_p!$, $t^\nu = t_1^{\nu_1}\ldots t_p^{\nu_p}$ and $\|t\|$ for the Euclidean norm of t. We shall prove Lemmas 5.2 and 5.3 under the set up of Section 4. The corresponding assertions under the stationarity of X_i's will then follow as special cases.

Lemma 5.2. Suppose that $\ell\to\infty$ and $\ell=O(n^{(1-\delta)/4})$ for some $\delta > 0$. Then a.s.(P)

(i) $\quad \ell\|\hat{\mu}_n-\bar{\mu}_n\| = o(1)$

(ii) $\quad \|\ell\hat{M}_n-M_n\| = o(1)$

(iii) $\quad \ell^2 E_*(U_1^*-\hat{\mu}_n)^\nu - n^2\, E(\bar{X}_n-\bar{\mu}_n)^\nu = o(1) \qquad$ for all $|\nu|=3$

(iv) $\quad \|\sqrt{n}(\bar{X}_n-\hat{\mu}_n)\| = o(1)$.

where $\bar{\mu}_n = n^{-1}\sum_{i=1}^n \mu_i$ and $\mu_i = EX_i$.

Proof of Lemma 5.2(i). Let $U_t = \ell^{-1}\sum_{i=1}^{\ell} X_{t+i-1}$, $t=1,\ldots,b$. Then

$$\hat{\mu}_n-\bar{\mu}_n = b^{-1}\sum_{t=1}^b (U_t-m_t) + \sum_{t=1}^b (m_t-\bar{\mu}_n)/b. \tag{5.3}$$

Using the condition $\sup\{E\|X_j\|^4 : j\ge1\} < \infty$ and the definition of m_t, one can show that there exists a constant $C > 0$ such that

$$\|b^{-1}\sum_{t=1}^b m_t-\bar{\mu}_n\| \le C\ell b^{-1}.$$

For the other term in (5.3), without loss of generality (w.l.g.), we may assume that $\mu_j = 0$ for all $j \geq 1$. Then,

$$b^{-1} \sum_{t=1}^{b}(U_t - m_t) = (b\ell)^{-1}[\sum_1 \ell X_j + \sum_2 j(X_j+X_{n-j+1})]$$

where $\sum_1$ extends over $j=\ell,\ldots,b$ and $\sum_2$ extends over $j=1,2,\ldots,\ell-1$. By Condition (C.1) and Borel-Contelli Lemma, it follows that $\|X_n\| = o(n^{(1+\delta)/4})$ a.s.(P). One can use the above fact to show that a.s.(P)

$$\|(b\ell^{-1}\sum_2 j(X_j+X_{n-j+1})\| = o(n^{-(3-\delta)/4}\ell)$$
$$\text{and} \tag{5.4}$$
$$\|b^{-1}\sum_1(X_j-T(X_j))\| = O(n^{-1})$$

where $T(\cdot) = T(n;(1+\delta)/4,\cdot)$.

Next, note that for all $x,y \in \mathbb{R}^p$, there exists a constant C, independent of n, such that

$$\|T(x) - T(y)\| \leq C. \quad \|x-y\|.$$

Hence, by Borel-Cantelli Lemma and condition (C.2) (with $m=a_n$), it follows that

$$\|b^{-1}\sum_1(T(X_j)-T(Y_{j,m}))\| = O(e^{-dm/2}) \quad \text{a.s.(P)}$$
$$\text{and} \tag{5.5}$$
$$\|b^{-1}\sum_1 E(T(X_j)-T(Y_{j,m}))\| = O(e^{-dm}).$$

Finally use Lemma 3.1 and condition (C.2) to conclude that for any $a \in [0,1/4)$

$$\|b^{-1}\sum_1 T(Y_{j,m}) - ET(Y_{j,m})\| = O(b^{-a}) \quad \text{a.s.}$$

This proves Lemma 3.2(i).

<u>Proof of (ii)</u> By part (i), it is enough to show that the matrix $(\ell E_n(U_1^*-\bar{\mu}_n)(U_1^*-\bar{\mu}_n)'-M_n)$ goes to the zero matrix almost surely (P). To that end, note that

$$\ell E_*(U_1^*-\bar{\mu}_n)(U_1^*-\bar{\mu}_n)'$$
$$= \ell b^{-1}\Sigma_{t=1}^{b}[(U_t-m_t)(U_t-m_t)' + (m_t-\bar{\mu}_n)(U_t-m_t)'$$
$$+ (U_t-m_t)(m_t-\bar{\mu}_n)' + (m_t-\bar{\mu}_n)(m_t-\bar{\mu}_n)'].$$

Clearly, by assumption (C.6), the last term above goes to zero. Hence, it is enough to show that a.s.(P),

(a) $\|\ell b^{-1}\Sigma_{t=1}^{b}(U_t-m_t)(U_t-m_t)' - M_n\| \to 0$

(b) $\|\ell b^{-1}\Sigma_{t=1}^{b}(m_t-\bar{\mu}_n)(U_t-m_t)'\| \to 0$ and

(c) $\|\ell b^{-1}\Sigma_{t=1}^{b}(U_t-m_t)(m_t-\bar{\mu}_n)'\| \to 0$.

We shall give a proof of part (a) only. Parts (b) and (c) can be proved using similar arguments. As in the proof of lemma 5.2(i), w.l.g. assume that $\mu_j=0$ for all $j\geq 1$. Fix $1\leq i_o, j_o\leq p$. With $m=a_n$ and T as in the proof of Lemma 5.2(i), denote the i_o and the j_o components of $T(Y_{t,m}) - ET(Y_{t,m})$ by Z_t and V_t respectively. Then, by (5.5), (i_o,j_o)th element of $\ell\hat{M}_n$

$$= (\ell b)^{-1} \Sigma_{t=1}^{b} (\Sigma_{i=1}^{\ell} Z_{t+i-1})(\Sigma_{i=1}^{\ell} V_{t+i-1}) + o(1) \quad \text{a.s.}$$

$$= (\ell b)^{-1}(\Sigma_{k=0}^{\ell-1} \Sigma_{j=1}^{n-k} a(k,j)(Z_jV_{j+k} + Z_{j+k}V_j)) + o(1) \quad \text{a.s.}$$

where $a(k,j) = (j\wedge b) - (j-\ell+k+1)\vee 1$, for $k\neq 0$

$= \text{Min}\{j, n-j+1, \ell\}/2$ for $k=0$.

The term k=0 can be shown to converge to its expectation a.s. (P) as in the proof of Lemma 5.2(i). Hence, it is enough to put a bound on

$$\Delta_n=E\,(\Sigma_{k=1}^{\ell-1} \Sigma_{j=1}^{n-k} a(k,j)\,(Z_jV_{j+k} - EZ_jV_{j+k}))^4.$$

Note that Lemma 3.1 is not directly applicable in this case. However, one can use a similar argument and write

$$\Delta_n = E\sum_{j=1}^{4}\sum_1 C(\alpha_1..\alpha_j)\sum_2 \Pi_{i=1}^{j}\{a(j_i,k_i)(V_{j_i+k_i}Z_{j_i} - EV_{j_i+k_i}Z_{j_i})\}^{\alpha_i}$$

where $\sum_1$ extends over all j tuples $(\alpha_1,\ldots\alpha_j)$ with $\alpha_i > 0$ and $\alpha_1 +\ldots+ \alpha_j = 4$ and $\sum_2$ extends over j distinct pairs of indices (j_i,k_i), $1\le i\le j$.

For $j \le 2$, one uses the trivial bound $n^2\ell_n^2$ on the number of summands under $\sum_2$. The other values of j require more careful estimates. First we consider the case when j=4. Note that in this case it is enough to find a bound on

$$\ell_n^4 \sum_2'\{E\, \Pi_{i=1}^{4}(V_{j_i+k_i}Z_{j_i} - EV_{j_i+k_i}Z_{j_i})\} = \Delta_{4n} \text{ (say)},$$

where $\sum_2'$ extends over all distinct pairs (j_i,k_i) under $\sum_2$ with $j_1\le j_2\le j_3\le j_4$.

Write $\sum_2' = \sum_3+\sum_4$ where $\sum_3$ runs over all distinct pairs of indices (j_i,k_i), i=1,...,4 such that

$$\text{Max}\{j_2 - (j_1+k_1),\ j_4 - t_o\} \ge 3m+1$$

where $t_o = \max\{j_i+k_i\colon j_i+k_i \le j_4,\ i=1,2,3\}$. Then, by weak dependence, each term under $\sum_3$ is at most $\alpha(m)n^2$ and the number of terms under $\sum_4$ is less than $n^2(\ell+m)^6$. Next expand the product of the differences to write

$$\sum_4 E\, \Pi_{i=1}^{4}\ (V_{j_i+k_i}Z_{j_i} - EV_{j_i+k_i}Z_{j_i})$$

$$= (\sum_4 E\, \Pi_{i=1}^{4}\ V_{j_i+k_i}Z_{j_i}) + R_n$$

$$= (\sum_5 + \sum_6)\{(\ \sum_{i=1}^{4} \Pi\ \ V_{j_i+k_i}Z_{j_i})\} + R_n$$

where $\sum_5$ extends over all indices (j_i,k_i), i=1,..,4 such that $j_1\le j_2\le j_3\le j_4$ and at least one of $\{j_1,\ldots,j_4;\ k_1,\ldots,k_4\}$ differ from both the neighbouring indices by 3m+1 or more. Then, by weak dependence, the sum $\sum_5$ is negligible. To estimate $\sum_6$, consider the number of subsequences of $\{j_1,..j_4, k_1,...k_4\}$ where the adjacent indices differ by at most 3m. Then, under $\sum_6$, there can be at most four such subsequences, each having length two or more. Consequently, the number of summands under $\sum_6$ is less than $C.n^2\ell_2 m^4$. By similar arguments, one can show that $R_n < Cn^2\ell^2m^4$. Hence, the bound on Δ_{4n} is $O(n^2\ell^6m^4)$.

The case j=3 also admits the same bound. Hence, it follows that (i_0,j_0)th element of $\ell\hat{M}_n$ converges to $(\ell b)^{-1}\sum_{k=0}^{\ell-1}\sum_{j=1}^{n-k}EZ_jV_{j+k}$ a.s. (P) for all $1\le i_0,j_0\le p$. The proof of Lemma 5.2(ii) can now be completed by expressing the elements of M_n as a similar sum and then using the weak dependence condition.

Proof of (iii) Similar to part (ii). In this case, the bound on $E(L.H.S.)^4$ is of the order $O(n^{-2}\ell^4m^6)$ which necessarily forces $\ell = O(n^a)$ with $a < 1/4$.

Proof of (iv) Follows from Borel Cantelli lemma and the moment conditions.

For the next lemma, define $w_n(t) = E_* \exp(i\sqrt{\ell}t'U_1^*)$, $v_n(t) = b^{-1}\sum_{j=1}^{b}\exp(-t'M_{nj}t/2)$, and $u_n(t) = b^{-1}\sum_{j=1}^{b}E\exp(i\sqrt{\ell}t'U_j)$, $t \in \mathbb{R}^p$.

Lemma 5.3. Let $\ell = O(n^a)$ for some $a < 1/2$. Then for any real number $\lambda>0$, $\sup\{|w_n(t)-v_n(t)|:\|t\|\le\ell^\lambda\} = o(1)$ a.s.(P).

Proof of Lemma 5.3. Fix $\delta>0$, $\lambda>0$. With $T(\cdot) = T(n;(1-a)/2,\cdot)$, define $\mathbb{U}_t=\ell^{-1}\sum_{j=1}^{\ell} T(X_{j+1-t})$, t=1,...,b. Then, uniformly in $t \in \mathbb{R}^p$,

$$|w_n(t) - b^{-1} \sum_{j=1}^{b} \exp(it\sqrt{\ell}\, \bar{U}_j)| \le 2b^{-1} \sum_{j=1}^{b} I(U_j \ne \bar{U}_j) \tag{5.6}$$

$= O(n^{-1})$, a.s. (P).

Also, by condition (C.2), it follows that

$$L = \operatorname{Sup}\{E\|T(Y_{n,m})\| : m > d^{-1},\ n \ge 1\} < \infty.$$

Let $\underset{\sim}{U}_{j,m} = \ell^{-1} \sum_{i=1}^{\ell} T(Y_{i+j-1,m})$ and $\underset{\sim}{W}_{j,m}(t) = E \exp(it\sqrt{\ell}\, \underset{\sim}{U}_{j,m})$. Using arguments similar to those in the proof of Lemma 5.2(i) and taking $m = O(\ell + a_n)$, one can show that uniformly over all t in $\mathbb{R}^p$,

$$b^{-1} \sum_{j=1}^{b} |\exp(it\sqrt{\ell}\, \underset{\sim}{U}_{j,m}) - 1| \le \|t\|\sqrt{\ell}(L + O(1)) \text{ a.s. (P) and}$$

$$b^{-1} \sum_{j=1}^{b} |\underset{\sim}{W}_{j,m}(t) - 1| \le L\|t\|\sqrt{\ell}.$$

Now using Lemma 5.1 above and a discretizing argument similar to the one given in the proof of Lemma 4.2 of Babu and Singh (1984) (hereafter referred to as [BS]), one can show that

$$\operatorname{Sup}\{|b^{-1} \sum_{j=1}^{b} (\exp(it\sqrt{\ell}\, \underset{\sim}{U}_{j,m}) - \underset{\sim}{W}_{j,m}(t))| : \|t\| \le \ell^{\lambda}\} = o(1) \text{ a.s. (P)}.$$

This, together with equations (5.5) and (5.6), yields

$$\sup\{|w_n(t) - u_n(t)| : \|t\| \le \ell^{\lambda}\} = o(1) \quad \text{a.s. (P)}.$$

Next, recall that $M_{nj} = \operatorname{Cov}(\sqrt{\ell} U_j)$. Using the arguments in the proof of Theorem 2.8 of [GH], one can show that uniformly in $1 \le j \le b$,

$$\operatorname{Sup}\{|E \exp(i\sqrt{\ell} t' U_j) - \exp(-t' M_{nj} t/2)| : \|t\| < \ell^{\lambda}\} = o(1).$$

Hence, Lemma 5.3 follows.

<u>Lemma 5.4</u>. Under the hypotheses of Theorem 4.1, there exists a constant $C > 0$ such that $E_*\|\hat{A}_n(U_1^* - \hat{\mu}_n)\|^4 < C < \infty$ a.s. (P).

Proof of Lemma 5.4: By Lemma 5.2(ii) and condition (C.1)′, $\ell^{-1/2}\hat{A}_n$ converges almost surely (P) to $M^{-1/2}$ and hence, is bounded in norm. Therefore it is enough to estimate $E_*\|\sqrt{\ell}(U_1^* - \hat{\mu}_n)\|^4$. By Lemma 5.2(i) and condition (C.6), there exists a constant $C > 0$ such that

$$E_*\|\sqrt{\ell}(U_1^*-\hat{\mu}_n)\|^4 \le C + C\ell^2 b^{-1}\Sigma_{t=1}^b \|U_t-m_t\|^4.$$

W.l.g., assume that $\mu_j=0$ for all $j\ge1$. Next define the truncated variables $\underset{\sim}{X}_j = T(j;(1-2\beta)/2, X_j)$, $\underset{\sim}{Y}_j = T(j;(1-2\beta)/2, Y_{j,m})$ and $Y_j = T(j;\theta,Y_{j,m})$, for $j\ge1$, where $m=a_j$, $\theta=(1/12) + (1-4\beta)/48$ and $0<\beta<1/4$ (as in the hypothesis of Theorem 4.1). Then, using conditions (C.1), (C.2) and the arguments in (5.5), it can be shown that

(a) $\|EY_j\| \le C.j^{-11\theta}$ for all $j\ge1$,

(b) $X_j=\underset{\sim}{X}_j$ and $Y_{j,m}=\underset{\sim}{Y}_{j,m}=Y_j$, eventually, a.s. (P)

(c) $\ell^2 b^{-1}\Sigma_{t=1}^b\|U_t-m_t\|^4 \le C\ell^2 b^{-1}\Sigma_{t=1}^b\|V_t\|^4 + o(1)$ a.s. (P)

where $V_t = \ell^{-1}\Sigma_{j=1}^{\ell}(Y_{t+j-1}-EY_{t+j-1})$, $1\le t\le b$. Next, a series of long and tedious arguments as in the proof of Lemma 5.2(ii) show that

$$\ell^2 b^{-1}\Sigma_{t=1}^b(\|V_t\|^4-E\|V_t\|^4) = o(1) \quad \text{a.s. (P)}.$$

Let Z denote a p-dimensional standard normal random vector. Then, using the arguments in the proof of Theorem 2.8 of [GH], one can show that

$$b^{-1}\Sigma_{t=1}^b(\ell^2 E\|V_t\|^4-E\| \ M_n{}^{1/2}Z\|^4) = o(1).$$

Since $\|M_{nj}{}^{1/2}\| = \|M_{nj}\|^{1/2} \le C \sup\{(E\|X_j\|^4)^{1/4}: j\ge1\}$ for all $1\le j\le n$, the lemma is proved.

For the next lemma, we need to introduce some more notation. For $\nu\in(Z^+)^p$, $|\nu| \le 3$, let $\tilde{\chi}_{\nu,n} = E_*(\hat{A}_n(U_1^*-\hat{\mu}_n))^\nu$. Let ψ_n denote the finite signed measure corresponding to the Fourier transform

$$\hat{\psi}_n(t) = (1 + \Sigma_{|\nu|=3} \ \frac{\tilde{\chi}_{\nu,n}}{\nu!}\cdot(it)^\nu)\exp(-\|t\|^2/2), \ t\in\mathbb{R}^p.$$

Lemma 5.5. Assume that the hypotheses of Theorem 4.1 hold. Let $t>0$ be a real number. Then, there exist a set $\Gamma_t \subseteq \Omega$ with $P(\Gamma_t)=1$ such that for every sample point $\omega \in \Gamma_t$, there exists a constant $C \equiv C(t,\omega) > 0$ satisfying

$$|E_* f(T_{1n}^*) - \int f d\psi_n| \le C[M_4(f) \cdot n^{-1}\ell + \bar{w}(f, n^{-t}; \Phi)]$$

for every $n \ge 1$ and any Borel measurable function $f: \mathbb{R}^p \to \mathbb{R}$ with $M_4(f) < \infty$.

Proof of Lemma 5.5. The main steps in the proof of Lemma 5.5 are similar to those in the proof of Theorem 20.1 of [BR]. Therefore, we only mention the differences. For $1 \le j \le k = [n/\ell]$, define:

$Z_{1,j} = \hat{A}_n(U_j^* - \hat{\mu}_n)$; $Z_{2,j} = Z_{1,j}\ I(\|Z_{1,j}\|^2 < k)$ and $Z_j = Z_{2,j} - E_* Z_{2,j}$.

Then, by Lemma 5.4,

$$E_n\|Z_{2,1}\|^4 < C < \infty \quad \text{a.s. (P)}. \tag{5.7}$$

Using (5.7), Lemma 5.2 above and Lemma 14.1 of [BR] one can show that, almost surely (P),

$$\|E_* Z_{2,1}\| = O(k^{-3/2}),\ \|E_n Z_1 Z_1' - I\| = O(k^{-1}) \text{ and}$$
$$|E_n Z_1^\nu - E Z_{1,1}^\nu| = O(k^{-1/2}) \text{ for any } \nu \in (\mathbb{Z}^+)^p \text{ with } |\nu| = 3 \text{ a.s. (P)}. \tag{5.8}$$

Now fix a sample point for which (5.7), (5.8) and the conclusions of Lemma 5.2 and Lemma 5.3 (for a suitable λ depending only on α and t) hold. Clearly, it is enough to prove the assertion of the lemma for this sample point. Fix a function f with $M_4(f) < \infty$. Using the arguments in the proof of Theorem 20.1 of [BR], one can show that

$$|E_* f(T_n^*) - \int f d\psi_n| \le C.M_4(f) k^{-1} + |\int f_n\, dH_n|$$

where $f_n(x) = f(x + \sqrt{k}\ E_* Z_{2,1})$, H_n is the signed measure with Fourier transform

$$\hat{H}_n(t) = \{E_* \exp(itZ_1)\}^k - \sum_{s=0}^{p+1} k^{-s/2} \underset{\sim}{P}_s(it; \{\hat{\chi}_{\nu,n}\}) e^{-t' D_n t},$$

$2D_n = (E_* Z_1 Z_1')^{-1}$ and $\hat{\chi}_{\nu,n} = \nu$th cumulant of Z_1 under E_*, $\nu \in (Z^+)^p$ and $\underset{\sim}{P}_s(\cdot,\cdot)$ are polynomials (see page 51, [BR] for definitions of $\underset{\sim}{P}_s$). As noted in [BS], the proof of Theorem 20.1 of [BR] has to be modified for estimating the last term above. Instead of Lemma 11.2, one has to use Lemma 24.1 of [BR]. Rest of the proof can be completed following the arguments in the proof of Theorem 20.1 of [BR] and noting that, by condition (C.7), $\sup\{|v_n(t)|: \delta \le \|t\| \le \ell^\lambda\} < 1$ for every $\delta > 0$.

Proof of Theorem 2.1: Note that $P_*(T_n^* \in B) = E_* f_n(T_n^*)$ where $f_n(x) = 1_B(x - B_n)$, $x \in \mathbb{R}^p$ and 1_B denotes the indicator function of the set B. Hence $\bar{w}(f_n, \eta; \Phi) = \Phi(B_n + (\partial B)^\eta) \le C\|B_n\| + \Phi((\partial B^\eta))$ for any $\eta > 0$. Consequently, by Lemma 5.2 and the definition of ψ_n,

$$\begin{aligned} P_*(T_n^* \in B) &= \int_B f_n d\psi_n + O(\|B_n\|) \\ &= \Phi(B + B_n) + O(\|B_n\| + n^{-1/2}) \quad \text{a.s.}(P). \end{aligned}$$

Also, by Theorem 2.8 of [GH],

$$P(T_n \in B) = \Phi(B) + O(n^{-1/2}).$$

Hence, it follows that a.s. (P)

$$P_*(T_n^* \in B) - P(T_n \in B) = \Phi(B + B_n) - \Phi(B) + O(\|B_n\| + n^{-1/2}).$$

This proves part (a) of Theorem 2.1.

Part (b) follows by a similar argument noting that there exists a constant $C_0 > 0$ such that $\mathrm{Sup}\{\bar{w}(1_{(-\infty,x]}(\cdot - y), \eta; \Phi) : x \in \mathbb{R}^p\} = C_0 \eta$ for all $\eta > 0$, and $y \in \mathbb{R}^p$.

As for part (c), note that by condition (C.1) (w.l.g., assuming $\mu_j = \mu = 0$ for all $j \ge 1$),

$$\bar{X}_n - \hat{\mu}_n = -(nb)^{-1}(\ell - 1)\sum_{n=1}^{j} X_j + \sum_{j=1}^{\ell-1} a_{nj}(X_j + X_{n-j+1})$$

$$= o_p(n^{-5/4}) + (\ell b)^{-1}\sum_{j=1}^{\ell-1}(\ell-j)(X_j+X_{n-j+1}).$$

where $a_{nj} = (\ell-j)(\ell b)^{-1}$, $1\le j\le \ell-1$.

Write $S_{n1} = \sum_{j=1}^{\ell-1} jX_j$, $S_{n2} = \sum_{j=1}^{\ell-1} jX_{n-j+1}$. Then by stationarity, $(\ell^3/3)^{-1}\mathrm{Cov}(S_{ni}) = M+o(1)$, for i=1,2 and S_{n2} has the same distribution as $S_{n3} = \sum_{j=1}^{\ell-1} jX_{\ell-j}$. By Lemma 27.2 of Billingsley (1986), conditions (C.2), (C.4) and arguments similar to those used in the derivation of (5.5), one has

$$|E\exp(it'(S_{n1}+S_{n2})) - \Pi_{i=1}^{2}E(\exp(it'S_{ni}))| = o(1) \qquad (5.9)$$

for every $t\in\mathbb{R}^p$. A 'big block - little block' argument together with conditions (C.2), (C.4), the Cramer-Wald device and the classical Lindeberg-Feller Theorem implies that $(\ell^3/3)^{-1/2}S_{ni}$ converges in distribution to $N_p(0,M)$ for i=1,3. Also, by equation (5.9), $(\ell^3/3)^{-1/2}S_{n1}$ and $(\ell^3/3)^{-1/2}S_{n2}$ are asymptotically independent. Hence it follows that $(\ell^3/3)^{-1/2}\sum_{j=1}^{\ell-1} j(X_j+X_{n-j+1})$ converges in distribution to $N_p(0,2M)$. Therefore, the assertion of part (c) follows from Lemma (5.2)(ii) and condition (C.1).

Proof of Theorem 2.2: Note that the signed measure ψ_n in Lemma 5.5 has density

$$(1 + k^{-1/2}\sum_{|\nu|=3} C(\nu)\,\tilde{\chi}_{\nu,n}\, q_\nu(x))\phi(x)$$

where $q_\nu(x)\phi(x) = (-D)^\nu\phi(x)$, and $\tilde{\chi}_{\nu,n} = E_*[\hat{A}_n(U_1^*-\mu_n)]^\nu$.

By Lemma 5.2, $\tilde{\chi}_{\nu,n} = \ell^{-1/2}n^2E(A_n(\bar{X}_n-\mu))^\nu + o(1)$ a.s. (P). Hence, Theorem 2.2 follows from Theorem 2.8 of [GH], Lemma 5.5 above and equation (2.1).

Proof of Theorem 3.1: Following the proofs of Theorem 1 of Bhattacharya (1985) or Theorem 2.2 of Bhattacharya and Ghosh (1978), one can show that the distribution of $\sqrt{n}(H(\bar{Y}) - H(\alpha))$ has the following expansion:

$$\sup_{B\in\mathcal{B}}|P(\sqrt{n}\Sigma_n^{-1/2}(H(\bar{Y}_n)-H(\alpha))\in B)-\int_B(1+n^{-1/2}g(x:\{\chi_{\nu,n}\}))\phi(x)dx|=o(n^{-1/2})$$

where g is a polynomial and its co-efficients are continuous functions of partial derivatives of H at α and of the cumulants $\{\chi_{\nu,n};|\nu| \le 3\}$ of $\bar{Y}_n$.

Also, with an observation as above, one can show that the bootstrapped statistic admits a similar expansion:

$$\sup_{B\in\mathcal{B}} |P_*(\sqrt{k}\ \hat{\Sigma}_n^{-1/2}(H(\bar{Y}_n^*) - H(\underset{\sim}{\alpha}_n))\in B) - \int_B (1 + n^{-1/2}g(x;\{\bar{\chi}_{\nu,n})\phi(x)dx|$$

$$= o(n^{-1/2}) \quad \text{a.s. } (P)$$

where $\bar{\chi}_{\nu,n}$ = νth cumulant of $\bar{Y}_n^*$. Theorem 3.1 follows from Lemma 5.2 (stated in terms of Y_i's).

<u>Proof of Theorem 3.2</u> W.l.g. assume that $\lambda(X_1,\ldots,X_{m+1})$ = $(X_1,X_1^2,\ldots,X_1X_{1+m})$. Define a function $H: \mathbb{R}^{m+2} \to \mathbb{R}$ by $H(y) = (y_1-\mu)/[f_0(y)]^{1/2}$, $y=(y_1,\ldots,y_{m+2}) \in \mathbb{R}^{m+2}$ with f_0 as in Section 3. Then, H has partial derivatives of all orders on the open set Q = $\{y\in \mathbb{R}^{m+2}: \sum_{r=0}^{m} a_r y_{r+2} > (2m+1)y_1^2\}$ containing $\underset{\sim}{\alpha}$ = $(\mu, EX_1^2, \ldots, EX_1X_{1+m})'$ and the vector D of first order partial derivatives at $\underset{\sim}{\alpha}$ is given by $D = [\tau,0\ldots0]$. Let $T_{4n} = \sqrt{n}H(\bar{Y}_n)$ where $\bar{Y}_n = n^{-1}\sum_{i=1}^{n}Y_i$ and $Y_i = (X_i, X_1^2,\ldots, X_iX_{i+m})$, $i\ge1$.

Then, using condition (C.2) and the m-dependence of X_i's, one can show that

$$P(|T_{3n} - T_{4n}| > n^{-7/12}) < Cn^{-7/12}$$

for some constant $C > 0$. Hence, by condition (2.1) and Theorem 3.1

(together with the m-dependence of X_i's), it follows that uniformly over $B \in \mathcal{B}$,

$$P(T_{3n} \in B) = P(T_{4n} \in B) + O(n^{-7/12})$$
$$= \int_B (1 + n^{-1/2} p(x, \{\mu_\nu\})) \phi(x) dx + o(n^{-1/2}) \tag{5.10}$$

where $p(x,\{\mu_\nu\})$ is a polynomial and its co-efficients are continuous functions of $\mu_\nu = EY_1^\nu$ for $|\nu| \le 3$, $\nu \in (Z^+)^{m+2}$. Since $\mathrm{Sup}\{|p(x,\{\mu_\nu\})\phi(x)| : x \in \mathbb{R}\} < \infty$ and $n^{-1/2} - (n-m)^{-1/2} = O(n^{-3/2})$, with $N=n-m$, relations (5.10) yields

$$\sup_{B \in \mathcal{B}} |P(T_{3n} \in B) - P(T_{4N} \in B)| = o(n^{1/2}). \tag{5.11}$$

Next, recall the definition of the block averages $\underset{\sim}{U}_i^*$, $1 \le i \le k_1 = [N/\ell]$. For notational similarity, set $\bar{Y}_N^* = \bar{U}_n^* = k_1^{-1}(\underset{\sim}{U}_1^* + \ldots + \underset{\sim}{U}_{k_1}^*)$.

Then it is obvious that the conclusions of Lemma 5.2 hold with X_i's replaced by Y_i's and U_i^*'s replaced by $\underset{\sim}{U}_i^*$'s. Next, define the function $H_n : \mathbb{R}^{m+2} \to \mathbb{R}$ by

$$H_n(y) = (y_1 - \mu_n)/[f_0(y)]^{1/2}$$

for $y = (y_1, \ldots, y_{m+2})' \in \mathbb{R}^{m+2}$. Then, for every $n > m$, H_n is infinitely differentiable on Q. Since $\underset{\sim}{\alpha}_n \to \underset{\sim}{\alpha}$ a.s. (P), it follows that a.s. (P), for sufficiently large n, $\underset{\sim}{\alpha}_n \in Q$ and H_n has partial derivatives of all orders at $\underset{\sim}{\alpha}_n$. Let D_n denote the vector of first order partial derivatives of H_n at $\underset{\sim}{\alpha}_n$. Then, trivially,

$$\hat{\Sigma}_n = D_n \, \mathrm{Disp}(\sqrt{k_1 \ell}\; \bar{Y}_n^*) D_n' = \tilde{\tau}_n^2$$

and $$\Sigma_n = D \, \mathrm{Disp}(\sqrt{n}\; \bar{Y}_n) D' = \tau_n^2/\tau^2, \text{ where } \tau_n^2 = \mathrm{Var}(\sqrt{n}\; \bar{X}_n).$$

Note that $T_{4N} = \sqrt{N}(H(\bar{Y}_N) - H(\underset{\sim}{\alpha}))$ and $T_{3n}^* = \sqrt{k_1 \ell}\, \hat{\Sigma}_n^{-1/2}(H_n(\bar{Y}_N^*) - H_n(E_* \bar{Y}_N^*))$.

Using arguments similar to those needed in the proof of Theorem 3.1, one can show that a.s. (P)

$$\sup_{B\in\mathcal{B}} |P(\Sigma_N^{-1/2}T_{4N}\in B) - P_*(T_{3N}^*\in B)| = o(n^{-1/2}). \tag{5.12}$$

Next observe that the m-dependence of X_i's implies $\tau_n^2-\tau^2 = O(n^{-1})$. Hence, by condition (2.1) and relation (5.10), one gets

$$\sup_{B\in\mathcal{B}} |P(T_{4N}\in B) - P(\Sigma_N^{-1/2}T_{4N}\in B)| = o(n^{-1/2}).$$

Theorem 3.2 now follows from (5.11) and (5.12).

Proof of Theorem 4.1: Follows from Lemmas 5.2–5.5.

References

Athreya, K.B. and Fuh, C.D. (1989): Bootstrapping Markov Chains: Countable Case. Preprint, Dept. of Stat., Iowa State University.

Babu, G.J. (1986). Bootstrapping Statistics with Linear Combinations of Chi-Squares as Weak Limit. Sankhya Ser A., **56**, 85-93.

Babu, G.J. and Singh, K. (1984). One-term Edgeworth Correction by Efron's Bootstrap. Sankhya Ser. A, **46**, 219-232.

Basawa, I.V., Green, T.A., McCormick, W.P. and Taylor, R.L. (1990) Asymptotic bootstrap validity for finite Markov chains. Preprint. Dept. of Statistics, University of Georgia.

Bhattacharya, R.N. (1985). Some Recent results on Cramer-Edgeworth Expansions with Applications. Multivariate Analysis - VI (Ed. P.R. Krishiah.), Elsevier Science Publishers B.V., 57-75.

Bhattacharya, R.N. and Ghosh, J.K. (1978). On the Validity of the Formal Edgeworth Expansion. Ann. Statist., **6**, 436-451.

Bhattacharya, R.N. and Ranga, Rao R. (1986). Normal Approximation and Asymptotic Expansions. R.E. Krieger Publishing Company, Malabar, Florida.

Billingsley, P. (1986). Probability and Measure, Wiley, New York.

Datta, S. and McCormick, W.P. (1990). Asymptotic accuracy of bootstrap for finite Markov chains. Technical Report, Dept. of Statistics, Univ. of Georgia.

Efron, B. (1979). Bootstrap Methods: Another Look at the Jackknife. Ann. Statist. **7**, 1-26.

Efron, B. and Tibshirani, R. (1986). Bootstrap Methods for Standard Errors, Confidence Intervals and other Measures of Statistical Accuracy (with discussion). Statist. Sci. **1**, 56-77.

Freedman, D.A. (1984). On Bootstrapping Two Stage Least Squares Estimates in Stationary Linear Models. Ann. Statist. **12**, 827-842.

Götze, F. and Hipp, C. (1983). Asymptotic Expansions for Sums of Weakly Dependent Random Vectors. Z. Wahrscheinlichkeitstheor. Verw. Geb. **64**, 211-240.

Götze, F. and Künsch, H.R. (1990). Blockwise bootstrap for dependent observation: Higher order approximations for studentized statistics. (Abstract). Institute of Mathematical Statistics bulletin. 19, 443.

Kulperger, R.J. and Prakasa Rao, B.L.S. (1987). Bootstrapping a finite state Markov chain. To appear in Sankhya.

Künsch, H.R. (1989). The Jackknife and the Bootstrap for General Stationary Observations. Ann. Statist. **17**, 1217-1241.

Lahiri, S.N. (1989). Bootstrap Approximation to the Distributions of M-estimators. Ph.D. dissertation, Michigan State University.

Lahiri S.N. (1990). Second Order Optimality of Stationary Bootstrap. To appear in Statist. Prob. Letters.

Liu, R.Y. (1988). Bootstrap Procedures Under Some Non-iid Models. Ann. Statist. 16, 1696-1708.

Liu, R.Y. and Singh, K. (1989). Using iid Bootstrap Inference for some Non-iid Models. Preprint. Department of Statistics, Rutgers University.

Liu, R.Y. and Singh, K. (1988). Moving Blocks Jackknife and Bootstrap Capture Weak Dependence. Preprint. Dept. of Statistics, Rutgers University.

Moore, M. and Rais, N. (1990). A bootstrap procedure for finite state stationary uniformly mixing discrete time processes. Preprint. Ecole Polytechnique de Montreal and Universite de Montreal.

Shi, X. and Shao, J. (1988). Resampling estimation when observations are m-dependent. Comm. Statist. A, 17 (11), 3923-3934.

Singh, K. (1981). On the Asymptotic Accuracy of Efron's Bootstrap. Ann. Statist. 9, 1187-1195.

Bootstrapping Signs

Raoul LePage
Michigan State University

Abstract. *One of the known limitations of the basic bootstrap is its sensitivity to heavy tails in the error distribution. Athreya (1987) studies bootstrap's failure to recover the sampling distribution of the sum of i.i.d. non-normal stable errors. Giné and Zinn (1989) prove that the same failure occurs essentially whenever i.i.d. errors are not subject to a central limit theorem (CLT) having a normal limit. This poses a practical difficulty, for how is one to know whether data meet such a requirement? Interestingly, in the case of symmetric errors, and probably much more generally, we can resolve this difficulty through a somewhat different application of Efron's (1979) bootstrap.*

In the case of symmetric error distributions, without the assumption of finite variance, we find that a simple bootstrapping of signs, instead of bootstrapping the observations themselves, approximately recovers the **conditional** *distribution of* $(\hat{\mu} - \mu)$ ***given*** $|\epsilon|$, *where* μ *denotes population location parameter,* $\hat{\mu}$ *is the average of the sample, and* $|\epsilon|$ *is the vector of the absolute values of the actual errors present in the data.*

Even for a seemingly routine application, where the symmetric error distribution is concentrated on an arbitrarily small interval around zero, the central limit theorem and bootstrapping residuals can be sub-optimal. Confidence intervals produced by bootstrapping signs can for some distributions be far narrower for small to moderate sample sizes.

1. Introduction.

Bootstrap has trouble recovering the distribution of a sum of i.i.d. long tailed errors ϵ_i based on observations $\mu + \epsilon_i$ (Athreya 1987, Giné and Zinn 1989). There is an idea which seems missing from previous attempts to

*Department of Statistics and Probability, Michigan State University, E. Lansing, MI 48824. Research partially supported by ONR grant N00014-91-J-1087.

Exploring the Limits of Bootstrap. Edited by Raoul LePage and Lynne Billard.
 ISBN: 0471-53631-8.

deal with this difficulty. I have not found it, for example, in Athreya (1987) or Knight (1989). Nor does it appear in Wu (1986) among the many topics relating to resampling methods in regression. This idea is that the presumed goal of bootstrap, recovery of the unconditional limit law of a statistic of interest, may be too modest. Perhaps we should seek to recover the conditional limit law given some characteristic of the errors.

Bootstrap confidence intervals, in cases where bootstrap appears to fail, are actually narrower than their unconditional counterparts for the same unconditional confidence level. Kinateder (1990) established an invariance principle linking bootstrap to the order statistics of the data, and presented simulations which show that for long tailed errors bootstrap confidence intervals based on the sample average tend to be smaller than those obtained from the unconditional limit law. The following histogram, taken from Figure 6.2 of Kinateder, is illustrative. It depicts the sampling distribution of $\log_e(\Delta)$ where 2Δ is the bootstrap confidence interval width (calibrated to have 0.95 unconditional confidence level) for samples of $n = 50$ from a symmetric distribution attracted to the Cauchy distribution. The vertical line in the figure points to the log-width of a .95-confidence interval based on the Cauchy limit law. The horizontal scale is about [-0.7, 7.0] (logarithmic). It can be seen that the sampling distribution of Δ places most of its mass far below the confidence interval width obtainable from the unconditional limit law (in this case Cauchy).

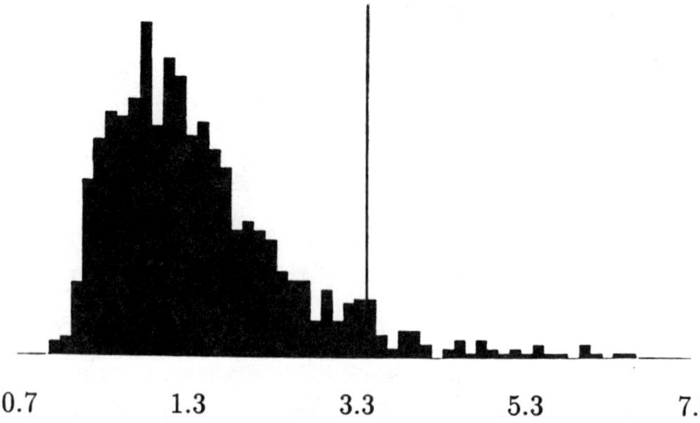

One is tempted merely to calibrate bootstrap by estimating the unconditional confidence level. This lacks simplicity but, more seriously, would appear to mask an unusual conditioning for which there is no sensible inferential interpretation.

Any effort to establish a truly conditional interpretation for bootstrap meets with one fundamental difficulty. The order statistics of a sample are approximately a Poisson point process (Lévy 1925; LePage, Woodroofe, and Zinn, 1981). Bootstrap, through its multinomial sampling of the data, approximately places i.i.d. mean-one Poisson counts on the order statistics of sample data points (Athreya 1987, Knight 1989, Kinateder 1990). Except in special circumstances, bootstrap resampling of a Poisson process is quite unlike the original, which is why bootstrap will give incorrect results for long tailed errors (see Proposition 3 below, comparing bootstrap to sign-bootstrap).

2. Bootstrapping signs.

Our problem is to use i.i.d. observations $Y_i = \mu + \epsilon_i$, $i \leq n$, to estimate the sampling distribution of $\hat{\mu} - \mu$, where $\hat{\mu} = \overline{Y}$. We consider the case of errors whose signs are randomly assigned by coin flips.

Assumption 1. The distribution of the error vector ϵ is identical to that of $\delta\epsilon$ defined by $\delta\epsilon = (\delta_1\epsilon_1, \ldots, \delta_n\epsilon_n)$, where $\delta = (\delta_1, \ldots, \delta_n)$ are i.i.d., independent of ϵ , and take the values ± 1 each with probability 0.5 . □

In may be convenient to interpret expressions like $\delta\epsilon$ as the product of a diagonal matrix, whose diagonal is δ, by a column vector ϵ . What we propose is very simple:

a) Find the conditional distribution $[\ \overline{\delta(Y-\overline{Y})}\ \mid Y\ ]$, of $\overline{\delta(Y-\overline{Y})}$ given Y.
b) Result (a) estimates the conditional distribution $[\ (\hat{\mu} - \mu) \mid |\epsilon|\ ]$.

Regarding (a) it can be noted that the Y-conditional characteristic function $E\ [\exp(\sqrt{-1}\ r\ \overline{\delta(Y-\overline{Y})}) \mid Y]$, involving only a linear expression in random signs, may be written as the product $\Pi_{i=1}^{n} \cos(r\ (Y_i-\overline{Y})/n)$. Fast Fourier transforms can be effectively used to determine the conditional distribution by inverting this characteristic function, addressing one of the Six Questions Raised by the Bootstrap (B. Efron, in this volume).

3. Examining the conditional distributions.

Proposition 1. If errors ϵ_i satisfy Assumption 1 (symmetry) then

$$[\,\overline{\delta(Y\text{-}\overline{Y})} \mid Y\,] = [\,\overline{\delta(Y\text{-}\overline{Y})} \mid \epsilon\,] = [\,\overline{\delta(\epsilon\text{-}\overline{\epsilon})} \mid \epsilon\,] \tag{3.1}$$

$$\approx [\,\overline{\delta\epsilon} \mid \epsilon\,] = [\,\overline{\epsilon} \mid |\epsilon|\,] = [\,(\hat{\mu} - \mu) \mid |\epsilon|\,], \tag{3.2}$$

where the approximation $\approx$ is justified if $\overline{\epsilon}^2$ is small compared with $\overline{\epsilon^2}$.

Proof. All relations are obvious, except perhaps the approximation $\approx$ in (3.2) which is justified as $n \to \infty$ provided $\overline{\delta}\ \overline{\epsilon}$ is small compared with $\overline{\delta\epsilon}$, conditionally on ϵ. Both $\overline{\delta}\ \overline{\epsilon} = (\ \overline{\delta(Y\text{-}\overline{Y})} - \overline{\delta\epsilon}\ )$ and $\overline{\delta\epsilon}$ have conditional mean zero, conditional on ϵ. Their conditional variances given ϵ are respectively $\overline{\epsilon}^2/n$ and $\overline{\epsilon^2}/n$. □

Often (see Lemma 1 below) a scaling $n^{-2/a}$ with $\alpha \leq 2$ will be required for distributional convergence of each of $(n\overline{\epsilon})^2$ and $n\overline{\epsilon^2}$. In such a case $\overline{\epsilon}^2/\overline{\epsilon^2}$ is of order $1/n$, which guarantees the relative smallness of the difference between bootstrapping signs of residuals versus bootstrapping signs on the actual errors present in the data.

Notice that one need not assume i.i.d. errors ϵ in Proposition 1. With the additional assumption of i.i.d. errors a far simpler comparison can be obtained.

Proposition 1.a. Define F to be the sigma algebra generated by the order statistics of the absolute values $|\epsilon_i|$. If errors ϵ_i satisfy Assumption 1 and are also i.i.d. then

$$E[\,(\ \overline{\delta(Y\text{-}\overline{Y})} - \overline{\delta\epsilon}\ )^2 \mid F\,] = \overline{\epsilon^2}/n^2, \tag{3.1.a.}$$

$$E[\,(\overline{\delta\epsilon})^2 \mid F\,] = \overline{\epsilon^2}/n, \tag{3.2.a.}$$

$$E[\,(\ \overline{\delta(Y\text{-}\overline{Y})}\ )\,(\overline{\delta}\overline{\epsilon}) \mid F\,] = 0.$$

Proof. These relations are a simple consequence of the fact that, conditional on *F*, the vectors δ and ϵ are independent; δ_i are conditionally i.i.d. signs; ϵ_i are conditionally uniform over all $2^n \times n!$ signed permutations of $|\epsilon_i|$. □

The relevance of relations (3.1.a.-3.2.a.) may be explained as follows. When attempting to assess the sampling variation of $\overline{Y}$ by bootstrapping signs, we would exactly recover the conditional distribution of the error term $\overline{\epsilon}$ given the vector $|\epsilon|$ except that we have thrown away a term $\overline{\delta}\ \overline{\epsilon}$. Relative to the conditioning *F*, that missing term has conditional mean zero and mean square (3.1.a.). On the other hand, the target $\overline{\epsilon}$ (equivalently $\overline{\delta\epsilon}$) has conditional mean zero and conditional mean square given by (3.2.a.). Thus the *F*-relative average magnitude of our error in using the sign-bootstrap approximation to the conditional sampling distribution of $\overline{\epsilon}$ given $|\epsilon|$ can be determined by taking the ratio of (3.1.a.) to (3.2.a.), which does not depend on ϵ and equals $1/n$.

Conditioning on *F*, as in Proposition 1.a., is particularly important for the regression and time-series counterparts of these results, since Proposition 1 does not extend to sums involving unequal coefficients on the errors, while Proposition 1.a. does have an extension to such sums.

Assumption 2. $Y = X\ \beta + \epsilon$ (a finite non-singular matrix regression model) in which the errors ϵ_i satisfy Assumption 1 (symmetry) and are i.i.d., and β is an unknown vector. □

Denote by $\hat{Y}$ the values fitted by linear least squares; $\hat{\epsilon}$ = linear least square fit to the errors; $\epsilon^{\perp} = Y - \hat{Y} = \epsilon - \hat{\epsilon}$. As before, let *F* denote the sigma algebra generated by the order statistics of the absolute values $|\epsilon_i|$. If v is any vector in R^n we denote by v/X its projection to the column space of X.

Proposition 2. Let Y, X, β, ϵ satisfy Assumption 2. For every vector v define (v, Y) = Cartesian inner product of Y with v in R^n. Then for every complete ortho-normal basis $\{u_k\}$ for the column space of X,

$$E\,[\,(\,(v,\,\delta\epsilon^{\perp}\,) - (v,\,\delta\epsilon)\,)^2\,|\,F\,] \;=\; \overline{\epsilon^2}\,(u^2,\,v^2), \tag{3.3}$$

$$E\,[\,(v,\,\delta\epsilon)^2 \mid F\,] \;=\; \overline{\epsilon^2}\,\|\,v\,\|^2, \tag{3.4}$$

$$E\,[\,(v,\,\delta\epsilon^{\perp})\,(v,\,\delta(\epsilon/X)) \mid F\,] \;=\; 0, \tag{3.5}$$

where $u^2 = \Sigma_k\, u_k^2$ does not depend upon the choice of c.o.n. $\{u_k\}$; the vectors u_k^2, v^2 being formed by squaring each coordinate.

Proof. Use the conditioning argument of Proposition 1.a. Let $\{u_k\}$ be any complete orthonormal basis for the column space of X. We make use of the fact that for every α, β,

$$E\,[\,(\epsilon,\,u_\alpha)\,(\epsilon,\,u_\beta) \mid F\,] \;=\; \Sigma_i\, u_{\alpha i}\, u_{\beta i}\, E\,[\,\epsilon_i^2 \mid F\,] \;=\overline{\epsilon^2}\,(\alpha=\beta), \tag{3.6}$$

where $(\alpha = \beta)$ denotes the indicator function. By expanding ϵ/X in terms of $\{u_k\}$ we obtain from (3.5) that for every i,

$$E\,[\,(\epsilon/X)_i^2 \mid F\,] \;=\; \Sigma_{\alpha\beta}\, u_{\alpha i}\, u_{\beta i}\, \overline{\epsilon^2}\,(\alpha=\beta) \;=\; \overline{\epsilon^2}\,\Sigma_k\, u_{ki}^2. \tag{3.7}$$

From (3.7) we obtain (3.3),

$$E\,[\,(\,(v,\,\delta\epsilon^{\perp}) - (v,\,\delta\epsilon)\,)^2 \mid F\,] = E\,[\,(\,(v,\,\delta(\epsilon/X)\,)^2 \mid F\,]$$

$$=\; \Sigma_i\, v_i^2\, E\,[\,(\epsilon/X)_i^2 \mid F\,] \;=\; \overline{\epsilon^2}\,\Sigma_k(u_k^2,\,v^2) \;=\; \overline{\epsilon^2}\,(u^2,\,v^2). \tag{3.8}$$

Expressions (3.4)-(3.5) have a similar proof. □

The main point of Proposition 2 is that, provided the ratio (3.3) to (3.4) (which only depends upon the *condition vector* u^2 associated with the design X) is small, the Y-conditional distribution of $(v,\,\delta\epsilon^{\perp})$ will, on F-average, be close to the conditional distribution of the sampling error $(v,\,\epsilon)$ given the absolute values $|\epsilon|$. The same method may be used to prove that,

$$E\,[\,\|\,(\delta\epsilon^{\perp})/X - (\delta\epsilon)/X\,\|^2 \mid F\,] \;=\; \overline{\epsilon^2}\,\Sigma_\alpha\,\Sigma_\beta\,(u_\alpha^2,\,u_\beta^2), \tag{3.9}$$

$$E\,[\,\|\,\delta\epsilon\,\|^2 \mid F\,] \;=\; \overline{\epsilon^2}\,d, \tag{3.10}$$

where $d > 0$ is the number of columns of X. Relations (3.3)-(3.10) also hold for bootstraping over all $2^n \times n!$ signed permutations of residuals.

To illustrate Proposition 2, consider using least squares linear regression to fit a cubic polynomial to $n = 101$ points with equally spaced x-values in the interval [0, 1], i.e. $x_1 = 0, \ldots, x_{101} = 1$. The ratios (3.3)/(3.4), computed for each of the four coefficients, are 0.0863, 0.0719, 0.0665, 0.0666. Suppose the actual "errors from the regression" are $\epsilon_i = (-1)^i\, i^2$, $0 \le i \le 101$. In the case of the first coefficient, the y-intercept, we find by computer calculations that bootstrapping signs gives a 0.95 confidence interval less than half the width of a confidence interval determined by either the CLT or by bootstrapping the residuals. Bear in mind that the relative widths of these three confidence interval types (bootstrap-signs of residuals, CLT, bootstrap residuals) are unchanged by scale change in the errors, so we could just as well be working with errors having the same distributional shape but concentrated arbitrarily close to the origin.

4. Performance on randomly signed powers of uniforms.

It is instructive to compare ordinary vs sign bootstraps for the case of the sum of i.i.d. symmetric errors having a distribution whose tail is a power. Such distributions are actually the prototype for errors attracted to a symmetric stable law. A simple Poisson process construction of such errors (correct for each n, but not for all n simultaneously) allows us to track and compare the performance of both bootstraps using almost sure arguments.

Lemma 1. Suppose $\{U_i\}$ are i.i.d. uniform on [0, 1]; $\epsilon_i = \delta_i' |\epsilon_i|$ with $\epsilon_i = U_i^p$ for a real number p, and δ_i' i.i.d as δ_i. All sequences of r.v. are independent of one another. Let γ_j = time of j-th arrival in a homogeneous Poisson process on $[0,\infty]$ having a mean of one arrival per unit time. Then for every n (but not for different n simultaneously) there is a standard construction of the increasing order statistics $U_{(i)} = \gamma_i / \gamma_{n+1}$, $i \le n$, in terms of which,

$$\bar{\epsilon} = \gamma_{n+1}^{-p}\, \overline{\delta' \gamma^p} \qquad \overline{\epsilon^2} = \gamma_{n+1}^{-2p}\, \overline{\gamma^{2p}}\,. \tag{4.1}$$

For $p < -1/2$ we have a.s. , as $n \to \infty$,

$$[\, n^{1+p}\, \overline{(\delta\text{-}\bar{\delta})Y} \mid Y\,] \text{ and } [\, n^{1+p}\, (\hat{\mu} - \mu) \mid |\epsilon|\,]$$
$$\to [\sum_{i=1}^{\infty} \delta_i\, \delta_i'\, \gamma_i^p \mid \{\delta_j'\, \gamma_j\}\,]\,. \tag{4.2}$$

Outline of Proof. Breiman (1968) is a good reference for the Poisson construction of uniform order statistics. The limits (4.2) are obtained by applying the three series theorem conditional on $\{\delta_j'\, \gamma_j\}$, or a martingale argument. For additional details see (LeP, 1980; L-W-Z, 1981). □

Only the case $p \le -1/2$ corresponds to a symmetric stable law, yet the other powers are of some interest for finite samples, before the central limit theorem takes hold. They are best examined from the conditional viewpoint of Proposition 1.a. since the limit law is uninteresting.

5. Symmetric errors attracted to a stable law. It is not surprising that the result (4.2) applies also to i.i.d. symmetric errors belonging to the domain of attraction of a stable law.

Proposition 3. The result (4.2) (with a scale constant on the right side) holds for the corresponding Poisson construction of symmetric errors $\{\epsilon_i,\ i \le n\}$ belonging to the domain of attraction of a stable law of index $0 < \alpha < 2$ (i.e. $p = -1/\alpha$ in the above) . Basic bootstrap (Kni, 1989) on the other hand gives,

$$[\, n^{1+p}\, (\overline{Y^*} - \overline{Y}) \mid Y\,] \to \text{const} \times [\sum_{i=1}^{\infty} \pi_i\, \delta_i'\, \gamma_i^p \mid \{\delta_j'\, \gamma_j\}\,]\,, \tag{5.1}$$

where, as is usual, $\overline{Y^*}$ denotes the bootstrap average, and $\{\pi_i\}$ are i.i.d. Poisson r.v., having expectation one, and independent of δ, etc. .

In the case $p \ge -1/2$ the conditioning vanishes in the limit.

Remarks on the Proof. The proofs follow by application of an invariance principle (Kinateder, Theorem 3., 1990; also in this volume), or by direct arguments along the lines of lemma 1. The form of the limit law in (5.1) is due to Knight (1989), who does not address the problem of developing a consistent bootstrap. Athreya (1987) obtained a characterization of the unconditional limit law (5.1) as a random distribution. □

The following figure is a plot of confidence interval half-widths for different samples of n = 30 from a standard normal distribution. The CLT half-width is 0.358. The pairs being plotted are y = confidence interval half-width obtained from bootstrapping signs; x = confidence interval half-width obtained by the same monte-carlo simulation applied instead to the signs of the actual errors in the data. Sampling variation is biased upward due to the use of only 500 passes in each bootstrap, but the point being made is only that the approximation of lemma 1 is operating very well. That is, bootstrapping signs is approximately recovering the conditional sampling distribution given the absolute values of the residuals.

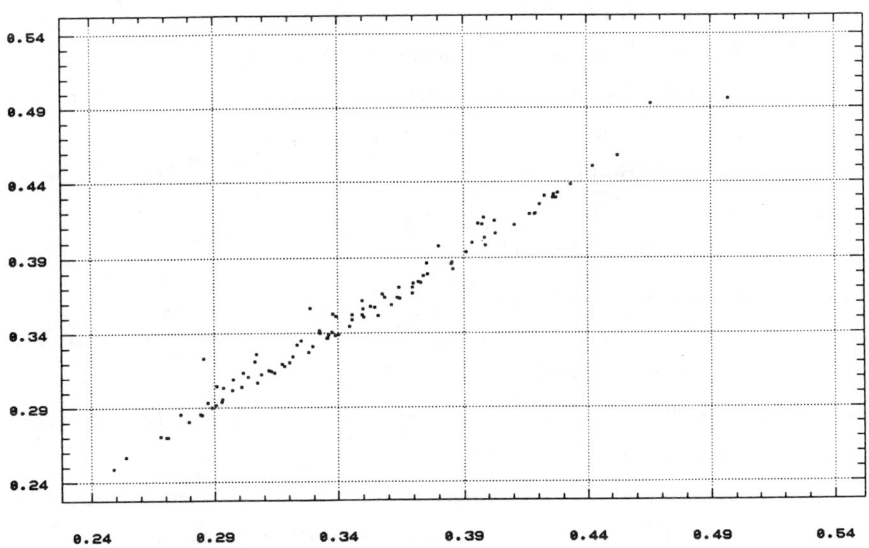

6. **Comments.** Symmetry does not appear to be essential for the basic principle. Rather, for symmetric errors we more easily see how to use resampling to change the object of our inference from the unconditional limit law to a conditional limit law likely to have narrower confidence limits

since E [(v, Y) | F] ≡ (v, E Y) (conditionally unbiased). This idea, appropriately modified, seems to extend to time series and other settings.

The conditional interpretation that we offer for bootstrapping signs appears to be related to the concept of second-order efficiency of bootstrap introduced in (Singh, 1981) for the case of finite second moments, since pivoting adjusts bootstrap to the variance of the sample. Bootstrapping signs can produce dramatic results, even in problems such as the regression cited earlier where one might expect the CLT to do reasonably well. In cases where consistent bootstrap methods, such as bootstrapping signs, are available, the way is open to compare different estimators based on bootstrap estimates of their respective sampling errors.

References

Athreya, K. B. (1987), Bootstrap of the mean in the infinite variance case. *Annals of Statistics,* **15**:724-731.

Breiman, L. (1968), *Probability.* Addison-Wesley, New York.

Bickel, P. J. and Freedman, D. A. (1981), Some asymptotic theory for the bootstrap. *Annals of Statistics,* **9**:1196-1217.

Efron, B. (1979), Bootstrap methods: Another look at the jackknife. *Annals of Statistics,* **7**:1-26.

Gine´, E. and Zinn, J. (1989), Necessary conditions for the bootstrap of the mean. *Annals of Statistics,* **17**:684-691, 1989.

Knight, K. (1989), On the bootstrap of the sample mean in the infinite variance case. *Annals of Statistics,* **17**:1168-1175.

Kinateder, J. (1990), An invariance principle applicable to the bootstrap. *Thesis,* Michigan State University.

LePage, R., Woodroofe, M. and Zinn, J. (1981), Convergence to a stable distribution via order statistics. *Annals of Probability,* **9**:624-632.

LePage, R. (1980), Reprinted in: Conditional moments for coordinates of stable vectors. *Lecture Notes in Mathematics,* **1391**:148-163.

Singh, K. (1981), On the asymptotic accuracy of Efron's bootstrap. *Annals of Statistics,* **9**:1187-1195.

Wu, C. F. (1986), Jackknife, bootstrap, and other resampling methods in regression analysis. *Annals of Statistics,* **14**:1261-1343.

Moving Blocks Jackknife and Bootstrap Capture Weak Dependence*

Regina Y. Liu and Kesar Singh
Rutgers University**

Abstract

Let $X_1,\ldots,X_n$ be a sequence of stationary m–dependent random variables with the common distribution F for each X_i. Let T_F be the parameter of interest and T_n be its considered estimator based on $(X_1,X_2,\ldots,X_n)$. To study the sampling distribution of T_n, we introduce two resampling procedures — the moving blocks jackknife (MBJ) and the moving blocks bootstrap (MBB). These two procedures resample from the moving blocks $B_1,\ldots,B_{n-b+1}$, where b is the size of each block and B_j stands for the block consisting of b consecutive observations starting from X_j, i.e., $B_j = \{X_j,X_{j+1},\ldots,X_{j+b-1}\}$. For the MBJ, pseudo–values are generated by deleting each of such blocks. As for the MBB, k i.i.d. block samples are drawn from $B_1,\ldots,B_{n-b+1}$. All elements in the k sampled blocks are then regarded as the bootstrap sample. Both MBJ and MBB tend to make partial correction for the m–dependence if b is a fixed positive integer. Furthermore, they both achieve the actual consistency under m–dependence if b is allowed to grow to infinity with n.

*This research was supported by NSF grants DMS–85–0294, DMS–88–02558.

**Department of Statistics, Hill Center, Rutgers University, New Brunswick NJ 08903.

Exploring the Limits of Bootstrap. Edited by Raoul LePage and Lynne Billard.
 ISBN: 0471-53631-8.

1. Introduction.

A sequence of time dependent observations tend to be dependent, unless the time indices are sufficiently far apart. The notion of m–dependence is probably the most basic model which takes into account such dependence. Let $\{X_1,X_2,...\}$ be a sequence of r.v.s (random variables). Let A and B be two events such that A depends on $\{X_1,...,X_k\}$ and B on $\{X_{k+m+1},X_{k+m+2},...\}$. The sequence $\{X_i\}$ is called m–dependent if any such pair of events A and B are independent. The sequence $\{X_i\}$ is called stationary if the distribution of any subcollection of X_i's does not change if each of the indices are shifted by the same amount. For instance, (X_1,X_3,X_6) has the same distribution as (X_{10},X_{12},X_{15}) under the stationarity. Consider the variance of the sum $\Sigma_{i=1}^n X_i$. If $\{X_i\}$ is stationary and m–dependent, $m \leq n$, then

$$\frac{1}{n} \operatorname{Var}\left(\sum_{i=1}^{n} X_i\right) = \operatorname{Var}(X_1) + 2\sum_{j=1}^{m}\left(1-\frac{j}{n}\right)\operatorname{cov}(X_1,X_{1+j}).$$

Thus, $\sigma_n^2 \equiv \frac{1}{n}\operatorname{Var}(\Sigma_{i=1}^n X_i) \to \sigma_\infty^2 \equiv \operatorname{Var}(X_1) + 2\Sigma_{j=1}^m \operatorname{cov}(X_1,X_{1+j})$ as $n \to \infty$. It is well known that $\sqrt{n}(\bar{X}-\mu)$ is asymptotically normal with mean 0 and variance σ_∞^2 (see p.174 of Billingsley (1968) for a more general result). Here $\bar{X} = \frac{1}{n}\Sigma_{i=1}^n X_i$ and $\mu = EX_1$. In general, if a statistic T_n is asymptotically $N(0, \operatorname{Var} g_F(X_1))$ under i.i.d. models, then it is $N(0, \sigma_\infty^2(T_n))$ under stationary m–dependent models, where

$$\sigma_\infty^2(T_n) = \operatorname{Var}(g_F(X_1)) + 2\sum_{i=1}^{m}\operatorname{cov}(g_F(X_1), g_F(X_{1+i})).$$

Here F is the c.d.f. of X_1. For instance, for the sample median, $g_F(X_1) = I(X_1 \leq F^{-1}(\frac{1}{2}))/F'(F^{-1}(\frac{1}{2}))$, and for the sample variance $g_F(X_1) = (X_1-\mu)^2$.

More general models of time dependence are φ–mixing and α–mixing. In these two models the dependence, in two different senses, is assumed to decrease to zero as time gap tends to infinity. All these models are generally referred to as weak dependence models. The asymptotics on weakly dependent r.v.s have been extensively studied. (See Billingsley (1968) for results on φ–mixing r.v.s.)

In this paper we propose a jackknife procedure and a bootstrap procedure for estimating sampling errors of general statistics which are consistent under m–dependent models. First of all, we examine why the classical jackknife and bootstrap procedures are inconsistent. Consider the simplest case of the sample mean $\bar{X}_n$. Let F_n be the empirical distribution based on $X_1,\ldots,X_n$. The classical bootstrap procedure is to draw a random sample of size n from F_n. Denote the bootstrapped sample by $Y_1,\ldots,Y_n$. The sampling distribution of $\sqrt{n}\,(\bar{X}_n - \mu)$ is then estimated by that of $\sqrt{n}(\bar{Y}_n - \bar{X}_n)$ under F_n, where $\bar{Y}_n$ is the mean of the bootstrap sample. The variance of $\sqrt{n}(\bar{Y}_n - \bar{X}_n)$ under F_n is $s_n^2 = \frac{1}{n}\Sigma_{i=1}^n(X_i - \bar{X}_n)^2$ which converges to $\mathrm{Var}_F(X_1)$. This can be seen by writing

$$s_n^2 = \frac{1}{n}\sum_{i=1}^n X_i^2 - (\bar{X}_n)^2$$

and using the law of large numbers for stationary m–dependent r.v.s. This inconsistency of the classical bootstrap was also mentioned in Remark of Singh (1981). Since $Y_1,\ldots,Y_n$ are i.i.d. observations from F_n, it is obvious that this bootstrap completely ignores the dependence structure of X_i's. Consequently under m–dependent models, the achieved asymptotic variance is incorrect. As for the classsical jackknife procedure, the estimate of the variance of $\sqrt{n}(\bar{X}_n - \mu)$ is $\frac{1}{n}\Sigma_{i=1}^n(J_i - \bar{X}_n)^2$, where J_i's are pseudo–values; i.e., $J_i = n\bar{X} - (n-1)\bar{X}_{n,-i} = X_i$. Here $\bar{X}_{n,-i} = \frac{1}{n-1}\sum_{j=1,\, j\neq i}^{n} X_j$ and $\bar{J}_n$ is the mean of $J_1,\ldots,J_n$. This estimate equals s_n^2 and hence also converges to $\mathrm{Var}_F(X_1)$. Thus the classical delete–1 jackknife is also inconsistent.

Now we describe the proposed bootstrap and jackknife procedures. Let B_i denote the block of b consecutive sample observations beginning from the ith one, i.e., $B_i \equiv \{X_i,X_{i+1},\ldots,X_{i+b-1}\}$, $i = 1,2,\ldots,n-b+1$. Let T_F indicate the parameter of interest and T_n the estimate of T_F based on $X_1,X_2,\ldots,X_n$.

A) <u>Moving Blocks Jackknife (MBJ)</u>.

We delete one of the moving blocks, say B_i, and then compute the functional T based on the remaining observations $\{X_1,\ldots,X_n\}\backslash B_i$. The latter is denoted by $T_{n,-i}$. For each $i = 1,\ldots,n-b+1$, the following pseudo–value is formed.

$$J_i = b^{-1}[nT_n - (n-b)T_{n,-i}]. \tag{1.1}$$

The proposed estimate of the variance of $\sqrt{n}T_n$ is

$$\hat{V}_{J,b}(T_n) \equiv b[\frac{1}{n-b+1} \sum_{i=1}^{n-b+1} (J_i - T_n)^2]. \tag{1.2}$$

At the first glance the MBJ procedure proposed here may seem somewhat to resemble the so–called delete–d jackknife. (See Miller (1974) for a general review on the jackknife.) However, they are intrinsically different. In fact, a delete–d version of MBJ would be to delete d of the moving blocks at a time, for all combinations of d such blocks.

B) Moving Blocks Bootstrap (MBB).

Under the MBB scheme we resample with replacement from the moving blocks $B_1,\ldots,B_{n-b+1}$ with B_i's equally likely to be drawn. Suppose we draw k blocks and denote the resulting sampled blocks by $\xi_1,\ldots,\xi_k$. Note that each ξ_j contains b elements which may be expressed as $\xi_j \equiv (\xi_{j1},\ldots,\xi_{jb})$. Let $\ell = kb$. We shall refer to $(Y_1,\ldots,Y_\ell) \equiv (\xi_1,\xi_2,\ldots,\xi_k)$ as the bootstrap sample under the MBB scheme. Here $Y_a \equiv \xi_{ji}$, if $a = (j-1)b + i$. For instance, with $b = 2$, $i = 3$, if $\xi_1 = (6,8)$, $\xi_2 = (4,7)$ and $\xi_3 = (7,2)$, then the bootstrap sample of size 6 is (6,8,4,7,7,2). Let F_ℓ^* denote the e.d.f. based upon $(Y_1,\ldots,Y_\ell)$, and let T_ℓ^* be the statistic computed from the bootstrap sample. In order to estimate the sampling distribution of $\sqrt{n}[T_n - T_F]$ under the MBB scheme, we may obtain a large number of bootstrap values $\sqrt{\ell}[T_\ell^* - T_n]$ to form a histogram. This histogram is the proposed bootstrap approximation of the sampling distribution of $\sqrt{n}[T_n - T_F]$.

To understand how MBJ and MBB reflect the weak dependence structure, we first focus on the case of sample mean with block size $b = 2$. The jackknife variance with $b = 2$ can be expressed as

$$\hat{V}_{J,2} = \frac{1}{n-1} \sum_{j=1}^{n-1} (\frac{X_j+X_{j+1}}{\sqrt{2}} - \sqrt{2}\,X_n)^2.$$

Let $U_j = (X_j+X_{j+1})/\sqrt{2}$, $j = 1,\ldots,n-1$, then U_j's are stationary and (m+1)–dependent. The average of $U_1,\ldots,U_{n-1}$ turns out to be $\sqrt{2}\,\bar{X} + O_p(n^{-1})$ since

$$\frac{1}{n-1}\sum_{j=1}^{n-1}\frac{X_j+X_{j+1}}{\sqrt{2}} = \sqrt{2}\,\bar{X} - \frac{1}{\sqrt{2}\,(n-1)}\,(X_1 + X_n).$$

This shows that $\hat{V}_{J,2}(\bar{X})$ is essentially the sample variance of $U_1,\ldots,U_{n-1}$. Consequently, the jackknife variance converges to $\mathrm{Var}(\frac{X_1+X_2}{\sqrt{2}})$. Now, we examine the estimated variance of $\bar{X}$ under the MBB scheme with $b = 2$. Let $S(\xi_i)$ indicate the sum of the components of the ith sampled block ξ_i. Then

$$\mathrm{Var}^*\,(\sqrt{\ell}\,\bar{Y}_\ell) = \ell \cdot \mathrm{Var}^*\,(\ell^{-1}[S(\xi_1) +\ldots+ S(\xi_k)]).$$

Here $\mathrm{Var}^*(\,\cdot\,)$ stands for the variance of $\cdot$ under the MBB procedure. Since $S(\xi_i)$, $i = 1,2,\ldots,k$ are i.i.d. under the MBB probability, the expression above reduces to $\mathrm{Var}^*\,(S(\xi_1)/\sqrt{2})$ which is the sample variance of U_j's. Therefore, this estimated variance also converges to $\mathrm{Var}\,((X_1+X_2)/\sqrt{2})$. By extending these arguments, one immediately concludes that for a general fixed b, both $\hat{V}_{J,b}(\bar{X}_n)$ and $\mathrm{Var}^*\,(\sqrt{\ell}\,\bar{Y}_\ell)$ converge to

$$\mathrm{Var}\,((X_1 + X_2 + \ldots + X_b)/\sqrt{b}).$$

The latter variance converges to the true asymptotic variance of $\sqrt{n}\,\bar{X}_n$ as $b \to \infty$.

The above discussion suggests that, for the case of sample mean, MBJ and MBB will make partial correction for the weak dependence when b is considered fixed. To clearly demonstrate this point, we consider the m–dependent model with minimum m, i.e., $m = 1$. The true asymptotic variance in this case is

$$\mathrm{Var}\,(X_1) + 2\,\mathrm{cov}\,(X_1,X_2).$$

The achieved asymptotic variance by MBJ and MBB with $b = 2$, 3 and 4 are, respectively,

$$\text{Var}(X_1) + \text{cov}(X_1,X_2),$$
$$\text{Var}(X_1) + \tfrac{4}{3}\,\text{cov}(X_1,X_2),$$
and
$$\text{Var}(X_1) + \tfrac{3}{2}\,\text{cov}(X_1,X_2).$$

Thus, with $b = 2$ we capture 50% of the dependence factor, namely $2\,\text{cov}(X_1,X_2)$. The fraction rises to 66% and 75% with $b = 3$ and 4, respectively. Therefore for practical purposes, replacing the classical jackknife or bootstrap by the moving blocks ones will amount to robustizing these procedures against suspected serial dependence. As for asymptotic theory, it is shown that the actual consistency is achieved under m–dependent models by the proposed moving blocks resampling procedures if b is allowed to grow to infinity with n.

Turning to general statistics, usually if a statistic T_n is asymptotically normal, it can be represented as

$$\frac{1}{n}\sum_{1}^{n} g_F(X_i) + o_p(n^{-1/2}),$$

where F is the c.d.f. of X_1 and $E\, g_F(X_1) = 0$. This comment applies under i.i.d. models as well as weak dependence models. When X_i's are i.i.d. the pseudo– values of the standard delete–1 jackknife are $g_F(X_i)$, $i = 1,2,\ldots,n$, with a remainder of order $o(1)$ uniformly in i (see Parr (1985) and Singh and Liu (1990)). It turns out that the ith pseudo–value of the MBJ procedure is

$$b^{-1}\sum_{j=i}^{i+b-1} g_F(X_j)$$

plus a remainder term of order $o(1)$ uniformly in i. Thus the foregoing discussion on the variance estimate of the sample mean by the MBJ carries over to general statistics, with X_j replaced by $g_F(X_j)$. Similar reasoning applies to the MBB. It follows from asymptotic representations that the classical bootstrap on T_n is equivalent, up to the first order limit, to bootstrapping the sample mean $n^{-1}\Sigma_1^n g_F(X_i)$ (see Liu, Singh and Lo (1989)). Evidently, the classical bootstrap gives the asymptotic variance $\text{Var}(g_F(X_1))$, which is incorrect under m–dependent models. We establish in Section III some representations which assert that the MBB on a "smooth" T_n is equivalent to that on the mean $\frac{1}{n}\Sigma g_F(X_i)$. This leads to the validity of the MBB for general statistics. In the course of obtaining these representations, we prove an extension of the Dvoretzky–Kiefer–Wolfowitz inequality for the empirical d.f. F^*_ℓ of the MBB sample. This inequality in turn implies that $\|F^*_\ell - F_n\|_\infty = O_p(k^{-1/2})$, in the MBB probability. This particular fact enables us to neglect second order terms in pursuing the asymptotic result.

We present the results for MBJ and MBB separately in Sections 2 and 3. For each scheme we shall discuss successively the case of sample mean and the case of more general statistics T_n. In each case, b starts out as a fixed number and later on it is allowed to tend to infinity as n tends to infinity. The case when b is a fixed number is technically more complete than the case when b tends to infinity with n. Indeed, the arguments are much more involved in the latter case and we have not attempted to include the most general results in order not to obscure the main ideas. Section 4 contains some concluding remarks and possibilities for further investigations.

After this manuscript was written (March, 1988), we were informed that the paper Hans Kunsch (1989) was in the process of publication in the Annals of Statistics. Though the same resampling methods were proposed in both papers, the techniques used differ substantially.

2. Moving Blocks Jackknife (MBJs).

Case (i) $T_n = \bar{X}_n$.

In this case, (1.1) and (1.2) can be expressed explicitly as

$$J_i = b^{-1} \sum_{j=i}^{i+b-1} X_j \qquad \text{for } i = 1,\dots,n-b+1 \tag{2.1}$$

and

$$\hat{V}_{J,b}(\bar{X}_n) = b\left[\frac{1}{n-b+1} \sum_{i=1}^{n-b+1} \left(b^{-1} \sum_{j=i}^{i+b-1} X_j - \bar{X}_n\right)^2\right]. \tag{2.2}$$

The following two theorems provide the asymptotic limit of $\hat{V}_{J,b}(\bar{X}_n)$ in the two situations: b is fixed and b tends to infinity.

Theorem 1. Let $X_1,\dots,X_n$ be a stationary m–dependent sequence of r.v.s with $EX_i^2 < \infty$. For any fixed integer $b \geq 1$, we have

$$\hat{V}_{J,b}(\bar{X}_n) \to \sigma_b^2 \equiv \frac{1}{b} \operatorname{Var}\left(\sum_{j=1}^{b} X_j\right) \quad \text{a.s.} \quad \text{as } n \to \infty.$$

Proof: (with $b = 2$) Without loss of generality, we assume $E X_1 = 0$. Note that

$$\hat{V}_{J,b}(\bar{X}_n) = \frac{1}{n-1}\sum_{i=1}^{n-1}\left(\frac{X_i + X_{i+1}}{\sqrt{2}} - \sqrt{2}\,\bar{X}_n\right)^2$$

$$= \frac{1}{n-1}\sum_{i=1}^{n-1}\frac{(X_i + X_{i+1})^2}{2} + 2(\bar{X}_n)^2$$

$$- 2\bar{X}_n \frac{1}{n-1}\sum_{i=1}^{n-1}(X_i + X_{i+1}).$$

It is easy to see that $(X_i + X_{i+1})$, $i = 1,2,\ldots$, is a stationary (m+1)–dependent sequence of r.v.s. Therefore, the claim of the theorem follows using the SLLN for m–dependent r.v.s (see the Appendix).

If b is treated as a sequence depending on n and if it is allowed to tend to infinity at some proper rate, then the following theorem shows that MBJ provides a consistent estimate of the variance of $\sqrt{n}\,\bar{X}_n$.

Theorem 2. Let $X_1,\ldots,X_n$ be a stationary m–dependent sequence of r.v.s with $EX_1^4 < \infty$. Let $b \equiv b_n$, $b \to \infty$ and $b/n \to 0$ as $n \to \infty$. Then

$$\hat{V}_{J,b}(\bar{X}_n) \to \sigma_\infty^2 \equiv \mathrm{Var}\,(X_1) + 2\sum_{j=1}^{m}\mathrm{cov}\,(X_1, X_{1+j}), \quad \text{in prob. as } n \to \infty.$$

Note that σ_∞^2 is the asymptotic variance of $\sqrt{n}\,\bar{X}_n$.

Proof: First we express $\hat{V}_{J,b}(\bar{X}_n)$ as

$$\frac{1}{n-b+1}\sum_{i=1}^{n-b+1}(\tilde{B}_i - \sqrt{b}\,\bar{X}_n)^2 \tag{2.3}$$

where $\tilde{B}_i = b^{-1/2}\sum_{j=i}^{i+b-1} X_j$. Since $E\,\tilde{B}_i = b^{1/2}\mu$, expression (2.3) can be rewritten as

$$\frac{1}{n-b-1}\sum_{i=1}^{n-b+1}[\tilde{B}_i - E\,\tilde{B}_i - \sqrt{b}\,(\bar{X}_n - \mu)]^2. \tag{2.4}$$

Clearly, $(\tilde{B}_i - E\,\tilde{B}_i)$, $i = 1,2,\ldots,n-b+1$ is a stationary (m+b)–dependent sequence of r.v.s. We observe that the splitting technique mentioned in the

Appendix for m–dependent r.v.s is not applicable here for getting the law of large numbers when $(m+b) \to \infty$ as $n \to \infty$. Next we express (2.4) further as $A_n + D_n - 2C_n$ with

$$A_n = \frac{1}{n-b+1} \sum_{i=1}^{n-b+1} (\breve{B}_i - E\breve{B}_i)^2,$$

$$D_n = b(\bar{X}_n - \mu)^2,$$

$$C_n = b^{1/2}(\bar{X}_n - \mu) \cdot \frac{1}{n-b+1} \sum_{i=1}^{n-b+1} (\breve{B}_i - E\breve{B}_i).$$

The limit of each of the above terms is established separately to prove the claim of the theorem. We show that $A_n \to \sigma_\infty^2$ in probability as $n \to \infty$. The fact that C_n converges to zero in probability follows from similar arguments. The term D_n converges to zero in probability in view of the bound $\bar{X}_n - \mu = O_p(n^{-1/2})$. (In fact, this bound also comes into use in showing that $C_n \to 0$ in probability.)

Note that

$$\text{Var}(A_n) = (n-b+1)^{-2}\{(n-b+1)\,\text{Var}\,[(\breve{B}_1 - E\breve{B}_1)^2]$$

$$+ 2 \sum_{i=1}^{m+b} (n-b+1-i)\,\text{cov}\,((\breve{B}_1 - E\breve{B}_1)^2, (\breve{B}_{1+i} - E\breve{B}_{1+i})^2)\}.$$

The last expression is bounded above by

$$(n-b+1)^{-2}[(n-b+1) + 2(m+b)(n-b+1)]\,\text{Var}\,((\breve{B}_1 - E\breve{B}_1)^2). \qquad (2.5)$$

Now,

$$\text{Var}\,(\breve{B}_1 - E\breve{B}_1)^2) \le E((\breve{B}_1 - E\breve{B}_1)^4) = b^{-2}\, E[\sum_{i=1}^{b} (X_i - \mu)]^4$$

which is of order $O(1)$ as $n \to \infty$, provided $EX_1^4 < \infty$. (See the Appendix.) Applying this particular observation and the condition $b/n \to 0$ in (2.5), we obtain that $\text{Var}\,(A_n) \to 0$ as $n \to \infty$. Using Chebyshev inequality it is easy to show that $\|A_n - \text{Var}(\tilde{B}_1)\| \to 0$ in probability. The result now clearly follows since $\text{Var}(\tilde{B}_1) \to \sigma_\infty^2$ in probability.

Case (ii) General Statistics T_n.

In order to study the asymptotic behavior of $\hat{V}_{J,b}(T_n)$ given in (1.2), we first establish a representation theorem for the pseudo–values J_i (given in (1.1)) when $T_F \equiv T(F)$ is a "sufficiently smooth" functional, and T_n is simply $T(F_n)$. The representation enables us to view T_n as a sample mean, so the arguments used for Case (i) can be adapted readily. As far as for the smoothness of $T(\cdot)$ is concerned, we choose to discuss the consistency under the so–called strong Frechét differentiability and twice Frechét differentiability. These conditions are satisfied by many commonly used statistics and they turn out to be "just right" in the theoretical development of the jackknife (see Parr (1985) for examples on the strong Frechét differentiability).

Definition 1. A functional $T(\cdot)$ is strongly Frechét differentiable at F if, for some function g_F,

$$T(G) - T(H) = \int g_F \, d(G - H) + o(\|G - H\|)$$

as G and H both converge to F in the sup norm, $\|\cdot\|$.

Definition 2. A functional $T(\cdot)$ is twice Frechét differentiable at F if

$$T(G) - T(F) = \int g_F \, d(G - F) - \int\int h_F(x,y) \, d[G(x) - F(x)] \, d[G(y) - F(y)] + o(\|G - F\|^2)$$

as G converges to F in the sup norm, $\|\cdot\|$, for some functions $g_F(\cdot)$ and $h_F(\cdot,\cdot)$. (W.L.O.G., we may assume $\int g_F \, dF = 0$.)

In both the definitions stated above, F, G and H may be restricted to some suitable class of distributions in order to achieve the differentiability in some restricted sense. For instance, for technical reasons one may restrict oneself to the distributions with compact supports.

We are now ready to state the representations for $T_n = T(F_n)$ under above differentiability conditions. Note that $T_{n,-i} = T(F_{n,-i})$ where $F_{n,-i}$ is the empirical d.f. based on the data $\{X_1, X_2, \ldots, X_n\} \backslash B_i$.

Theorem 3. Let $X_1,...,X_n$ be a stationary m–dependent sequence of r.v.s with the common distribution F. If the functional $T(\cdot)$ is strongly Frechét differentiable at F with the differential $g_F(\cdot)$, then

$$R_n \equiv \max_{1 \le i \le n-b+1} |J_i - T(F) - b^{-1} \sum_{j=i}^{i+b-1} g_F(X_j)|$$

tends to zero a.s. as $n \to \infty$. Here b is assumed to be a fixed positive integer.

To prove Theorem 3 we note that if we take $G = F_n$ and $H = F_{n,-i}$ in Definition 1, then $\|G - H\| = O(b/n)$. This is the key element in the proof. The rest of the proof is straightforward and it is therefore omitted.

Theorem 3 immediately implies the following corollary regarding the estimation of the variance of T_n under MBJ procedure.

Corollary 1. Under the same condition as in Theorem 3 and the additional moment condition $Eg_F^2(X_1) < \infty$, we have

$$\hat{V}_{J,b}(T_n) \to b^{-1} \operatorname{Var}(g_F(X_1) + ... + g_F(X_b)) \quad \text{a.s. as } n \to \infty.$$

When the block size b is allowed to go to infinity, we are able to establish Theorem 4 below. The strongly Frechét differentiability falls short of the required smoothness for the sake of achieving the representation similar to the one stated in Theorem 4 for the case $b \to \infty$ as $n \to \infty$. This is simply because in this situation $\|F_{n,-i} - F_n\|$ is of the order $O(b/n)$, and hence the difference

$$[T(F_{n,-i}) - T(F_n)] - \int g_F(x)\, d[F_{n,-i}(x) - dF_n(x)]$$

is of the order $o(b/n)$. To yield a useful representation, this remainder should rather be $o(\sqrt{b}/n)$. Therefore, when b is allowed to tend to infinity, we instead work with the twice Frechét differentiability (as stated in Definition 2). Note that these two differentiability notions are not directly comparable with each other (see the example on p.8 of Singh and Liu (1990)). We also remark here that Theorem 4 itself includes the case of b being a fixed integer. It is possible to relax the moment conditions to some extent if b is restricted to be a fixed integer specifically.

Theorem 4. Let $T(\cdot)$ be twice Frechét differentiable at F as stated in Definition 2 with $\int g_F \, dF = 0$. Assume

$$\int h_F^{4/3}(x,x)\, dF(x) < \infty \text{ and}$$
$$\int\int h_F^4(x,y)\, dF(x)\, dF(y) < \infty.$$

If $b^3/n \to 0$ as $n \to \infty$, then

$$J_i = T(F) + b^{-1} \sum_{j=i}^{i+b-1} g_F(X_j) + \gamma_i,$$

where $\max_{1 \le i \le n-b+1} \sqrt{b}\,|\gamma_i| \to 0$ in probability as $n \to \infty$.

A direct consequence of Theorems 2 and 4 is the following.

Corollary 2. Under the conditions of Theorem 4 and, additionally, $Eg_F^4(X_1) < \infty$, one has

$$\hat{V}_{J,b}(T_n) \to \mathrm{Var}\,(g_F(X_1)) + 2 \sum_{i=1}^{m} \mathrm{cov}\,(g_F(X_1), g_F(X_{1+i})), \text{ in probability}$$

which is the actual asymptotic variance of $\sqrt{n}\, T_n$.

Proof of Theorem 4: We write

$$nT(F_n) - (n-b)T(F_{n,-i}) = n[T(F_n) - T(F)]$$
$$- (n-b)[T(F_{n,-i}) - T(F)] + bT(F).$$

Then, we expand each of $[T(F_n) - T(F)]$ and $[T(F_{n,-i}) - T(F)]$ as a linear term plus a quadratic term utilizing the assumed twice Frechét differentiability. Since $\|F_{n,-i} - F_n\| = O(b/n)$ which in turn is $o(n^{-1/2})$ due to the condition $b/\sqrt{n} \to 0$, the remainders in both expansions mentioned above are of the order $o_p(n^{-1})$, uniformly in i. Thus, the claim of the theorem follows if we show that the pseudo–values in the quadratic terms are of appropriate order.

To this end, we define

$$Q_{n,-i} = b^{-1}[n\int\int h_F(x,y)\, d[F_n(x) - F(x)]\, d[F_n(y) - F(y)]$$
$$- (n-b)\int\int h_F(x,y)\, d[F_{n,-i}(x) - F(x)]\, d[F_{n,-i}(y) - F(y)].$$

We would like to prove that

$$b^{1/2} \max_{1 \leq i \leq n-b+1} |Q_{n,-i}| \to 0, \quad \text{in probability as } n \to \infty.$$

We rewrite $b\, Q_{n,-i}$ as

$$[n^{-1} - (n-b)^{-1}] \sum_j \sum_k u_{jk} + (n-b)^{-1}[\sum_j \sum_k u_{jk} - \sum_{j,k \neq i,\ldots,i+b-1} \sum u_{jk}]$$

$$= A_n + B_{ni} \quad \text{(say)}$$

where $u_{jk} = h_F(X_j,X_k) - h_F(X_j,*) - h_F(*,X_k) + h_F(*,*)$ with $h_F(x,*) = E\, h_F(x,X)$ and $h_F(*,*) = E\, h_F(X_j,X_k)$. Next, we express

$$A_n = \frac{b}{n(n-b)} \sum_{j=1}^{n} q_j$$

and

$$B_{ni} = \frac{1}{n-b} [\sum_{j=i}^{i+b-1} q_j + \sum_{j=i}^{i+b-1} q_{ij}]$$

where

$$q_j = \sum_{k=1}^{n} u_{jk}, \quad \text{and} \quad q_{ij} = \sum_{k \neq i,\ldots,i+b-1} u_{jk}.$$

It is clear that $A_n + B_{ni}$ cannot exceed

$$\frac{b}{n-b} [\max_{(*)} |q_{ij}| + 2 \max_{1 \leq j \leq n} |q_j|],$$

where (*) stands for $i \leq j \leq i+b-1$ and $1 \leq i \leq n-b+1$.
Thus, it only remains to show that

$$\max_{1 \leq j \leq n} |q_j| = o(n/\sqrt{b}),$$

and

$$\max_{(*)} |q_{ij}| = o(n/\sqrt{b}). \tag{2.6}$$

Following the argument in the proof of Lemma of Singh and Liu (1990) and using the splitting technique mentioned in the Appendix, it is shown that $Eq_j^4 = O(n^2)$ uniformly in j and $Eq_{ij}^4 = O(n^2)$ uniformly in (*). Finally by using Bonferroni inequality, we obtain

$$P(\max_{1\leq j\leq n} |q_j| > \epsilon \frac{n}{\sqrt{b}}) \leq \epsilon^4 n \cdot \frac{n^2 b^2}{n^4} = O(b^2/n) \to 0.$$

Exactly similar argument proves (2.6) which in fact uses the condition $b^3/n \to 0$. The proof is complete.

3 . Moving Blocks Bootstrap (MBB).

Recall that $\xi_1,\ldots,\xi_k$ are i.i.d. sampled blocks from the moving blocks $B_1,\ldots,B_{n-b+1}$; for each i, $\xi_i \equiv (\xi_{i1},\ldots,\xi_{ib})$; $Y_1,\ldots,Y_\ell \equiv (\xi_1,\ldots,\xi_k)$ stands for the bootstrap sample of size ℓ; and F_ℓ^* is the empirical d.f. based on $Y_1,\ldots,Y_\ell$. Throughout this section we assume ℓ and n are of the same order, namely ℓ/n is bounded below and above.

Case (i) $T_n = \bar{X}_n$.

For the bootstrap approximation of the sampling distribution of $\bar{X}_n$, based on the MBB procedure, with b fixed, we obtain Theorem 5 below.

Remark. The true asymptotic distribution of $\sqrt{n}(\bar{X}_n - \mu)$ is $N(0,\sigma_\infty^2)$. As mentioned in the Introduction $\sigma_b^2 \to \sigma_\infty^2$, as $b \to \infty$. Furthermore σ_b^2 may be viewed as an intermediate stage value between Var (X_1) (which is achieved by the classical bootstrap) and σ_∞^2 (which is the true asymptotic variance). Thus the discrepancy between the MBB approximation with b fixed and the actual distribution of $\sqrt{n}(\bar{X}_n - \mu)$ is $\|\Phi(x/\sigma_b) - \Phi(x/\sigma_\infty)\|$ in limit.

Theorem 5. Let $X_1,\ldots,X_n$ be a sequence of stationary m–dependent r.v.s. If $EX_1^2 < \infty$, then

$$\|P^*(\sqrt{\ell}(\bar{Y}_\ell - \bar{X}_n) \leq x) - \Phi(x/\sigma_b)\| \to 0 \text{ a.s. as } n \to \infty, \tag{3.1}$$

where P* stands for the bootstrap probability (i.e., under the MBB scheme), $\bar{Y}_n = \frac{1}{n}\Sigma_{i=1}^n X_i$, $\|\cdot\|$ stands for the sup norm over all x, and $\sigma_b^2 =$

$b^{-1}\mathrm{Var}\,(\Sigma_{j=1}^{b} X_j)$. Here b is a fixed integer.

Proof: (with $b = 2$) The proof hinges on the observation that under the MBB procedure the effective sample is $\tilde{\xi}_1,\ldots,\tilde{\xi}_k$, where $\tilde{\xi}_i \equiv (\xi_{i1} + \xi_{i2})/\sqrt{2}$.

Let $\bar{\tilde{\xi}}_k = \frac{1}{k}\Sigma_{i=1}^{k}\tilde{\xi}_i$; then $\sqrt{\ell}\,Y_\ell = \sqrt{k}\,\bar{\tilde{\xi}}_k$. Note that $\tilde{\xi}_1,\ldots,\tilde{\xi}_k$ are i.i.d. r.v.s under the MBB scheme. Hence,

$$E\bar{\tilde{\xi}}_k = E^*\tilde{\xi}_1 = (2X_n)/\sqrt{2} + o(n^{-1/2}) \text{ a.s.}$$

Here E^* (or Var^* as seen later) is the mean (variance) under P^*. Using the same arguments as in Theorem 1, we see that

$$\mathrm{Var}^*\,(\tilde{\xi}_1) = \frac{1}{n-1}\sum_{i=1}^{n-1}\left[\frac{X_i+X_{i+1}}{\sqrt{2}} - E^*\tilde{\xi}_1\right]^2$$

$$= \frac{1}{n-1}\sum_{i=1}^{n-1}\left(\frac{X_i+X_{i+1}}{\sqrt{2}} - \sqrt{2}\,X_n\right)^2 + o(1) \quad \text{a.s.}$$

$$\to \sigma_2^2 \quad \text{a.s.}$$

Now we shall use the Berry–Esseen theorem to bound the normed difference in (3.1) which can be rewritten as $\|P^*(\sqrt{k}(\bar{\tilde{\xi}}_k - E^*\tilde{\xi}_1) \le x) - \Phi(x/\sqrt{\mathrm{Var}^*(\tilde{\xi}_1)})\|$ $+ o(1)$ by $2E^*|\tilde{\xi}_1 - E^*\tilde{\xi}_1|^3/\sqrt{k}\,(\mathrm{Var}^*\,(\tilde{\xi}_1))^{3/2} + o(1)$ a.s. It only remains to be shown that $k^{-1/2}E^*|\tilde{\xi}_1|^3 \to 0$ a.s., i.e.,

$$k^{-1/2}\,\frac{1}{n-1}\sum_{i=1}^{n-1}|X_i + X_{i+1}|^3 \to 0 \text{ a.s.}$$

When $EX_1^2 < \infty$, the above statement follows from Marcinkiewicz–Zygmund SLLN, which does extend to m–dependent r.v.s, using the splitting technique. Thus, the theorem is proved.

Consider now the case that $b \equiv b_n$ and $b_n \to \infty$ as $n \to \infty$.

Theorem 6. $X_1,\ldots,X_n$ is a stationary m–dependent sequence of r.v.s with $EX_1 = \mu$ and $E|X_1|^{4+\delta} < \infty$, for some $\delta > 0$. If $b/n \to 0$ as $n \to \infty$, then

$$\|P^*(\sqrt{\ell}\,(Y_\ell - E^*Y_\ell) \le x) - P(\sqrt{n}\,(\bar{X}_n - \mu) \le x)\| \to 0 \qquad (3.2)$$

in probability. If the condition is strengthened to be $b/\sqrt{n} \to 0$ as $n \to \infty$, then (3.2) holds with E^*Y_ℓ replaced by $\bar{X}_n$.

Proof: Note first that under the MBB procedure

$$E^*Y_\ell = E^*[(\xi_{11} + \xi_{12} + \dots + \xi_{1b})/b]$$

$$= \frac{1}{n-b+1} \sum_{i=1}^{n-b+1} (X_i + X_{i+1} + \dots + X_{i+b-1})/b.$$

We may rewrite $\sqrt{\ell}\,(Y_\ell - E^*Y_\ell)$ as $k^{-1/2}\, \Sigma_{i=1}^{k}\, (\tilde{\xi}_i - E^*\tilde{\xi}_i)$, where $\tilde{\xi}_i = (\xi_{i1}+\dots+\xi_{ib})/\sqrt{b}$. Now, the result (3.2) will follow from Katz theorem (see Katz (1963)) on the Berry–Esseen bound provided that

i) $\mathrm{Var}^*\,(\tilde{\xi}_1)$ converges to σ_∞^2 in probability, and

ii) $E^*|\tilde{\xi}_1 - E\tilde{\xi}_1|^{2+\delta/2}$ is bounded in probability.

In order to replace E^*Y_ℓ by $\bar{X}_n$ in (3.2) legitimately, we need to show that

iii) $E^*Y_\ell = \bar{X}_n + o_p^*(n^{-1/2})$.

Here o_p^* stands for o_p under the bootstrap probability. We proceed to prove statements i), ii) and iii) in that order to complete the proof of the theorem.

Proof of i) $\mathrm{Var}^*\,(\tilde{\xi}_1) = \frac{1}{n-b+1}\, \Sigma_{i=1}^{n-b+1} (\tilde{B}_i - E^*\tilde{\xi}_1)^2$, where $\tilde{B}_i = (X_i + X_{i+1}+\dots+X_{i+b-1})/\sqrt{b}$. Let us rewrite $\mathrm{Var}^*\,(\tilde{\xi}_1)$ as

$$\frac{1}{n-b+1} \sum_{i=1}^{n-b+1} [(\tilde{B}_i - \sqrt{b}\mu) - (E^*\tilde{\xi}_1 - \sqrt{b}\mu)]^2.$$

Clearly $E^*\tilde{\xi}_1$ is a sample mean of (m+b)–dependent r.v.s whose mean is $\sqrt{b}\mu$. Therefore it follows from the arguments of Theorem 2 that $(E^*\tilde{\xi}_1 - \sqrt{b}\mu) \to 0$ in probability as $n \to \infty$, and consequently $Var^*(\tilde{\xi}_1)$ converges to σ_∞^2 in probability.

Proof of ii) We first use Minkowski inequality to obtain

$$\|\tilde{\xi}_1 - E^*\tilde{\xi}_1\|_{2+\delta/2} \le \|\tilde{\xi}_1 - \sqrt{b}\mu\|_{2+\delta/2} + \|E^*\tilde{\xi}_1 - \sqrt{b}\mu\|_{2+\delta/2}$$

where $\|Z\|_p = (E^*|Z|^p)^{1/p}$. Then, we apply the argument of Theorem 2 to show that

$$E^*|\tilde{\xi}_1 - \sqrt{b}\mu|^{2+\delta/2} - E|\tilde{B}_1 - \sqrt{b}\mu|^{2+\delta/2} \to 0 \text{ in probability.}$$

To prove this we need the result that $E|\Sigma_{i=1}^{b}(X_i - \mu)|^{4+\delta} = O(b^{2+\delta/2})$. This is a well–known fact for i.i.d. r.v.s and it can be extended easily to m–dependent r.v.s by the splitting technique mentioned in the Appendix.

Proof of iii) Note that

$$\begin{aligned}
E^*Y_\ell &= E^*[(\xi_{11} + \dots + \xi_{1b})/b] \\
&= \frac{1}{b(n-b+1)} \sum_{i=1}^{n-b+1} (X_i + \dots + X_{i+b-1}) \\
&= [b(n-b+1)]^{-1} \\
&\quad \cdot \{b \sum_{i=1}^{n} X_i - [(b-1)X_1 + (b-2)X_2 + \dots + 2X_{b-2} + X_{b-1}] \\
&\quad - [(b-1)X_n + (b-2)X_{n-1} + \dots + 2X_{n-(b-3)} + X_{n-(b-2)}]\} \\
&= \bar{X}_n + [n/(n-b+1) - 1]\bar{X}_n + O_p(b/n) \\
&= \bar{X}_n + O_p(b/n) \quad \text{in probability.}
\end{aligned}$$

To be able to replace $E^*(Y_\ell)$ by $\bar{X}_n$, we require

$$E^*(Y_\ell) = \bar{X}_n + o(\ell^{-1/2}).$$

Since ℓ and n are assumed to grow at the same rate, this requirement will be met in view of the above bound, provided $b/\sqrt{n} \to 0$ as $n \to \infty$.

Case (ii) General statistic T_n.

Let us investigate the asymptotic behavior of the MBB approximation when $T_F \equiv T(F)$, $T_n = T(F_n)$ and $T(\cdot)$ is a smooth functional. Our main goal is to establish representation theorems, for both b being fixed and $b \to \infty$, similar to the ones given in Liu, Singh and Lo (1989). The key elements in this task are probability inequalities on $\|F_\ell^* - F_n\|$ and $\|F_n - F\|$ which are extensions of the standard Dvoretzky–Kiefer–Wolfowitz inequality. These extensions not only help us obtain the representations, they also clarify how F_ℓ^* approximates F_n and F which does not manifest itself in the description of MBB.

Lemma 1. For any $\delta > 0$, there exists a universal constant c such that

i) $P^*(\|F_\ell^* - E^*F_\ell^*\| > \delta) \leq c\, b \exp(-2k\delta^2)$,

ii) $P^*(\|F_\ell^* - F_n\| > \delta) \leq c\, b \exp[-2k(\delta - \frac{2b}{n-b})^2]$.

Proof: Let us define $F_{\ell,i}^*$ as the empirical d.f. based on the k i.i.d. univariate observations consisting of all the ith coordinates from the sampled blocks $\xi_1, \xi_2, \ldots, \xi_k$. Thus

$$F_\ell^*(x) = b^{-1}[F_{\ell,1}^*(x) + \ldots + F_{\ell,b}^*]$$

and

$$\|F_\ell^* - F_n\| \leq \frac{1}{b} \sum_{i=1}^{b} \|F_{\ell,i}^* - F_n\|.$$

Consequently, by Bonferroni inequality we have

$$P^*(\|F_\ell^* - F_n\| > \delta) \leq \sum_{i=1}^{b} P(\|F_{\ell,i}^* - F_n\| > \delta). \tag{3.3}$$

Next we shall show that, for any i, $1 \leq i \leq b$,

$$\|E^*F_{\ell,i}^* - F_n\| \leq 2b/(n-b). \tag{3.4}$$

The claim ii) follows from (3.3) and (3.4) using the standard Dvoretzky–Kiefer–Wolfowitz inequality.

To prove (3.4), we first write

$$\begin{aligned} E^*(F_{\ell,i}^*(x)) &= [I(X_i \leq x) + \ldots + I(X_{n-b+i} \leq x)]/(n-b+1) \\ &= nF_n(x)/(n-b+1) + \gamma_n \end{aligned}$$

where $$|\gamma_n| \le \frac{i-1}{n-b+1} + \frac{b-i}{n-b+1} = \frac{b-1}{n-b+1}.$$

Therefore,

$$|E^*F^*_{\ell,1}(x) - F_n(x)| \le \left(\frac{n}{n-b+1} - 1\right) + \frac{b-1}{n-b+1} \le 2b/(n-b+1).$$

The claim i) is clearly contained in the above arguments.

This lemma immediately lends itself to the following corollary concerning the order of $\|F^*_\ell - F_n\|$.

Corollary 3. Assume $b/n \to 0$ as $n \to \infty$. We obtain

$$\|F^*_\ell - F_n\| = \begin{cases} O^*_p(n^{-1/2}) & \text{if } b \text{ is fixed,} \\ O^*_p(b^{1/2}n^{-1/2}(\log n)^{1/2}) & \text{if } b \to \infty. \end{cases}$$

Next, we extend the Dvoretzky–Kiefer–Wolfowitz inequality to the m–dependence case.

Lemma 2. Let $X_1,\dots,X_n$ be a sequence of stationary m–dependent r.v.s. For any $\delta > 0$, there exists a universal constant c such that

$$P(\|F_n - F\| > \delta) \le (m+1)c \exp\left[-2\frac{n}{m+1}\left(\delta - \frac{m}{n}\right)^2\right].$$

The proof of this lemma first uses the splitting technique mentioned in the Appendix to split the m–dependent sequence into (m+1) i.i.d. sequences and then the standard Dvoretzky–Kiefer–Wolfowitz inequality on each of the (m+1) i.i.d. empiricals.

We are now ready to present a representation theorem for $T(F^*_\ell) - T(F_n)$ under the MBB procedure.

Theorem 7. Assume b is a fixed positive integer. Let $T(\cdot)$ be (once) Frechét differentiable at F in the usual sense, i.e.,

$$T(G) - T(F) = \int g_F \, d(G - F) + o(\|G - F\|).$$

Then

$$T(F^*_\ell) - T(F_n) = \frac{1}{\ell}\sum_{j=1}^{\ell} g_F(Y_j) - \frac{1}{n}\sum_{j=1}^{n} g_F(X_j) + o^*_p(n^{-1/2}).$$

Proof: The basic idea is to write

$$T(F^*_\ell) - T(F_n) = [T(F^*_\ell) - T(F)] - [T(F_n) - T(F)].$$

The right–hand side of the above is equal to

$$[\tfrac{1}{\ell}\sum_{j=1}^{\ell} g_F(Y_j) - \tfrac{1}{n}\sum_{i=1}^{n} g_F(X_i)] + o(\|F^*_\ell - F\|) + o(\|F_n - F\|). \qquad (3.5)$$

Now the two exponential bounds of Lemmas 1 and 2 and the arguments of Theorem 1 of Liu, Singh and Lo (1989) can be used to show that the remainder term of (3.5) is of the order $o^*_p(n^{-1/2})$.

The following result regarding the asymptotic distribution of $\sqrt{\ell}(T(F^*_\ell) - T(F_n))$ can be viewed as a direct consequence of Theorem 5 above, the representation of Theorem 7 and the proposition in Section 1 of Liu, Singh and Lo (1989).

Theorem 8. Assume the conditions of Theorem 7 above and $E\, g^2_F(X_1) < \infty$. Then

$$\|P^*(\sqrt{\ell}(T(F^*_\ell) - T(F_n)) \leq x) - \Phi(x/\sigma_b(T_n))\| \to 0$$

in probability as $n \to \infty$, where $\sigma^2_b(T_n) = b^{-1}\,\mathrm{Var}\,[g_F(X_1) + \dots + g_F(X_b)]$.

Analogous to Theorems 7 and 8, we have the following results in the case of $b \equiv b_n \to \infty$.

Theorem 9. Assume $T(\cdot)$ admits the following expansion at F for some g_F:

$$T(G) - T(F) = \int g_F\, d(G - F) + O(\|G - F\|^2).$$

If $b \log n/\sqrt{n} \to 0$, as $n \to \infty$, then

$$T(F^*_\ell) - T(F_n) = \tfrac{1}{\ell}\sum_{i=1}^{\ell} g_F(Y_i) - \tfrac{1}{n}\sum_{i=1}^{n} g_F(X_i) + o^*_p(n^{-1/2}) \text{ a.s.}$$

Before providing a proof of this theorem, we remark that if the requirement on the remainder term is relaxed to $O(\|G - F\|^{1+\delta})$, $0 < \delta \leq 1$, then the theorem will still hold though the conditions on b will be more stringent.

Proof: To establish the claim, we essentially need to prove that for any $\epsilon > 0$, $P^*(\|F_\ell^* - F\|^2 + \|F_n - F\|^2 > \epsilon n^{-1/2}) \to 0$ a.s. Clearly $\|F_\ell^* - F\|^2 + \|F_n - F\|^2 \leq 2\|F_\ell^* - F_n\|^2 + 3\|F_n - F\|^2$. It follows from Lemma 2 that

$$\|F_n - F\|^2 = O(n^{-1/2} \log n) \text{ a.s.}$$

With each fixed sequence of X_i's for which $\|F_n - F\| = O(n^{-1/2} \log n)$ a.s., we have

$$P^*(2\|F_\ell^* - F_n\|^2 + 3\|F_n - F\|^2 > \epsilon n^{-1/2}) \leq P^*(\|F_\ell^* - F_n\| \geq \epsilon_1 n^{-1/4}). \quad (3.6)$$

Then ii) of Lemma 1 implies that (3.6) $\to 0$.

The representation of Theorem 9 immediately leads to the following:

Theorem 10. Under the condition of Theorem 9 and the condition $b \log n/\sqrt{n} \to 0$ as $n \to \infty$, we have

$$\|P^*(\sqrt{\ell}(T(F_\ell^*) - T(F_n)) \leq x) - P(\sqrt{n}(T(F_n) - T(F)) \leq x)\| \to 0$$

in probability as $n \to \infty$.

It may be appropriate to mention here that our theorems of this section do not cover general sample quantiles since the quantiles are not Frechét differentiable. Instead they are compactly differentiable, which is a weaker concept. We do expect that our results can be extended to cover compactly differentiable functionals. In any case, the consistency of MBB (though not MBJ) can be achieved for quantiles by an alternative approach via Bahadur–Kiefer representations.

4. Concluding Remarks.

Remark 4.1. Extension to Linear Regression Model: On time series data which indicate the presence of a trend and a seasonal effect, the linear regression model with a m–dependence structure on the errors is a commonly accepted model. This suggests that the extensions of the resampling procedures proposed in this article to linear regression models will be quite useful in time series analysis. We delay the investigation of this extension to a follow–up study.

Remark 4.2. Extension to More General Dependence Models: Our discussion throughout the paper is restricted to the m–dependence model which is the most basic weak dependence model. We expect that the MBJ and MBB procedures are consistent under much more general dependence models, such as the α–mixing model, under reasonable conditions. A study on this line should be worthwhile.

Remark 4.3. σ_b^2 Increases to σ_∞^2 if the Covariances are Non–Negative: Recall that

$$\sigma_\infty^2(X_n) \equiv \text{Var}(X_1) + 2 \sum_{i=1}^{m} \text{cov}(X_1, X_{1+i}).$$

One can express $\sigma_b^2(X_n) = \frac{1}{b} \text{Var}(X_1 + \ldots + X_b)$ as

$$\text{Var}(X_1) + 2 \sum_{i=1}^{\min(b,m)} (1 - \tfrac{i}{b}) \text{cov}(X_1, X_{1+i}).$$

Thus if all the covariances are non–negative, then $\sigma_b^2(X_n)$ increases monotonically to $\sigma_\infty^2(X_n)$, as $b \to \infty$. This monotonicity means that even in the case when b is considered fixed, there is a definite improvement compared to the classical procedures; also the improvement increases as b increases. Consider the example: $X_i = a_0 e_i + a_1 e_{i-1} + \ldots + a_m e_{i-m}$ where $\{e_i\}$ is an i.i.d. sequence of r.v.s and the coefficients a_i's are non–negative. In this instance, all the covariances turn out to be non–negative and the above comment on $\sigma_b^2(X_n)$ applies. This remark extends to a general statistic T_n if the differential $g_F(x)$ is a monotonic function of x. For example, this is the case with the sample median.

On the other hand, if all the covariances are negative, then σ_b^2 decreases to σ_∞^2 and the above comments on MBB and MBJ still hold.

In the mixed situation when the covariances are not of the same sign it is not guaranteed that σ_b^2 is between σ_1^2 and σ_∞^2 for every b. However, the partial correction on the ith covariance (i.e., cov (X_1, X_{1+i})) is inversely proportional to i. Since the lower terms of covariances (and the variance term) typically dominate the overall asymptotic variance, such a correction is expected to be quite effective.

Remark 4.4. Optimal Choice of b and k: Is there an optimal choice of the block size b for MBJ or the pair (b,k) for MBB? For the classical bootstrap it follows from the second order theory (i.e., the one term Edgeworth expansions) that the bootstrap sample size ℓ should satisfy $\ell/n \to 1$, as $n \to \infty$. Such a second order theory for MBB with m–dependent r.v.s is yet unavailable. One way to choose (b,k) is to assume $\{X_i\}$ to be i.i.d. and choose (b,k) which minimizes the loss of efficiency in comparison with the classical bootstrap. This idea is motivated by the remark in the Introduction that replacing the classical bootstrap or jackknife by MBB or MBJ may be viewed as robustizing both procedures against possible serial dependence of the observations. Preliminary calculations show that this viewpoint leads to the condition $\ell/n \to 1$ for MBB. Further work would be of value.

Remark 4.5. Functions of Several Means: To demonstrate the applicability of our theorems proved for the functionals, we consider the following class of many commonly used statistics. Suppose that X_i's are p–variate, and on each X_i there are q functions: $h_j(X_i)$ for $j = 1,2,\ldots,q$. The functional $T(\cdot)$ can be expressed as a smooth function $H(\cdot)$ of the q means $Eh_j(X_1)$, $j = 1,\ldots,q$. This class includes, for example, the central moments $E(X - \mu_F)^r$ in the univariate case, the correlation and regression coefficients in the bivariate case. In the case of variance: $p = 1$, $q = 2$; $h_1(X) = X^2$, $h_2(X) = X$. In the case of the correlation coefficient: $p = 2$, $q = 5$; $h_1(X) = X^{(1)}$, $h_2(X) = X^{(2)}$, $h_3(X) = (X^{(1)})^2$, $h_4(X) = (X^{(2)})^2$ and $h_5(X) = X^{(1)}X^{(2)}$; here $X = (X^{(1)},X^{(2)})$.

For this class of functionals, the smoothness required in our theorems can be checked simply by expanding $H(\cdot)$ using Taylor's expansion, provided $H(\cdot)$ has sufficiently many derivatives (which is often the case) and the class of distributions is suitably restricted. For instance, in the simple case of the variance,

$$\int(x-\mu_G)^2\,dG - \int(x-\mu_F)^2\,dF = \int(x-\mu_F)^2\,d(G-F) - (\mu_G-\mu_F)^2.$$

The term $(\mu_G - \mu_F) = O(\|G - F\|^2)$ if the distributions considered have compact support.

In order to include other examples such as the correlation coefficient, our theorems need to be extended to multivariate settings. This should be a simple task in view of the multivariate DKW inequality (see, e.g., Serfling (1980)).

Appendix: The splitting technique for m–dependent sequence of r.v.s.

The sum of $\Sigma_{i=1}^n X_i$ of m–dependent r.v.s can be split into (m+1) sums of independent r.v.s. For examples, let $m = 1$ and n be even, then $\Sigma_{i=1}^{n} X_i = (X_1 + X_3 + \ldots + X_{n-1}) + (X_2 + X_4 + \ldots + X_n)$. Applying the triangle inequality and this splitting technique, many asymptotic bounds for independent r.v.s can be extended immediately. Several such extensions are used repeatedly in the text. Some of them are listed here for references: If $\{X_i\}$ is stationary and m–dependent, then

1) $n^{-1}\Sigma_{i=1}^{n} X_i \to EX_1$ (if $E|X_1| < \infty$);
2) $\Sigma_{i=1}^{n} X_i = O_p(n^{1/2})$ (if $EX_1^2 < \infty$);
3) $E|\Sigma_{i=1}^{n} X_i|^p = O(n^{p/2})$ (if $E|X_1|^p < \infty$);
4) $\Sigma_{i=1}^{n} X_i = O(n^{1/2}(\log \log n)^{1/2})$ a.s. (if $EX_1^2 < \infty$).

References

Billingsley, P. (1968). *Convergence of Probability Measures.* John Wiley & Sons.

Katz, M. L. (1963). " Note on the Berry–Esseen Theorem." *Ann. Math. Statist.* **34**, 1107–1108.

Kunsch, H. (1989). "The jackknife and the bootstrap for general stationary observations." *Ann. Stat.* **17**, 1217–1241.

Liu, R. (1988). "Bootstrap procedures under some non–i.i.d. models." *Ann. Stat.* **16**, 1696–1708.

Liu, R., Singh, K. and Lo, S. (1989). "On a representation of the bootstrap." *Sankhyā A.* **51**, 168–177.

Miller, R. G. (1974). "The jackknife–a review." *Biometrika* **6**, 1–15.

Parr, W. C. (1985). "Jackknifing differentiable statistical functionals." *JRSS*, B, **47**, 56–66.

Serfling, R. (1980). *Approximation Theorems of Mathematical Statistics.* John Wiley & Sons.

Singh, K. (1981). "On the asymptotic accuracy of Efron's bootstrap." *Ann. Statist.* **9**, 1187–1195.

Singh, K. and Liu, R. (1990). "On the validity of jackknife procedures." *Scand. J. Stat.* **17**, 11–21.

Bootstrap Bandwidth Selection

J. S. Marron
University of North Carolina

Abstract: Various bootstrap methodologies are discussed for the selection of the bandwidth of a kernel density estimator. The smoothed bootstrap is seen to provide new and independent motivation of some previously proposed methods. A curious feature of bootstrapping in this context is that no simulated resampling is required, since the needed functionals of the distribution can be calculated explicitly.

1. Introduction

This is a review of results concerning application of bootstrap ideas to bandwidth or smoothing parameter selection. The main ideas are useful in all types of nonparametric curve estimation settings, including regression, density and hazard estimation, and also apply to a wide variety of estimators, including those based on kernels, splines, orthogonal series, etc. However as much of the work so far has focused on perhaps the simplest of these, kernel density estimation, the discussion here will be given in this context.

The density estimation problem is often mathematically formulated by assuming that observations $X_1,...,X_n$ are a random sample from a probability density $f(x)$, and it is desired to estimate $f(x)$. The kernel estimator of $f(x)$ is defined by

$$\hat{f}_h(x) = n^{-1} \sum_{i=1}^{n} K_h(x - X_i),$$

*Department of Statistics, University of North Carolina, Chapel Hill, NC 27599–3260. Research Partially Supported by NSF Grant DMS–8902973

Exploring the Limits of Bootstrap. Edited by Raoul LePage and Lynne Billard.
 ISBN: 0471-53631-8.

where K_h denotes the rescaling $K_h(\cdot) = K(\cdot/h)/h$, of the "kernel function" K (assumed throughout to be a symmetric probability density) by the "bandwidth" h. See Silverman (1986) for early references, good practical motivation, and useful intuitive discussion concerning this estimator.

The most important hurdle in the practical implementation of $\hat{f}(x)$ (and indeed essentially any nonparametric curve estimator), is the choice of the bandwidth h. While many methods have been proposed and studied, see the survey Marron (1988) for example, this remains an important problem in the application of curve estimators in general.

For a mathematical approach to this problem, most workers consider error criteria. Here the focus will be on the expected L^2 norm, or Mean Integrated Squared Error,

$$\mathrm{MISE}(h) = E \int [\hat{f}_h(x) - f(x)]^2 \, dx.$$

Devroye and Györfi (1984) provide an array of arguments in favor of the L^1 norm, but for simplicity of presentation and development of ideas, extensions to this challenging case will only be briefly discussed at the end.

An important advantage of MISE(h) as an error criterion is that it admits the simple variance–bias^2 representation

$$\mathrm{MISE}(h) = V(h) + B^2(h)$$

where

$$V(h) = n^{-1}\{h^{-1}\int K^2 + \int (K_h * f)^2\}$$
$$B^2(h) = \int (K_h * f - f)^2$$

using $*$ to denote the convolution. Deep insight into the bandwidth selection problem comes from the asymptotic analysis, as $n \to \infty$, $h \to 0$, with $nh \to \infty$, assuming two uniformly continuous derivatives on f,

$$V(h) = n^{-1} h^{-1} \int K^2 + o(n^{-1}h^{-1}),$$
$$B^2(h) = h^4 \int (f'')^2 \left(\int x^2 K/2\right)^2 + o(h^4),$$

see Section 3.3.1 of Silverman (1986) for example. From this one can see why choice of h is not a simple matter, h small gives too much variability to the estimator (expected intuitively since there are not enough points in the effective local average when the window width is too small), h big introduces too much bias (again intuitively clear since a large window width introduces information from too far away).

Section 2 discusses bandwidth selection by minimization of bootstrap estimates of MISE(h). In particular it is seen why the smoothed bootstrap is very important here. An interesting and unusual feature of this case is that the bootstrap expected value can be directly and simply calculated, so the usual simulation step is unnecessary in this case.

Asymptotic analysis and comparison of these methods is described in Section 3. Connection to other methods, including Least Squares Cross Validation, is made in Section 4. Simulation experience and a real data example are given in Section 5.

2. Bootstrap MISE Estimation

The essential idea of bootstrapping in general is to find a useful approximation to the probability structure of the estimation process being studied, denoted in this case by $\mathcal{L}\{\hat{f}_h(x) - f(x)\}$. The usual simple means of doing this involves thinking about "resampling with replacement". One way of thinking about this probability structure is through random variables $I_1,\ldots,I_n$ which are independent of each other, and of $X_1,\ldots,X_n$, and are uniformly distributed on the integers $\{1,\ldots,n\}$. These new random variables contain the probability structure (conditioned on $X_1,\ldots,X_n$) of the "resample" $X_1^*,\ldots,X_n^*$ defined for $i = 1,\ldots,n$ by

$$X_i^* = X_{I_i}.$$

As a first attempt at using this new conditional probability structure to model the bandwidth trade–off in the density estimation problem, one might define the "bootstrap density estimator"

$$\hat{f}_h^*(x) = n^{-1} \sum_{i=1}^{n} K_h(x - X_i^*),$$

and then hope the approximation

$$\mathcal{L}^*\{\hat{f}_h^*(x) - \hat{f}_h(x) \mid X_1,\ldots,X_n\} \cong \mathcal{L}\{\hat{f}_h(x) - f(x)\},$$

is useful. Faraway and Jhun (1987) have pointed out that this approximation is in fact not useful for bandwidth selection because

$$E^*\{\hat{f}_h^*(x)\} = E^*\{K_h(x - X_{I_1})\} = \hat{f}_h(x),$$

where E^* is expected value "in the bootstrap world", in other words with respect to the bootstrap distribution $\mathcal{L}^*\{\cdot|X_1,...,X_n\}$, i.e. the average over all possible values of $I_1,...,I_n$. This shows that there is no bias in this bootstrap world, which is disastrous because as shown by the MISE(h) analysis in section 1, bias constitutes one of the two essential quantities to be balanced in bandwidth selection. Actually this is not too surprising because $\hat{f}_h$ is not in fact the density of the $\mathcal{L}^*\{\cdot|X_1,...,X_n\}$ distribution. Note moreover that this philosophical flaw can not be simply fixed because this is a discrete distribution (supported on $\{X_1,...,X_n\}$) and has no density.

This motivates finding another bootstrap probability structure to approximate $\mathcal{L}\{\hat{f}_h(x) - f(x)\}$. A natural candidate, proposed for bandwidth selection by Faraway and Jhun (1987) and Taylor (1989), which does have a density is the smoothed bootstrap, introduced in Efron (1979). An alternative clever idea based on subsampling is proposed in Hall (1990), but this will not be discussed further here because connections to other methods, and comparison with them have not yet been well understood.

A means of studying the smoothed bootstrap, using notation as above, is to define additional random variables $\epsilon_1,...,\epsilon_n$, independent of each other and of $X_1,...,X_n$ and $I_1,...,I_n$, having probability density $L_g(x)$, where for $g > 0$, L_g denotes the rescaling $L_g(\cdot) = L(\cdot/g)/g$. Now redefine the bootstrap sample by, for $i = 1,...,n$

$$X_i^* = X_{I_i} + \epsilon_i.$$

Observe that the distribution of $X_1^*|X_1,...,X_n$, i.e. the bootstrap distribution $\mathcal{L}^*\{\cdot|X_1,...,X_n\}$, has probability density

$$\hat{f}_g(x) = n^{-1}\sum_{i=1}^{n} L_g(x - X_i).$$

Hence it seems natural to study when the approximation

$$\mathcal{L}^*\{\hat{f}_h^*(x) - \hat{f}_g(x) | X_1,\dots,X_n\} \cong \mathcal{L}\{\hat{f}_h(x) - f(x)\},$$

is useful. This depends on the choices of L and g which are discussed in the next section.

This motivates the use of the bandwidth h^* which minimizes (break ties arbitrarily) the MISE in the bootstrap world,

$$\begin{aligned} \mathrm{MISE}^*(h) &= E^* \int \{\hat{f}_h^*(x) - \hat{f}_g(x)\}^2\, dx, \\ &= V^*(h) + B^{*2}(h), \end{aligned}$$

where $\hat{f}_h$ is defined as above but using the smoothed bootstrap data, and where

$$\begin{aligned} V^*(h) &= n^{-1}\{h^{-1}\int K^2 + \int (K_h * \hat{f}_g)^2\}, \\ B^{*2}(h) &= \int (K_h * \hat{f}_g - \hat{f}_g)^2. \end{aligned}$$

An interesting fact about this setup is that the desired functionals of the bootstrap distribution can be simply calculated, so the usual computationally expensive simulation step is unnecessary. This is usually the case only for very simple examples, see section Chapter 5 of Efron (1982), although an interesting exception is in quantile estimation, see Sheather and Marron (1990).

It is straightforward to compute $\mathrm{MISE}^*(h)$ because V^* and B^{*2} admit the simple representations

$$V^*(h) = n^{-1}[h^{-1}\int K^2 + n^{-2}\sum_i\sum_j \{K_h * K_h * K_g * K_g\}(X_i - X_j)],$$

$$B^{*2}(h) = n^{-2}\sum_i\sum_j \{K_h * K_h * K_g * K_g - 2K_h * K_g * K_g + K_g * K_g\}(X_i - X_j).$$

Hence calculation of h^* requires about the same computational effort as the least squares cross–validated bandwidth (discussed in section 3.4.3 of section Silverman 1986).

For a completely different approach to bootstrapping in density estimation, see Hall (1990).

3. Asymptotics

In this section, choice of g and L is considered. A sensible first attempt, see Taylor (1989), would be to try $g = h$ and $K = L$. This can be easily analyzed using the assumptions and asymptotics at the end of section 1, with the important part being

$$B^{*2}(h) \cong h^4 \int(\hat{f}_g{}'')^2 \left(\int x^2 K/2\right)^2.$$

This presents a problem, because for $g \sim n^{-1/5}$, which is the reasonable range for h see section 3.3.2 of Silverman (1986) for example, $\hat{f}_g{}''(x)$ does not even converge to $f''(x)$ (because the variance does not tend to 0). For this reason, Faraway and Jhun (1987) propose using $g > h$. However observe that $f''(x)$ is not what is needed here, instead we need the functional $\int(f'')^2$ which is a different problem. Indeed for $g \sim n^{-1/5}$, Hall and Marron (1987) show that

$$\int(\hat{f}_g{}'')^2 \to \int(f'')^2,$$

although this choice of the bandwidth g is quite inefficient in the sense that it gives a slower than necessary rate of convergence.

A means of quantifying this inefficiency, which is relevant to bandwidth selection, is to study its effect on the relative rate of convergence. In remark 3.6 of Hall, Marron and Park (1990), it is shown that

$$(h^*/h_0) - 1 \sim n^{-1/10},$$

when $g = h$, where h_0 denotes the minimizer of MISE(h). This very slow rate of convergence is the same as that well known to obtain for least squares cross–validation, and for the biased cross–validation method of Scott and Terrell (1987) (which uses $g = h$ in a slightly different way). For this reason, as well as the fact that the appropriate bandwidth for estimating $\int(f'')^2$ is different from that for estimation of $f(x)$, the choice $g = h$ does not seem appropriate.

Good insight for the problem of how to choose g has been provided by the main results of Hall, Marron and Park (1990). The minor modification of these results presented here is explicitly given in Jones, Marron and Park (1990), where it is seen that if f has slightly more than four derivatives, and L is a probability density, for C_1, C_2 and C_3 constants depending on f, K and L,

$$(h^*/h_0) - 1 \overset{d}{\to} C_1 n^{-1} g^{-9/2} Z + (C_2 g^2 + C_3 n^{-1} g^{-5}),$$

where Z is a standard normal random variable. Note $C_3 = 0$ and $g \sim n^{-1/5}$ gives the slow $n^{-1/10}$ rate in the above paragraph. This expansion is important because it quantifies the trade–off involved in the choice of g. In particular there is too much "variance" present if $g \to 0$ too rapidly, and a "bias" term that penalizes $g \to 0$ too slowly. This variance and bias can be combined them into an "asymptotic mean squared error" which can then be optimized over g to see that the best g has the form

$$g \sim C_4(f,K,L)\, n^{-1/7},$$

which gives

$$(h^*/h_0) - 1 \sim n^{-5/14}.$$

Data based methods for estimating C_4 are given in Jones, Marron and Park (1990). Note that this rate of convergence is much faster than $n^{-1/10}$.

A natural question at this point is: can the rate $n^{-5/14}$ be improved? As noted in remark 3.3 of Hall, Marron and Jones (1990), by taking L to be a higher order kernel, this rate can be improved all the way up to the parametric $n^{-1/2}$ (L needs to be of "order 6" for this fast rate). This rate has been shown to be best possible by Hall and Marron (1990). However there is a distinct intuitive drawback to this in that when L is a higher order kernel, it can no longer be thought of as a probability density, so h^* is no longer a bootstrap bandwidth, at least in the usual sense of the word.

A more intuitive way of achieving root n convergence is given in Jones, Marron and Park (1990), who consider factorizations of g, in terms of h, of the form

$$g = C\, n^p\, h^m.$$

In particular for $m = -2$ and suitable p they obtain

$$(h^*/h_0) - 1 \sim n^{-1/2},$$

in the much more natural case $K = L$.

4. Connection to Other Methods

Bandwidths that are essentially the same as h^* have been derived from considerations much different from bootstrapping. In particular note that the dominant part of the representation of $V(h)$ at the end of section 1 does not depend on f, so it is natural to estimate $V(h)$ by $n^{-1}h^{-1}\int K^2$, which is asymptotically equivalent to $V^*(h)$. The fact that $B^{*2}(h)$ provides a natural estimate of $B^2(h)$ can also be derived in a natural way by thinking about replacing the unknown f in B^2 by the pilot kernel estimator $\hat{f}_g$. Such non–bootstrap motivations for a bandwidth selector very close to h^* were developed independently in an unpublished paper by Ramlau–Hansen and in the related regression setting by Müller (1985). See Chiu (1990) for related ideas from the Fourier Transform point of view.

Hall, Marron and Park (1990) motivate a very similar bandwidth selector, but by a different method. They propose decreasing the variability of the least squares cross–validated bandwidth through a "pre–smoothing" of the pairwise differences of the data. Note that, using $\hat{f}_{hj}$ to denote the kernel estimator based on the sample with X_j excluded, the least squares cross–validation criterion can be written in the form

$$\begin{aligned} CV(h) &= \int \hat{f}_h^{\,2} - 2\, n^{-1} \sum_{j=1}^{n} \hat{f}_{hj}(X_j) \\ &\cong n^{-1}h^{-1}\int K^2 + n^{-1}(n-1)^{-1} \sum\sum_{i \neq j} \{K_h * K_h - 2K_h\}(X_i - X_j), \end{aligned}$$

where the approximation comes from replacing a factor of n^{-1} by $(n-1)^{-1}$.

Note that the first term provides the same sensible estimate of $V(h)$ discussed in the paragraph above, while the second term has features reminiscent of the representation of B^{*2} given at the end of Section 2. To make this connection more precise, note that when there are no replications among $X_1,\dots,X_n$, the second term is the limit as $g \to 0$ of

$$\hat{B}^2(h) = n^{-1}(n-1)^{-1}\sum_{i \neq j}\sum D_{hg}(X_i - X_j),$$

where

$$D_{hg} = \{K_h{*}K_h{*}K_g{*}K_g - 2K_h{*}K_g{*}K_g + K_g{*}K_g\}.$$

Note also that by the associative law for convolutions, one may view this as first plugging the differences into $K_g{*}K_g$, and then putting the result into the bias part of CV, which is why this idea was called smoothed cross–validation by Hall, Park and Marron (1990).

The important difference between $\hat{B}^2(h)$ and $B^{*2}(h)$ is whether or not the "terms on the diagonal" are included in the double sum. At first glance one may feel uncomfortable about these terms because they do not depend on the data. At second glance it is not clear that they will have a very large effect, unless $g <= h$, when they contribute a term of order $n^{-1}h^{-1}$. For this reason Taylor(1989) deleted these terms in his $g = h$ implementation of the smoothed bootstrap. However more careful analysis, in the case $g >> h$, shows a rather slight theoretical superiority of B^{*2} over $\hat{B}^2$ has been demonstrated by Jones, Marron and Park (1990) in terms of the relative rate of convergence. Simulation work has also indicated usually small superiority of the diagonals in approach, although the improvement is sometimes much larger because the diagonals out version is less stable. One possible explanation as to why this happens is that B^{*2} is the smoothed bootstrap estimate of B^2, while $\hat{B}^2$ does not seem to have any such representation.

Faraway and Jhun (1987) have pointed out that the bootstrap approximation can be used to understand the bandwidth trade–off entailed by other means of assessing the error in $\hat{f}_h$. For example one could replace the L^2 based MISE with the expected L^1 norm. A major drawback to this is that it seems that an exact calculation of the bootstrap expected value E^* is no longer realistically available. Hence this expected value will need to be evaluated by simulation, which will be far more (perhaps prohibitively?) expensive from the computational viewpoint. Another example is the replacement of MISE by the pointwise Mean Squared Error, where one focuses on estimation of f at one fixed location x. Here the exact calculation of E^* can be done, however this has not been explored carefully yet, mostly because it seems sensible to postpone investigation of this more challenging pointwise case until more is understood about the global MISE problem.

5. Simulations and an Application

To see how these methods worked in a simulation context, various versions of the bootstrap bandwidth selectors were tested. Several methods of choosing the pilot bandwidth g, as discussed in Jones, Park and Marron (1990), including immediate use of a $N(0,\hat{\sigma}^2)$ reference distribution and also estimation of the unknown functionals as suggested in Jones and Sheather (1990), were tried. The results were usually better when estimates were used, so one step estimators of this type were used for the following discussion. To speed the computations, a binned implementation of the type described in Scott and Härdle (1990) was employed.

For this, 500 pseudosamples of size $n = 100$ were generated from the normal mean mixture density described in Park and Marron (1990). The results are visually summarized in Figure 1, which is very similar to Figure 3b in that paper. The bandwidth selectors CV, BCV and OS there are not shown here, because as one would expect from the results of that paper they were inferior to these newer ones. PI is the main bandwidth selector discussed by Park and Marron. Note that Taylor's $g = h$ method performed quite poorly in comparison to the others, with a strong bias towards oversmoothing. This poor performance is not surprising in view of the theoretical results described above. The simple bootstrap, denoted BSS,

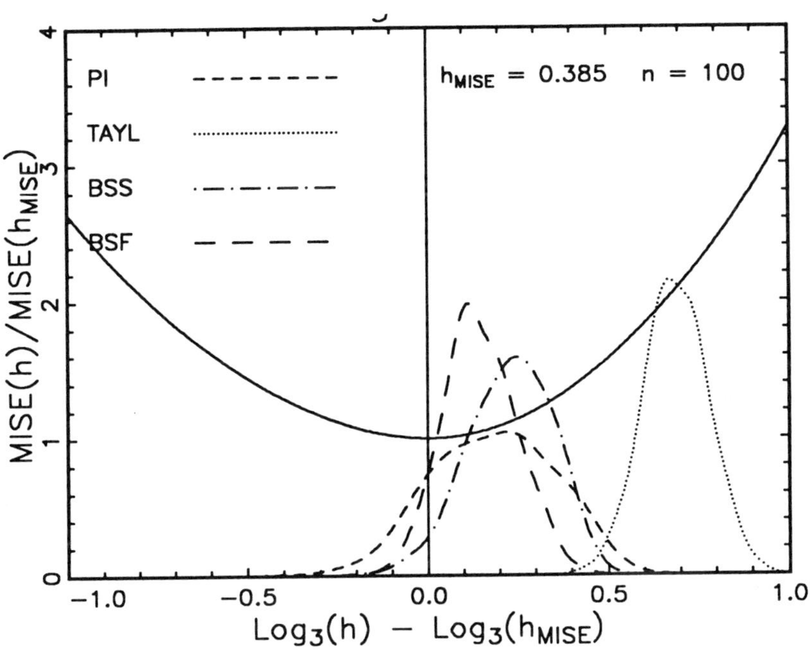

Figure 1: MISE and kernel density estimates of the distributions of the automatically selected bandwidths (log 3 scale). Based on 500 Monte Carlo replications of samples of size 100 from the model .5N(−1,4/9) + .5N(1,4/9).

which uses a data based g chosen independently of h, gave performance roughly comparable to the Park and Marron PI. It is not straightforward to compare these, because there is slightly more bias, but slightly less variability. However the bandwidth factorized bootstrap, i.e. the $n^{-1/2}$ method described at the end of Section 3, denoted BSF, gave much better performance, having less variability and also less bias than the others.

These selectors have also been tried for other sample sizes and other densities as well. For those densities not too far from the Normal in shape, the asymptotics describe the situation well, with larger sample sizes giving more rapid improvements in BSF than the others (as expected from its faster rate of convergence). For the N(0,1) BSF gave really superlative performance, in fact even beating out the Normal reference distribution bandwidth given at (3.28) of Silverman (1986). For densities which are still unimodal, but depart strongly from the normal in directions of strong skewness or kurtosis, the performance was not so good (in fact CV is typically the best in terms of MISE), but can be improved a lot by using a

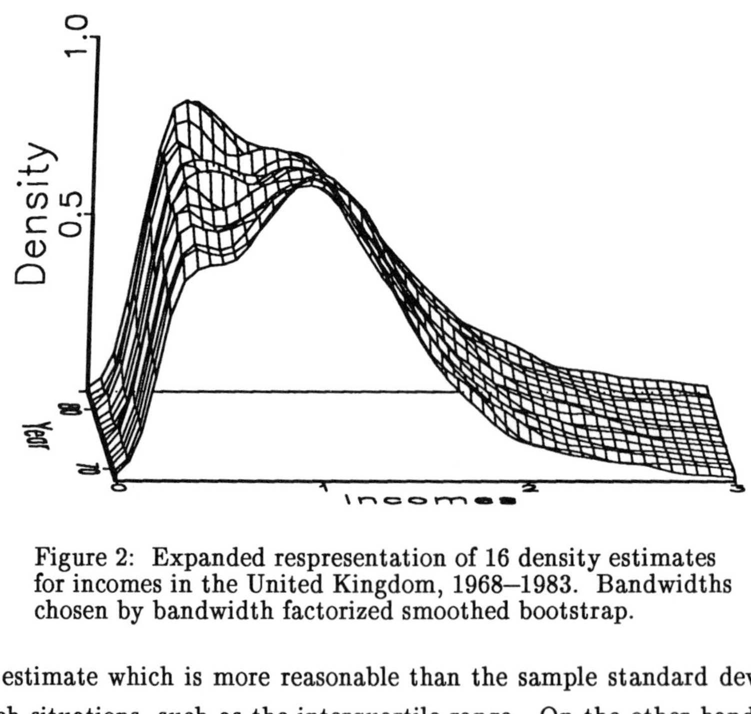

Figure 2: Expanded respresentation of 16 density estimates for incomes in the United Kingdom, 1968–1983. Bandwidths chosen by bandwidth factorized smoothed bootstrap.

scale estimate which is more reasonable than the sample standard deviation in such situations, such as the interquartile range. On the other hand when f is far from normal in the direction of heavy multimodality, again most of these newer bandwidth selectors were inferior to CV in the MISE sense, but the sample standard deviation was a more reasonable scale estimate than the IQR. A way to view both of the above situations, is that they are cases where it takes very large sample sizes before the effects described by the asymptotics take over. There is still work to be done in finding a bandwidth selector which works acceptably well in all situations.

To see how well these methods work on a real data set, they were tried on the income data shown in Figure 2 of Park and Marron (1990). The data and importance of that type of display are discussed there. Several of the bootstrap bandwidth selectors considered in this paper were tried on this data set. The best result was for SBF with the $N(0,\hat{\sigma}^2)$ reference distribution used immediately. Figure 2 here, which compares nicely to Figure 2 in Park and Marron shows the result. The other variants, involving estimation steps in the choice of g, tended to give smaller bandwidths, which are probably closer to the MISE value, but gave estimates that are too rough for effective presentation of this type.

REFERENCES

Chiu, S. T. (1990) Bandwidth selection for kernel density estimation, unpublished manuscript.

Devroye, L. and Györfi, L. (1984), Nonparametric density estimation: the L_1 view. Wiley, New York.

Efron, B. (1979) Bootstrap methods: another look at the jackknife, Annals of Statistics, 7, 1–26.

Efron, B. (1982) The jackknife, the bootstrap and other resampling plans, CBMS Regional Conference series in Applied Mathematics, Society for Industrial and Applied Mathematics, Philadelphia.

Faraway, J. J. and Jhun, M. (1987) Bootstrap choice of bandwidth for density estimation, unpublished manuscript.

Hall, P. (1990) Using the bootstrap to estimate mean squared error and select smoothing parameter in nonparametric problems, to appear in Journal of Multivariate Analysis.

Hall, P. and Marron, J. S. (1987) Estimation of integrated squared density derivatives, Statistics and Probability Letters, 6, 109–115.

Hall, P. and Marron, J. S. (1990) Lower bounds for bandwidth selection in density estimation, to appear in Probability Theory and Related Fields.

Hall, P., Marron, J. S. and Park, B. U. (1990) Smoothed cross–validation, unpublished manuscript.

Jones, M. C. and Sheather, S. J. (1990) Using nonstochastic terms to advantage in kernel–based estimation of integrated squared density derivatives, unpublished manuscript.

Jones, M. C., Marron, J. S. and Park, B. U. (1990) A simple root n bandwidth selector, unpublished manuscript.

Marron, J. S. (1988) Automatic smoothing parameter selection: a survey, Empirical Economics, 13, 187–208.

Müller, H.–G. (1985) Empirical bandwidth choice for nonparametric kernel regression by means of pilot estimators, Statistics and Decisions, Supplement no. 2, 193–206.

Scott, D. W. and Härdle, W. (1990) Weighted averaging using rounded points, to appear in Journal of the Royal Statistical Society, Series B.

Scott, D. W. and Terrell, G. R. (1987) Biased and unbiased cross–validation in density estimation, Journal of the American Statistical Association, 82, 1131–1146.

Sheather, S. J. and Marron, J. S. (1990) Kernel quantile estimation, to appear in Journal of the American Statistical Association.

Silverman, B. W. (1986) Density Estimation for Statistics and Data Analysis, Chapman and Hall, New York.

Taylor, C. C. (1989) Bootstrap choice of the smoothing parameter in kernel density estimation, Biometrika 76, 705–712.

A CIRCULAR BLOCK-RESAMPLING PROCEDURE FOR STATIONARY DATA

by

Dimitris N. Politis and Joseph P. Romano

Dimitris N. Politis
Department of Statistics
Purdue University
West Lafayette, IN 47907-1399

Joseph P. Romano
Department of Statistics
Stanford University
Stanford, CA 94305

ABSTRACT

A block-resampling bootstrap for the sample mean of weakly dependent stationary bservations has been recently introduced by Künsch (1989) and independently by Liu and ingh (1988). In Lahiri (1990) it was shown that the bootstrap estimate of sampling dis-ibution is more accurate than the normal approximation, provided it is centered around ie bootstrap mean, and not around the sample mean as customary. In this report, we itroduce a variant of this block-resampling bootstrap that amounts to 'wrapping' the data round in a circle before blocking them. This 'circular' block-resampling bootstrap, has ie additional advantage to be automatically centered around the sample mean. The con-stency and asymptotic accuracy of the proposed method are demonstrated in comparison ith the corresponding results for the block-resampling bootstrap.

MS 1980 subject classifications: Primary 62G05; secondary 62M10.

ey words and phrases: Resampling schemes, bootstrap, time series, mixing sequences, eak dependence, nonparametric estimation.

Exploring the Limits of Bootstrap. Edited by Raoul LePage and Lynne Billard.
 ISBN: 0471-53631-8.

1. Introduction

Suppose $X_1, \ldots, X_N$ are observations from the (strictly) stationary multivariate sequence $\{X_n, n \in \mathbf{Z}\}$, and the statistic of interest is the sample mean $\bar{X}_N = N^{-1}\sum_{i=1}^{N} X_i$. The sequence $\{X_n, n \in \mathbf{Z}\}$ is assumed to have a weak dependence structure. Specifically, the α-mixing (also called strong mixing) condition will be assumed, i.e. that $\alpha_X(k) \to 0$, as $k \to \infty$, where $\alpha_X(k) = \sup_{A,B} |P(A \cap B) - P(A)P(B)|$, and $A \in \mathcal{F}_{-\infty}^{0}, B \in \mathcal{F}_{k}^{\infty}$ are events in the σ-algebras generated by $\{X_n, n \leq 0\}$ and $\{X_n, n \geq k\}$ respectively.

The objective is to set confidence intervals for $\mu = EX_1$, for which an approximation to the sampling distribution of $\bar{X}_N$ is required. For this reason, a block-resampling bootstrap procedure has been introduced by Künsch (1989) and independently by Liu and Singh (1988). This method can be described as follows:

- Define B_i to be the block of b consecutive observations starting from X_i, that is $B_i = (X_i, \ldots, X_{i+b-1})$, where $i = 1, \ldots, q$ and $q = N - b + 1$. Sampling with replacement from the set $\{B_1, \ldots, B_q\}$, defines a (conditional on the original data) probability measure P^*. If k is an integer such that $kb \sim N$, then letting $\xi_1, \ldots, \xi_k$ be drawn i.i.d. from P^*, it is seen that each ξ_i is a block of b observations $(\xi_{i,1}, \ldots, \xi_{i,b})$. If all $l = kb$ of the $\xi_{i,j}$'s are concatenated in one long vector denoted by $Y_1, \ldots, Y_l$, then the block-resampling bootstrap estimate of the variance of $\sqrt{N}\bar{X}_N$ is the variance of $\sqrt{l}\bar{Y}_l$ under P^*, and the bootstrap estimate of $P\{\sqrt{N}(\bar{X}_N - \mu) \leq x\}$ is $P^*\{\sqrt{l}(\bar{Y}_l - \bar{X}_N) \leq x\}$, where $\bar{Y}_l = \frac{1}{l}\sum_{i=1}^{l} Y_i$.

It is obvious that taking $b = 1$ makes the block-resampling bootstrap coincide with the classical (i.i.d.) bootstrap of Efron(1979) which has well-known optimality properties (cf. Singh(1981)). It can be shown (cf. Lahiri(1990)) that a slightly modified bootstrap estimate of sampling distribution turns out to be more accurate than the normal approximation, under some regularity conditions, resulting to more accurate confidence intervals for μ. The modification amounts to approximating the quantiles of $P\{\sqrt{N}(\bar{X}_N - \mu) \leq x\}$ by the corresponding quantiles of $P^*\{\sqrt{l}(\bar{Y}_l - E^*\bar{Y}_l) \leq x\}$, where $E^*\bar{Y}_l$ denotes the expected value of $\bar{Y}_l$ under the P^* probability (conditional on the original data). The need for re-centering the bootstrap distribution so as to have mean zero can also be traced back to Künsch's (1989) short calculation of the skewness of his block-resampling bootstrap.

The reason that the re-centered bootstrap provides a more accurate approximation is that $E^*\bar{Y}_l = q^{-1}\sum_{i=1}^{q} b^{-1}\sum_{j=i}^{i+b-1} X_j = \bar{X}_N + O_p(b/N)$, where, for consistency of the bootstrap in the dependent data setting, $b \to \infty$ as $N \to \infty$. In other words, the distribution of $\bar{Y}_l$ under P^* possesses a random bias of significant order. This bias is associated with the block-resampling bootstrap that assigns reduced weight to X_i's with $i < b$ or $i > N-b+1$. In other words, if we let P_i^* be the limit (almost sure in P^*) of the proportion l^{-1}(number of the Y_j's that equal X_i) as $l \to \infty$, (and assuming no ties among the X_i's), although $P_i^* = b/R$, with $R = b(N-b+1)$, for any i such that $b \leq i \leq N-b+1$, this proportion drops to $P_i^* = i/R$, for any $i < b$, and $P_i^* = (N-i+1)/R$, for any $i > N-b+1$.

2. A circular block-resampling bootstrap

A simple and 'automatic' way to have an unbiased bootstrap distribution is to 'wrap' the X_i's around in a 'circle', that is, to define (for $i > N$) $X_i \equiv X_{i_N}$, where $i_N = i(\text{mod}N)$, and $X_0 \equiv X_N$. This idea is closely associated with the definition of the circular autocovariance sequence of time series models. The 'circular' block-resampling bootstrap amounts to resampling whole 'arcs' of the circularly defined observations, and goes as follows.

- Define the blocks B_j as previously, that is, $B_i = (X_i, \ldots, X_{i+b-1})$, but note that now for any integer b, there are N such B_j, $j = 1, \ldots, N$. Sampling with replacement from the set $\{B_1, \ldots, B_N\}$, defines a (conditional on the original data) probability measure P^*. If k is an integer such that $kb \sim N$, then letting $\xi_1, \ldots, \xi_k$ be drawn i.i.d. from P^*, it is seen that each ξ_i is a block of b observations $(\xi_{i,1}, \ldots, \xi_{i,b})$. If all $l = kb$ of the $\xi_{i,j}$'s are concatenated in one long vector denoted by $Y_1, \ldots, Y_l$, then the 'circular' block-resampling bootstrap estimate of the variance of $\sqrt{N}\bar{X}_N$ is the variance of $\sqrt{l}\bar{Y}_l$ under P^*, and the 'circular' block-resampling bootstrap estimate of $P\{\sqrt{N}(\bar{X}_N - \mu) \leq x\}$ is $P^*\{\sqrt{l}(\bar{Y}_l - \bar{X}_N) \leq x\}$, where $\bar{Y}_l = \frac{1}{l}\sum_{i=1}^{l} Y_i$.

The 'circular' construction is an integral part of a related resampling method in which blocks of random size are resampled (cf. Politis and Romano (1991)). It can also be immediately applied to bootstrapping general linear and nonlinear statistics, as in Künsch (1989), Liu and Singh (1988), and Politis and Romano (1989). Künsch's proposal of 'tapering' the observations in the B_j blocks can also be incorporated in the 'circular' construction without changing the first-order asymptotic results.

The following two theorems concern the consistency and asymptotic accuracy of the 'circular' block-resampling bootstrap. The theorems are stated for univariate sequences $\{X_n, n \in \mathbf{Z}\}$, although their extension to multivariate settings is straightforward.

Theorem 1 *Assume that* $E|X_t|^{6+\delta} < \infty$, *for some* $\delta > 0$, *and* $\sum_{k=1}^{\infty} k^2(\alpha_X(k))^{\frac{\delta}{6+\delta}} < \infty$. *As* $N \to \infty$, *let* $l/N \to 1$, *and let* $b \to \infty$, *but with* $bN^{-1} \to 0$. *Then* $\sigma_N^2 \equiv Var(\sqrt{N}\bar{X}_N)$ *has a finite limit* σ_∞^2, *and* $Var^*(\sqrt{l}\bar{Y}_l) \xrightarrow{P} \sigma_\infty^2$, *where* $Var^*(\sqrt{l}\bar{Y}_l)$ *is the variance of* $\sqrt{l}\bar{Y}_l$ *under* P^*

conditional probability, as well as

$$\sup_x |P^\star\{\sqrt{l}(\bar{Y}_l - \bar{X}_N) \le x\} - P\{\sqrt{N}(\bar{X}_N - \mu) \le x\}| \to 0 \quad (1)$$

for almost all sample sequences $X_1, \ldots, X_N$.

Proof. The proof of Theorem 1 is immediate in view of the proof of the consistency of the block-resampling bootstrap of Künsch(1989). If we let $E^\star, E^*, Var^\star, Var^*$, represent expectation and variance under the $P^\star$ and P^* probabilities, then it is easily calculated that $E^\star \bar{Y}_l = \bar{X}_N$, and that

$$Var^\star(\sqrt{l}\bar{Y}_l) = \frac{b}{N}\sum_{i=1}^{N}(b^{-1}\sum_{j=i}^{i+b-1} X_i - \bar{X}_N)^2$$

$$= \frac{b}{N}\{\sum_{i=1}^{N-b+1}(b^{-1}\sum_{j=i}^{i+b-1} X_i - \bar{X}_N)^2 + \sum_{i=N-b+2}^{N}(b^{-1}\sum_{j=i}^{i+b-1} X_i - \bar{X}_N)^2\} = Var^*(\sqrt{l}\bar{Y}_l) + O_p(b/N)$$

where it was used that $E^*\bar{Y}_l = \bar{X}_N + O_p(b/N)$, and

$$Var^*(\sqrt{l}\bar{Y}_l) = \frac{b}{N-b+1}\sum_{i=1}^{N-b+1}(b^{-1}\sum_{j=i}^{i+b-1} X_i - E^*\bar{Y}_l)^2$$

$$= \frac{b}{N-b+1}\sum_{i=1}^{N-b+1}(b^{-1}\sum_{j=i}^{i+b-1} X_i - \bar{X}_N)^2 + O_p(b/N)$$

Since $Var^*(\sqrt{l}\bar{Y}_l) \xrightarrow{P} \sigma^2_\infty$, it is seen that the first two moments of $\sqrt{l}\bar{Y}_l$ under the $P^\star$ probability are asymptotically correct.

Finally, the moment and mixing conditions assumed are sufficient (cf. Hall and Heyde(1980)) to imply that $\sqrt{N}(\bar{X}_N - \mu)$ is asymptotically normal $N(0, \sigma^2_\infty)$. Noting that $\sqrt{l}(\bar{Y}_l - \bar{X}_N)$ is also asymptotically normal (conditionally on the data $X_1, \ldots, X_N$), the proof is concluded. □

It is noteworthy that to make the bias of the block-resampling bootstrap distribution to be of smaller order than its standard deviation, Künsch (1989) imposed the additional condition $bN^{-1/2} \to 0$ which is unnecessarily strong in our setting.

Theorem 2 *Assuming that the sequence $\{X_n, n \in \mathbf{Z}\}$ is defined on the probability space $(\Omega, \mathcal{A}, P)$, denote $\mathcal{D}_n, n \in \mathbf{Z}$, a sequence of sub σ-fields of $\mathcal{A}$, and $\mathcal{D}_{n_1}^{n_2}$ the σ-field generated by $\mathcal{D}_{n_1}, \ldots, \mathcal{D}_{n_2}$. Also assume that $E|X_t|^{6+\delta} < \infty$, for some $\delta > 0$, and, as $N \to \infty$, let*

$l/N \to 1$, and $b \to \infty$, but with $bN^{-1/3} \to 0$. Under the following regularity conditions:

(a_0) $\alpha_X(k)$ decreases geometrically fast;

(a_1) $\exists d > 0$ such that for all $k, n \in \mathbf{N}$, with $n > 1/d$, there exists a $\mathcal{D}_{k-n}^{k+n}$ measurable random variable $Z_{k,n}$, for which $E|X_k - Z_{k,n}| \leq d^{-1}e^{-dn}$, and $E|Z_{k,n_k}|^{12}I(|Z_{k,n_k}| < k^{1/4}) < d^{-1}$, where n_k is a sequence of real numbers satisfying $\log k = o(n_k)$ and $n_k = O(\log k)^{1+d^{-1}}$, as $k \to \infty$;

(a_2) $\exists d > 0$ such that for all $k, n \in \mathbf{N}$, with $k > n > 1/d$, and all $t > d$,

$$E|E(e^{jt(X_{k-n}+X_{k-n+1}+\dots+X_{k+n})}|\mathcal{D}_i, i \neq k)| \leq e^{-d}$$

where j is used to denote the imaginary unit $\sqrt{-1}$;

(a_3) $\exists d > 0$ such that for all $k, n_1, n_2 \in \mathbf{N}$, and $A \in \mathcal{D}_{n_1-n_2}^{n_1+n_2}$,

$$E|P(A|\mathcal{D}_i, i \neq n_1) - P(A|\mathcal{D}_i, 0 < |n_1 - i| \leq k + n_2)| \leq d^{-1}e^{-dk};$$

the following is true (provided of course that $\sigma_\infty^2 > 0$),

$$\sup_x |P^\star\{\sqrt{l}\frac{\bar{Y}_l - \bar{X}_N}{\sqrt{Var^\star(\sqrt{l}\bar{Y}_l)}} \leq x\} - P\{\sqrt{N}\frac{\bar{X}_N - \mu}{\sigma_\infty} \leq x\}| = o_p(N^{-1/2}) \tag{2}$$

Proof. As the proof of Theorem 1 relied on comparing the first two moments of $\sqrt{l}(\bar{Y}_l - \bar{X}_N)$ under $P^\star$ with the corresponding ones under P^*, the proof of Theorem 2 follows by looking at the third order moment. Specifically, under the regularity conditions we have assumed, first order Edgeworth expansions for $P^\star\{\sqrt{l}\frac{\bar{Y}_l - \bar{X}_N}{\sqrt{Var^\star(\sqrt{l}\bar{Y}_l)}} \leq x\}$ and for $P\{\sqrt{N}\frac{\bar{X}_N - \mu}{\sigma_\infty} \leq x\}$ are valid (cf. Lahiri(1990) where an extensive discussion relative to these regularity conditions can be found). Furthermore, equation (2) would be true, provided $b^2E^\star(U_1^\star - \bar{X}_N)^3 - N^2E(\bar{X}_N - \mu)^3 = o_p(1)$, where $U_1^\star = b^{-1}\sum_{j=1}^b \xi_{1,j}$, and the $\xi_{1,1}, \dots, \xi_{1,b}$ are the elements of the first block-resample drawn from $P^\star$.

However, in Lahiri(1990) it was shown that $b^2E^*(U_1^* - E^*U_1^*)^3 - N^2E(\bar{X}_N - \mu)^3 = o_p(1)$, where $U_1^* = b^{-1}\sum_{j=1}^b \xi_{1,j}$, and the $\xi_{1,1}, \dots, \xi_{1,b}$ are the elements of the first block-resample drawn from P^*. Finally, it is easily seen that

$$E^\star(U_1^\star - \bar{X}_N)^3 = \frac{1}{N}\sum_{i=1}^{N}(b^{-1}\sum_{j=i}^{i+b-1} X_i - \bar{X}_N)^3$$

$$= \frac{1}{N}\{\sum_{i=1}^{N-b+1}(b^{-1}\sum_{j=i}^{i+b-1} X_i - \bar{X}_N)^3 + \sum_{i=N-b+2}^{N}(b^{-1}\sum_{j=i}^{i+b-1} X_i - \bar{X}_N)^3\} = E^*(U_1^* - E^*U_1^*)^3 + O_p(b/N)$$

where it was used that $E^*U_1^* = \bar{X}_N + O_p(b/N)$, and

$$E^*(U_1^* - E^*U_1^*)^3 = \frac{1}{N-b+1}\sum_{i=1}^{N-b+1}(b^{-1}\sum_{j=i}^{i+b-1} X_i - E^*\bar{Y}_l)^3$$

$$= \frac{1}{N-b+1}\sum_{i=1}^{N-b+1}(b^{-1}\sum_{j=i}^{i+b-1} X_i - \bar{X}_N)^3 + O_p(b/N)$$

Hence, $E^*(U_1^* - \bar{X}_N)^3 - E^*(U_1^* - E^*U_1^*)^3 = O_p(b/N) = o_p(b^{-2})$, and the proof is concluded.□

References

[1] Efron, B. (1979), Bootstrap Methods: Another Look at the Jackknife, *Ann. Statist.*, 7, 1-26.

[2] Hall, P. and Heyde, C. (1980), *Martingale Limit Theory and its Applications*, Academic Press, New York.

[3] Künsch H.R. (1989), The jackknife and the bootstrap for general stationary observations, *Ann. Statist.*, 17, 1217-1241.

[4] Lahiri, S.N.(1990), Second order optimality of stationary bootstrap, Technical Report No.90-1, Department of Statistics, Iowa State University, (to appear in *Statist. Prob. Letters*).

[5] Liu, R.Y. and Singh, K. (1988), Moving Blocks Bootstrap and Jackknife Capture Weak Dependence, unpublished manuscript, Department of Statistics, Rutgers University.

[6] Politis D.N. and Romano, J.P. (1989), A General Resampling Scheme for Triangular Arrays of α-mixing Random Variables with application to the problem of Spectral Density Estimation, Technical Report No.338, Department of Statistics, Stanford University.

[7] Politis D.N. and Romano, J.P. (1991), The Stationary Bootstrap, Technical Report No. 365, Department of Statistics, Stanford University.

[8] Singh, K.(1981), On the asymptotic accuracy of Efron's bootstrap, *Ann.Statist.*, 9, 1187-1195.

Some applications of the bootstrap in complex problems

Robert Tibshirani
Department of Preventive Medicine and Biostatistics
and
Department of Statistics
University of Toronto

1 Introduction

In this paper I give two examples of the application of the bootstrap in some complex problems. These problems arose as part of statistical consultations.

2 Prediction limits for exercise output

Dr. G. Canny of Toronto's Hospital for Sick Children collected data on the exercise performance for 118 normal children. Each child was given a series of increasing workloads on a stationary bicycle and the heart rate was recorded at each workload. The workloads were irregularly spaced and different for each child, and the maximum workload used depended somewhat on the observed heart rate. Sex and height of the children were recorded and were thought to be important factors. The objective was to construct heart rate prediction limits for each sex-height group for use in testing children who were suspected to have asthma.

In order to obtain heart rate values for a common set of workloads, an interpolating cubic spline was fit to the data for each child. Heart rate values for workloads 0 to 250 watts in increments of 25 watts were then extracted. For each sex-height group a cubic smoothing spline was then fit to the mean

Exploring the Limits of Bootstrap. Edited by Raoul LePage and Lynne Billard.
 ISBN: 0471-53631-8.

heart rates. The smoothing parameter was chosen so that the degrees of freedom of the fitted curve was about 4 (see Hastie and Tibshirani 1990 chapter 3). This fixed smoothing parameter was used throughout. Despite the correlation across workloads for a given individual, the mean heart rate value is the generalized least squares estimate of the within group mean; see for example Rice and Silverman, 1988. The results are the solid curves in Figure 1.

The construction of prediction bands for these curves seems very difficult analytically. Denote the estimated curve at workload w by $\hat{f}(w)$. Since the smoothing operation is linear, an estimate of the pointwise standard error of $\hat{f}(w)$ is easily obtained, say $s(w)$. Then if our nominal level is 80%,

$$\hat{f}(w) \pm 1.28(1 + \frac{1}{n})s(w) \tag{1}$$

represents an approximate central pointwise prediction interval for normal child's heart rate profile. However, what is wanted here is a global rather than pointwise band, so that there is a fixed probability of a new child's heart rate falling outside of the bands at any of the workloads $0, 25, \ldots 250$. For this reason, the pointwise bands need to be widened. In addition, there is correlation across workloads for a given child and this will also make the pointwise bands too narrow. Still another factor is the assumption of normality used in construction of (3).

The bootstrap can be used to estimate the amount by which the pointwise bands should be widened. Consider prediction bands of the form

$$\left(\hat{f}(w) - c_1(1 + \frac{1}{n})s(w), \hat{f}(w) + c_2(1 + \frac{1}{n})s(w)\right) \tag{2}$$

We allow c_1 and c_2 to differ to allow the bands to be different widths below and above the curve. Then we seek the value of c_1 so that the probability of a new curve falls outside of the lower band at any workload is 10%, and similarly for c_2. To apply the bootstrap, we sample children with replacement, sampling x_j times from the jth sex-height group, where x_j is the number of children in the jth group. For a fixed values of c_1 and c_2, prediction bands of the form (1) are constructed from the bootstrap data, using interpolating and cubic smoothing splines as described above. Then the number of times that the original heart rate profiles fall outside the bands (at least once) is

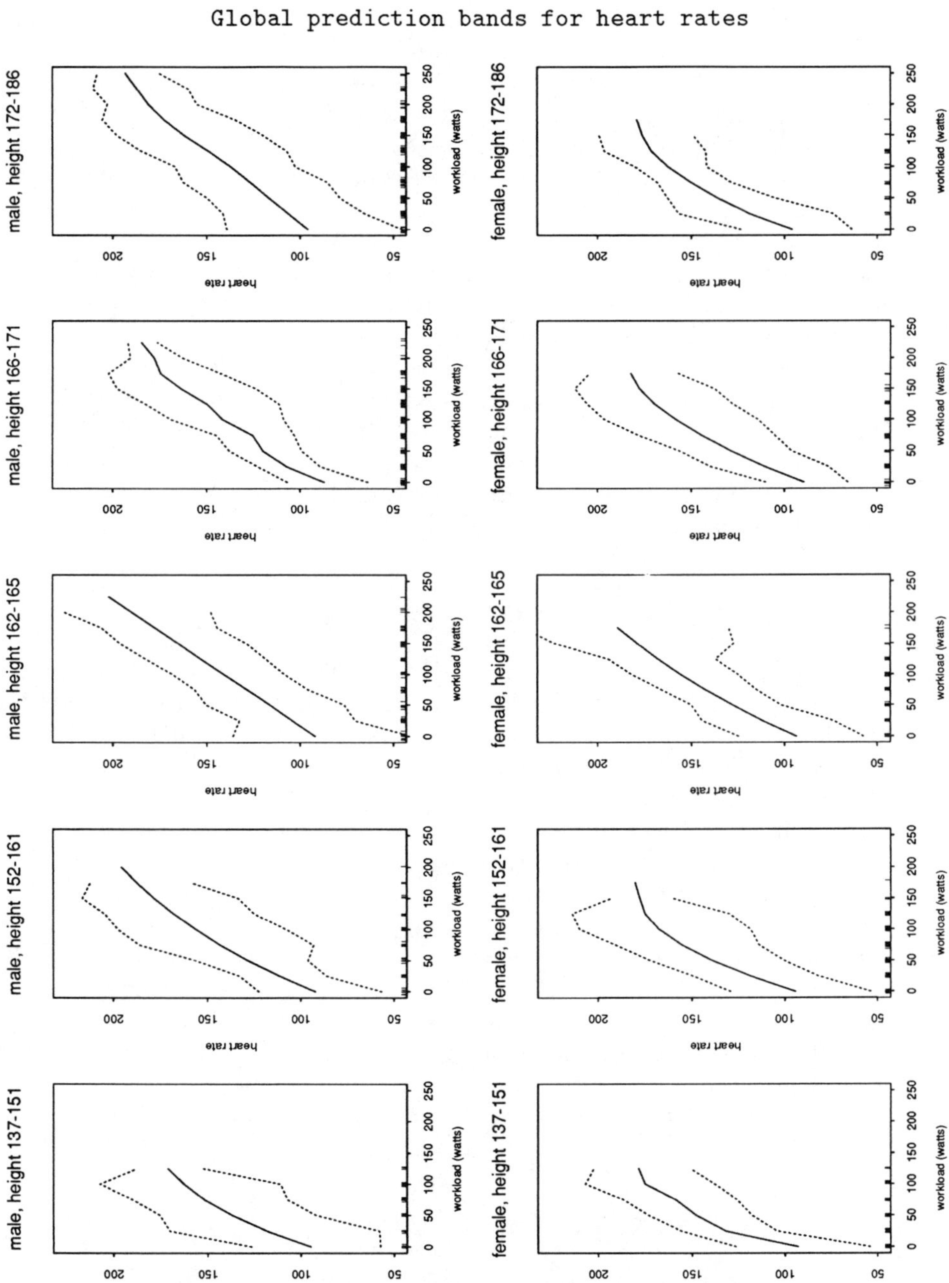

Fig. 1. *Solid curves are estimate of heart rates at each workload, for children in indicated sex-height group. Broken lines are 80% global prediction bands. Tick marks on the horizontal axis indicate workloads where measurements were taken.*

computed. This entire process is repeated, increasing c_1 and c_2 until the error rate, accumulated across all groups, is 10% below and above the curve.

This is an example of one of the prediction error estimators studied in Efron (1983).

The resulting values of c_1 and c_2 were 2.8 and 2.4, much larger than the pointwise values of 1.28. The global prediction bands are shown in Figure 1. There are clear differences between males and females, and between shorter and taller children. Since 11 workload values were used, a Bonferroni-type adjustment for the pointwise values would use $\Phi^{-1}(1-.1/11) = 2.36$, which is close to c_2 but not c_1. It is interesting to note that although the Bonferroni bounnds are usually conservative, the bootstrap bands, which account for other potential sources of error (e.g nonnormality) are wider in this problem.

A potential concern, pointed out by a referee, is the bias of a smoother in estimating the true underlying curve. In this particular problem, the curvature of the underlying functions does not seem to be substantial and hence bias is not a major concern. However, in general one might need to account for bias in forming the prediction bands.

3 Clustering of cortical cells

Dr. Derek Vanderkoy and Les Krushel of the Dept. of Anatomy, University of Toronto, conducted experiments to study the way in which cells evolve in the layers of the cortex from stem cells in the interior. The data consists of cell counts in 6 layers along "spokes" in a cortical cell of 17 animals. The layers are labelled 2 to 6. Each spoke represents descendents from a single stem cell; there are measurements on approximately 9 spokes per animal.

A certain theory suggests that the cells do not spread themselves uniformly through the layers but cluster in layers 2,3,4, or 5,6 in a given spoke.

Let x_{ij} represent the cell count in the jth layer for the ith spoke, x_{i+} be the total count in the ith spoke, and x_{++} be the total cell count. Figure 2 shows a plot of $(\sum_{j=2}^{4} x_{ij} - \sum_{j=5}^{6} x_{ij})/x_{i+}$ by animal. The plotting symbol is x_i.

There seems to be an abundance of points near -1 and 1, giving some credence to the clustering theory.

To formally test this, we consider the mixed multinomial model

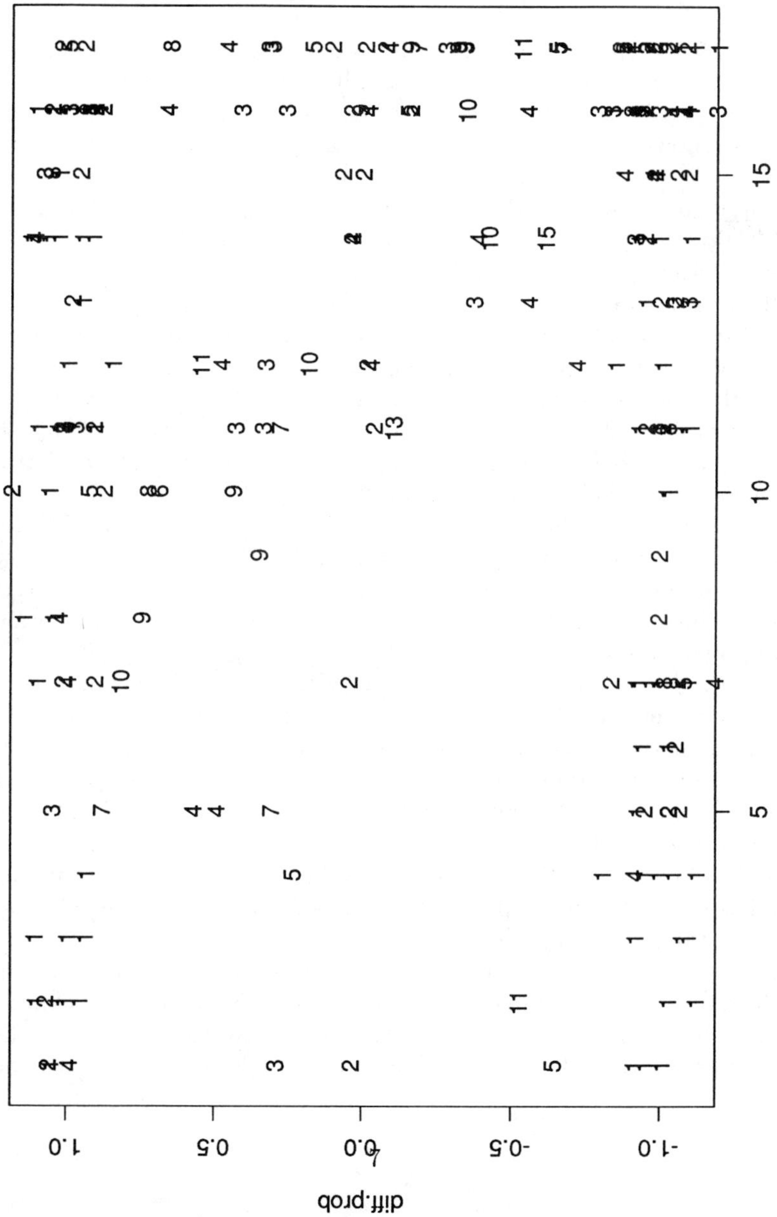

Fig. 2. *Plot of difference in proportion of cells in layers 2,3,4 vs 5,6, by animal. The plotting symbol is the total cell count in a given spoke.*

$$(x_{i2}\dots x_{i5}) \sim (1-\theta)M(x_{i+},\{p_1,p_1,p_1,q_1,q_1\}) + \theta M(x_{i+},\{p_2,p_2,p_2,q_2,q_2\}) \quad (3)$$

where $M(m,\{p_1,p_2,\dots p_5\})$ represents a five-category multinomial distribution with n draws and success probabilities $p_1,\dots p_5$, and $1/5 \le p_1 \le 1/3$, $0 \le p_2 \le 1/5$, and $3p_i + 2q_i = 1$, $i = 1,2$. The idea behind this model is as follows. Examination of the data revealed that $x_{+2} \approx x_{+3} \approx x_{+4}$, and $x_{+5} \approx x_{+6}$. Thus a reasonable "no-clustering model" would be $M(x_{i+},\{p_1,p_1,p_1,q_1,q_1\})$ for the counts in the *ith* spoke. The alternative model (2) (along with the restrictions on the p_is) says that the population is a mixture of two populations, one that clusters in layers 2,3,4 and the other in 5,6. The parameter θ determines the proportion of each type. The "no clustering" model results if $\theta = 0$ or 1 and is therefore nested in the mixture model. Note that at this point we are ignoring potential animal effects.

We fit this model to the data by maximum likelihood. The estimates were $\hat{\theta} = .556$, $\hat{p}_1 = .253$, $\hat{p}_2 = .060$, giving a log-likelihood value of $l_{max} = -769.97$. The model is suggesting an approximate 50-50 mixture of two multinomials with probabilities $(.253,.253,.253,.12,.12)$ and $(.06,.06,.06,.41,.41)$. The estimates for the no clustering model were $\hat{p} = .154$ and $l^0_{max} = -807.47$. The twice log-likelihood ratio statistic $R = 2[l_{max} - l^0_{max}]$ equals 75.0, a very large value considering that the mixture model has only 2 more parameters than the no clustering model. However, it is not clear that the distribution of R is approximately χ^2_2.

To investigate the distribution of R, we apply a simulation or bootstrap technique. Datasets are generated from the no clustering model, using $\hat{p}_1$, fixing the x_{i+} at their observed values, and computing R for each dataset. The maximum value of R over 200 simulations was about 6, suggesting that the observed value of 75.0 is highly unlikely to have occurred by chance if the no clustering model truly held.

Animal effects could still be inflating the value of R, so we tried sampling by animal as well. The maximum value of R was 35, so we still conclude that the observed value is highly significant.

There were numerous other questions that arose in this study. One involved the alternative clustering 2,3 vs. 4,5,6. The maximized log-likelihood value was -794.47 for this clustering, or 24.5 units lower than that for the (2,3,4 vs 5,6) clustering. The investigators wanted to know which of the two clusterings (2,3,4 vs 5,6) or (2,3 vs. 4,5,6) was better. The two models are not nested, so this is a difficult question to answer. To address it, I bootstrapped by sampling with replacement from the rows of the data matrix, and for each bootstrap sample, I computed the difference in maximized log-likelihood values for the two models. The standard deviation of this difference was 10.4, and so the observed difference of 24.5 is "significantly" large.

Acknowlegment:

I would like to thank William DuMouchel for helpful suggestions. This work was supported by the Natural Sciences and Engineering Research Council of Canada.

References:

Efron, B. (1983). Estimating the error-rate of a prediction rule: some improvements on cross-validation. J. Amer. Statist. Assoc., vol 78, pages 316-331.

Hastie, T. and Tibshirani, R. (1990) *Generalized Additive Models.* Chapman and Hall. London.

Rice, J.A and Silverman, B.W. (1988) Estimating the mean and covariance structure nonparametrically when the data are curves. Unpublished report.

APPROXIMATING THE DISTRIBUTION OF A GENERAL STANDARDIZED FUNCTIONAL STATISTIC WITH THAT OF JACKKNIFE PSEUDO VALUES

by

D. S. Tu

Center for Multivariate Analysis
The Pennsylvania State University

ABSTRACT

By randomly weighting the jackknifed pseudo values, a method to approximate the distribution of a general standardized differentiable functional statistic is proposed. The computation based on this method is simpler, the variance estimator is similar to that of jackknife variance estimator, and this approach also posesses the second order accuracy.

1. INTRODUCTION

The jackknife method, deleting one observation each time from the sample for generating pseudo values and then averaging those values to get the new estimate, was used mainly to reduce the bias of an estimator and construct the variance estimates for complex statistics. The properties of this method were investigated by many authors from the end of 1950's. For detailed studies and applications of this method, the reader may refer to review papers of Miller(1974) and Shi(1987) and the bibliography by Frangos (1987).

In this paper we will consider another use of jacknife pseudo

Exploring the Limits of Bootstrap. Edited by Raoul LePage and Lynne Billard.
 ISBN: 0471-53631-8.

values, that is, to give the approximation for the distribution of a general standardized functional statistic. Let $X_1, X_2, \ldots, X_n$ be independent identically distributed random variables with unknown cdf F. Suppose that $T_n=T(F_n)$ is an estimate of $\theta=T(F)$, where T is a specified real-valued functional and F_n is the empirical distribution of $X_1, X_2, \ldots, X_n$. Then our problem is to approximate the unknown distribution H_n of $(T_n-\theta)/[var(T_n)]^{1/2}$.

Since under some suitable conditions H_n is asymptotically the normal distribution with mean 0 and variance 1, we may simply use the normal distribution to approximate H_n. One of disadvantages of normal approximation lies in that it does not possess a higher order accuracy. To get an approximation with higher order accuracy, we may use the Edgeworth expansions which have been discussed in detail by Pfanzagl(1985). The exact form of Edgeworth expansions of H_n can be found in Takahashi(1988). An alternative but simpler approach, the bootstarp introduced by Efron(1979), can also give a second order accurate approximation with some superiorties over the one-term Edgeworth expansion approximation (Bhattacharya and Qumisiyeh, 1989). To use the bootstrap method, we need resampling from the original sample and recomputing T_n many times.

Our operation to approximate the distribution of a functional statistic is based on the idea of random weighting. We generate a group of random weights directly from the computer, use them to weight jackknife pseudo values, and then employ the conditional distribution of the weighted average of jackknife pseudo values (with a suitable transformation) given the original sample to approximate the distribution of the general functional statistic. The computation in this approach is simpler(which needs the computation of the original statistic n(n+1)/2 times) compared with the bootstrap(which needs the computation of the statistic 1,000~2,000 times(Efron, 1987)) for moderate and small n. The variance estimator deduced from this method is similar to that of the jackknife estimator. Moreover, this approach possesses also the second order accuracy and in some cases conditions guaranting this property are weaker than that for the bootstrap approach(see the remark in Section 4). By selecting weights properly, the random weighting approximation may even be more accurate than bootstrap approximation for tail probabilities (Zhang and Tu, 1990). For a detailed discussion about the motivation and general properties of the random weighting method, see Tu and Zheng (1989).

Beran(1984) and Hinkley and Wei(1984) also considered the use of the jackknife method to approximate H_n. The idea of their approaches is using the jackknife to estimate the standardized bias, variance and skewness of T_n and then pluging these jackknife estimates into the Edgeworth expansion of H_n to get a jackknife Edgeworth expansion estimate. The difference between their approach and ours lies in that the jackknife method is used before asymptotic expansion in our approach. The improvement for approximate accuracy via the transformation on statistics has also been considered by Hall(1983), Withers (1983) and Abramovitch and Singh(1985). The first two papers gave the modification only in a compact interval of real line. The improvement given in the later paper is over the whole line. Our work is closely related to that of Beran(1984), Hinkley and Wei(1984) and Abramovitch and Singh(1985). In fact, an exact form of transformation suggested in Abramovitch and Singh (1985) is given by combining methods of the jackknife used in Beran (1984) and Hinkley and Wei(1984) and the random weighting suggested in Tu and Zheng(1989) for the functional statistics. Recently, Konishi(1989) presented an improvement to normal approximation by finding the normalizing transformation for

T_n.

The rest of the paper is orginized as follows: In Secton 2 we give some preliminary notations and basic ideas, including the definition of a general functional statistic and its random weighting version. The main results of this paper, concerning the second order accuracy of our approach, are stated in Section 2 and proved in Section 3. Finally in Section 4 we apply these results to U-statistics and M-estimates.

2. SOME PRELIMINARY NOTATIONS AND BASIC IDEAS

Let Λ be the space of distribution functions containing a fixed distribution F and all discrete distributions with finite support and R be the real line. A mapping from Λ to R is called a functional on Λ. For this functional we introduce the definition of the strong Frechet differentiability at $F\in\Lambda$.

Definition 2.1 T is said to possess a strong Frechet differential at F if and only if there exists ψ: $R \longrightarrow R$ depending only on T and F such that

$$\frac{|T(G)-T(H)-\int\psi(x)d(G-H)(x)|}{\|G-H\|} \longrightarrow 0 \tag{2.1}$$

as $\|G-F\|+\|H-F\| \longrightarrow 0$ for G, $H\in\Lambda$, where $\|h\| \overset{\Delta}{=} \sup_x |h(x)|$. Without the loss of generality in this definition we can take $\int\psi(x)dF(x)=0$.

This definition of differentiability was used by Parr(1985) to discuss the consistency of jackknife variance estimates. It is stricter than that of Frechet differentiability or Hadamard differentiability. For the relationship between these definitions and examples satisfying this definition, refer to Parr(1985) and Reeds (1976).

Further we can give the definition of the second order

differentiability of T at F.

DEFINITION 2.2 T is said to possess a strong Frechet second order differential with remainder of type $\mathcal{G}_b^{2,\lambda}(\lambda>0)$ at F if and only if there exist ψ: $R \rightarrow R$ and φ: $R^2 \rightarrow R$ depending only on T and F and a bounded set $\tilde{B} \subset R$ such that

$$\frac{|T(G)-T(H)-\int\psi(x)d(G-H)(x)-\int\varphi(x,y)d(G-H)(x)d(G-H)(y)|}{||G-H||^{2+\lambda}} \in \tilde{B} \quad (2.2)$$

as $||G-H||+||H-F|| \longrightarrow 0$ for G, $H \in \Lambda$.

This definition was also given and discussed by Reeds(1976).

Let now $X_1, X_2, \ldots, X_n$ be n i.i.d. random variables with unknown distribution function F defined on R and $\theta=T(F)$ be a functional parameter of F. Denote

$$F_n(x) = \frac{1}{n}\sum_{i=1}^{n} I(X_i \le x), \quad (2.3)$$

where I(A) stands for the indicator function of the set A. Corresponding to the parameter θ, we consider the estimate $T_n = T(F_n)$, which is called the functional statistic. The main purpose of this paper is to estimate the unknown distribution of

$$L_n = (T_n-\theta)/[var(T_n)]^{1/2} \quad (2.4)$$

with that of jackknife pseudo values of T_n. First we introduce the jackknife pseudo values to T_n. Let

$$F_{n-1}^{(i)}(x)=(n-1)^{-1}\sum_{\substack{j\ne i\\ j=1}}^{n} I(X_j \le x) \qquad i=1, 2, \ldots, n, \quad (2.5)$$

$$T_{n-1}^{(i)}=T(F_{n-1}^{(i)}) \text{ and } T_{n,i}=nT_n-(n-1)T_{n-1}^{(i)} \quad (i=1, 2, \ldots, n). \quad (2.6)$$

Then $T_{n,i}(i=1, 2, \ldots, n)$ are called jackknife pseudo values of T_n and

$$T_n^* = \frac{1}{n}\sum_{i=1}^{n} T_{n,i} \quad \text{and} \quad U_n^* = (n-1)^{-1}\sum_{i=1}^{n}(T_{n,i}-T_n^*)^2 \tag{2.7}$$

are the classical jackknife estimate of θ and jackknife variance estimate of T_n, respectively. Next we give the random weighting statistic of $(T_n-\theta)$. Let $\varsigma_1, \varsigma_2, \ldots, \varsigma_n$ be n i.i.d. random variables with the common probability density

$$\rho(x) = \frac{1}{\Gamma(4)}(2x)^3 e^{-2x} I(x>0). \tag{2.8}$$

Then

$$W_n^* = \frac{1}{n}\sum_{=1}^{n}\varsigma_i(T_{ni}-T_n^*) \tag{2.9}$$

is definied as the random weighting statistic of $(T_n-\theta)$. The preliminary idea of random weighting is that we can use the conditional distribution H_n^* of $L_n^*=W_n^*/[\mathrm{var}*(W_n^*)]^{1/2}$ given X_1, $X_2, \ldots, X_n$ to approximate the distribution of L_n. The conditional variance of $\sqrt{n}W_n^*$,

$$s_n^2 = \mathrm{var}*(\sqrt{n}W_n^*) = \frac{1}{n}\sum_{i=1}^{n}(T_{ni}-T_n^*)^2, \tag{2.10}$$

where var^* stands for the conditional variance given $X_1, X_2, \ldots,$ X_n(Below we will use the symbol * to denote the conditional quantities given $X_1, \ldots, X_n$), can be used as an estimate for the variance of $\sqrt{n}T_n$. It is easily seen that s_n^2 is almost the same as the jackknife variance estimate U_n^*.

The consistency of above approximation can be proved by applying the Lindeberg Central Limit Theorem to L_n and L_n^* respectively. In comparing asymptotic expansions of H_n and H_n^* we find that, however, this approximation can not possess the desired second order accuracy(see the discussion in next section). Following the idea of Abramovitch and Singh(1985), Beran(1984) and Hinkley and Wei(1984) we can use a transformation on L_n^* via the jackknife method to get a second order accurate approximation. Let

$$h_n(y)=(y - \frac{\beta_{0n}}{\sqrt{n}}) - \frac{\beta_{2n}}{\sqrt{n}}(y - \frac{\beta_{0n}}{\sqrt{n}})^2 + \frac{\beta_{2n}^2}{3n}(y - \frac{\beta_{0n}}{\sqrt{n}})^3 \quad (2.11)$$

where

$$\beta_{0n}=\hat{\alpha}_{0n}- \frac{\sqrt{n}}{6} \sum_{j=1}^{n} r_{nj}^3, \quad \beta_{2n}=\hat{\alpha}_{2n}+ \frac{\sqrt{n}}{6} \sum_{j=1}^{n} r_{nj}^3,$$

$$\hat{\alpha}_{0n}= \frac{\hat{k}_{32}}{6} - \frac{\hat{k}_{31}}{2}, \quad \hat{\alpha}_{2n}= - \frac{\hat{k}_{32}}{6} \quad (2.12)$$

with the jackknife estimates

$$r_{ij}= \frac{(T_{n,j}-T_n^*)}{\left[\sum_{i=1}^{n} (T_{n,i}-T_n^*)\right]^{1/2}}, \qquad \hat{k}_{31}= \frac{(n-1) \sum_{i=1}^{n} [T_{n-1}^{(i)}-T_n]}{s_n},$$

$$\hat{k}_{32}= \frac{1}{s_n^3} \left\{\frac{(n-1)^3}{n} \sum_{i=1}^{n} (T_{n-1}^{(i)}-T_n)^3 \right.$$

$$+ \frac{3(n-1)^2}{n} \sum_{j\neq k}[T_{n-1}^{(j)}-T_n][T_{n-1}^{(k)}-T_n][nT_n-(n-1)(T_{n-1}^{(k)}+T_{n-1}^{(j)})$$

$$\left. +(n-2)T_{n-2}^{(k,j)}]\right\}$$

$$T_{n-2}^{(k,j)}=T(\frac{1}{n-2} \sum_{i\neq k,j} I\{X_i\leq x\}). \quad (2.13)$$

Then we can use the conditional distribution of $h_n(L_n^*)$ given X_1, X_2, ... , X_n to approximate the distribution of L_n. For the second order accuracy of this kind approximation we have a following theorem.

THEOREM 2.1 Assume that T is a twice strongly Frechet differentiable functional with remainder of type $\mathcal{G}_b^{2,\lambda}(\lambda>0)$ at F. Suppose also that the distribution of $\psi(X_1)$ is nonlattice and $E\{|T(F_n)|^4\}<\infty$. Then we have

$$\sqrt{n}\sup_x|P^*\{h_n(L_n^*)\leq y\}-P\{L_n\leq y\}|\longrightarrow 0 \quad \text{a.s.} \tag{2.14}$$

The proof of the theorem is deferred in next section. Next we make some general remarks for this theorem before closing this section.

REMARK 2.1 In the applications of jackknife and bootstrap, for example, to set a confidence interval for θ, we are mostly concerned about the approxiamation for the distribution of the studentized statistic

$$\tilde{L}_n=\sqrt{n}(T_n-\theta)/\tilde{s}_n,$$

where $\tilde{s}_n$ is an estimate of the variance of T_n, for example, the jackknife estimate s_n. The method of this paper can be used in principal in this case. The exact form of the transformation can be obtained using the jackknife estimates for the cumulants of $\tilde{L}_n$ given in Hinkley and Wei(1984). See also the discussion of Konishi (1989).

REMARK 2.2 In fact the method suggested in this paper can be used to approximate the distribution of the jackknifed statistic T_n^* of T_n and, even more, that of a functional jackknife statistic of T_n suggested by Sen(1988). The accuracy of the approximation to these statistics needs further investigation.

REMARK 2.3 The jackknife technique used in this paper has a little bit difference with that in Beran(1984). In fact, Beran used the add-one jackknife other than the ordinary delete-one jackknife used by most authors. The jackknife estimates $\hat{k}_{31}$ and $\hat{k}_{32}$ of this paper are essentially taken from Hinkley and Wei(1984).

REMARK 2.4 Numerical results for the comparision of the method suggested above with the bootstrap and some other related methods are presented in Zhang and Tu(1990).

3. THE SECOND ORDER ACCURACY OF THE RANDOM WEIGHTING APPROXIMATION

The purpose of this section is to prove Theorem 2.1, which shows that the random weighting approximation suggested in the above section is second order accurate. We first give some results about the Edgeworth expansions for a functional statistic and that for a weighted summation of iid random variables. The following theorem is taken from Takahashi(1988).

THEOREM 3.1 Let

$$G(x)=P\{[T(F_n)-T(F)]/\sqrt{VAR[T(F_n)]}\le x\} \tag{3.1}$$

and assume that T is a twice strongly Frechet differentiable functional with remainder of type $\mathcal{g}_b^{2,\lambda}(\lambda>0)$ at F. Suppose also that the distribution of $\psi(X_1)$ is nonlattice and $E\{|T(F_n)|^4\}<\infty$, then uniformly in x

$$G(x)=\Phi(x)-n^{-1/2}[k_{3,1}+(1/6)k_{3,2}(x^2-1)]\phi(x) +o(1/\sqrt{n}) \tag{3.2}$$

where $\Phi(x)$ and $\phi(x)$ are the distribution function and density function of standard normal, respectively,

$$\sigma^2=E\{\psi^2(X_1)\},\ k_{3,1}=\sigma^{-1}E\{\varphi(X_1,X_1)\} \tag{3.3}$$

and

$$k_{3,2}=\sigma^{-3}\{6E[\psi(X_1)\psi(X_2)\varphi(X_1,X_2)]\}+E\{\psi^3(X_1)\}. \tag{3.4}$$

REMARK 3.1 We note that the Edgeworth expansion given above has a little bit different form with that given in Takahashi(1988). The reason is that in Definition 2.2 the second Frechet derivative of T is one half of that given in Definition 2 of Takahashi(1988).

Next we discuss the expansion for the weighted summation of iid random variables.

THEOREM 3.2 Suppose that $\varsigma_1, \varsigma_2, \ldots, \varsigma_n$ are iid with density function as in (2.8), $\{a_{ni}\}$ is a sequence of triangle arrays satisfying

$$\sum_{j=1}^n a_{nj}=0,\quad 0<\frac{1}{n}\sum_{j=1}^n a_{nj}^2\to\sigma^2<\infty,\quad \frac{1}{\sqrt{n}}\max_{1\le j\le n}|a_{nj}|\to 0$$

$$\text{and } \overline{\lim_{n\to\infty}}\,\frac{1}{n}\sum_{j=1}^n|a_{nj}|^3\le\eta<\infty.$$

Let

$$K_n(y)=P\left\{\sum_{j=1}^n a_{nj}(\varsigma_j-2)/(\sum_{j=1}^n a_{nj}^2)^{1/2}\le y\right\} \tag{3.5}$$

then uniformly in x we have

$$K_n(y)=\Phi(y)-(1/6)\phi(y)(y^2-1)\sum_{j=1}^n r_{nj}^3+o(1/\sqrt{n}) \tag{3.6}$$

where

$$r_{nj}=a_{nj}/(\sum_{j=1}^n a_{nj}^2)^{1/2}.$$

To prove this result we need a lemma on the Edgeworth expansion for sums of independent random variables proved by Bai and Zhao (1986).

LEMMA 3.1 Suppose that $Y_1, Y_2, \ldots, Y_n$ is a sequence of independent random variables with

$$EY_j=0,\quad \sum_{j=1}^{n} EY_j^2>0 \text{ and } E|Y_j|^3<\infty.$$

Let

$$G_n(y)=P\left\{\sum_{j=1}^{n} Y_j/\left(\sum_{j=1}^{n} EY_j^2\right)^{1/2}\le y\right\} \tag{3.7}$$

then we have

$$|G_n(y)-\Phi(y)+(1/6)\phi(y)(y^2-1)\mu_3/(\mu_2)^{3/2}|$$

$$\le C\left\{\frac{1}{(\mu_2)^{3/2}}\sum_{j=1}^{n} E|W_{nj}|^3 + \frac{1}{(\mu_2)^2}\sum_{j=1}^{n} E|Y_j|^4\right.$$

$$\left.+n^6\left(\sup_{|t|\ge\delta_n}\frac{1}{n}\sum_{j=1}^{n}|\nu_j(t)| + \frac{1}{2n}\right)^n\right\} \tag{3.8}$$

where C is a absolute constant,

$$\mu_2=\sum_{j=1}^{n} EY_j^2,\quad \mu_3=\sum_{j=1}^{n} EY_j^3,\quad W_{nj}=Y_jI(|Y_j|>(\mu_2)^{1/2}),$$

$$\nu_j(t)=E\exp(itY_j) \text{ and } \delta_n = \frac{1}{12}\mu_2\left(\sum_{j=1}^{n} E|Y_j|^3\right)^{-1}.$$

THE PROOF OF THEOREM 3.2 . Let $Y_j = \frac{1}{\sqrt{n}}a_{nj}(\varsigma_j-2)$, then $\{Y_j\}_{j=1}^{n}$ is a sequence of independent random variables. Hence from Lemma 3.1 we have

$$|K_n(y)-\Phi(y)+(1/6)\phi(y)(y^2-1)\sum_{j=1}^{n} r_{nj}^3|$$

$$\le C\left\{\frac{1}{(\mu_2)^{3/2}}\sum_{j=1}^{n} E|W_{nj}|^3 + \frac{1}{(\mu_2)^2}\sum_{j=1}^{n} E|Y_j|^4\right.$$

$$\left.+n^6\left(\sup_{|t|\ge\delta_n}\frac{1}{n}\sum_{j=1}^{n}|\nu_j(t)| + \frac{1}{2n}\right)^n\right\}$$

$$\triangleq J_1+J_2+J_3$$

where

$$\mu_2 = \frac{1}{n}\sum_{j=1}^{n} a_{nj}^2, \quad W_{nj} = \frac{a_{nj}}{\sqrt{n}}(\varsigma_j - 2) I\left[\left|\frac{a_{nj}}{\sqrt{n}}(\varsigma_j-2)\right| > (\mu_2)^{1/2}\right],$$

$$\nu_j(t) = E\exp(itY_j) = (1-(1/2)it\frac{a_{nj}}{\sqrt{n}})^{-4}$$

$$\text{and } \delta_n = \frac{1}{12}\mu_2\left[\left(\sum_{j=1}^{n} a_{nj}^3\right)E(\varsigma_1-2)^3\right]^{-1}.$$

For J_1, since

$$\sqrt{n}\sum_{j=1}^{n} E|W_{nj}|^3 = \sum_{j=1}^{n}\left[\frac{1}{n}|a_{nj}|^3 E|\varsigma_1-2|^3 I\left(\frac{1}{\sqrt{n}}|a_{nj}||\varsigma_j-2| > (\mu_2)^{3/2}\right)\right]$$

$$\le \frac{1}{n}\sum_{j=1}^{n}|a_{nj}|^3 E|\varsigma_1-2|^3 I\left[|\varsigma_1-2| > \frac{\sqrt{n}(\mu_2)^{1/2}}{\max_{1\le j\le n}|a_{nj}|}\right] \text{ and}$$

$$\mu_2 = \frac{1}{n}\sum_{j=1}^{n} a_{nj}^2 \longrightarrow \sigma^2 \neq 0,$$

we have

$$\sqrt{n}\sum_{j=1}^{n} E|W_{nj}|^3 \longrightarrow 0,$$

that is $J_1 = o(1/\sqrt{n})$. Consider J_2. Since

$$\sqrt{n}\sum_{j=1}^{n} E|Y_j|^4 = \sqrt{n}\sum_{j=1}^{n}(1/n^2)a_{nj}^4 E|\varsigma_1-2|^4$$

$$\le \frac{1}{\sqrt{n}}\max_{1\le j\le n}|a_{nj}|\left(\frac{1}{n}\sum_{j=1}^{n}|a_{nj}|^3\right)E|\varsigma_1-2|^4$$

$$\longrightarrow 0$$

we have $J_2 = o(1/\sqrt{n})$. Finally we treat J_3. Observe that

$$\sup_{|t|\ge\delta_n}\frac{1}{n}\sum_{j=1}^{n}|\nu_j(t)| = \sup_{|t|\ge\delta_n}\frac{1}{n}\sum_{j=1}^{n}\left|(1-(1/2)it\frac{a_{nj}}{\sqrt{n}})^{-4}\right|$$

$$\le \frac{1}{n}\sum_{j=1}^{n}(1+\frac{1}{n}n^{-1}\delta_n^2 a_{nj}^2)^{-2}$$

$$\le \frac{1}{n}\sum_{j=1}^{n}(1+n^{-3/2}\delta_n^2 a_{nj}^2)^{-2}$$

$$\le \frac{1}{n}\sum_{j=1}^{n}[1-2n^{-3/2}\delta_n^2 a_{nj}^2+6n^{-3}\delta_n^4 a_{nj}^4]$$

$$=1-\frac{2}{\sqrt{n}}(\delta_n/\sqrt{n})^2(\frac{1}{n}\sum_{j=1}^{n}a_{nj}^2)+\frac{6}{n}(\frac{\delta_n}{\sqrt{n}})^4(\frac{1}{n}\sum_{j=1}^{n}a_{nj}^4)$$

$$\le 1-\frac{2}{\sqrt{n}}(\frac{\delta_n}{\sqrt{n}})^2\Big[\frac{1}{n}\sum_{j=1}^{n}a_{nj}^2$$

$$-(\frac{\delta_n}{\sqrt{n}})^2[\frac{3}{n}\max_{1\le j\le n}|a_{nj}|](\frac{1}{n}\sum_{j=1}^{n}|a_{nj}|^3)\Big]$$

$$\triangleq 1-\frac{1}{\sqrt{n}}a_n$$

so that

$$\log\Big[\sqrt{n}n^6(\sup_{|t|\ge\delta_n}\frac{1}{n}\sum_{j=1}^{n}|\nu_j(t)|+\frac{1}{2n})^n\Big]$$

$$\le 7\log n+n\log(1-\frac{1}{\sqrt{n}}a_n+\frac{1}{2n})$$

$$=7\log n+n(\frac{1}{\sqrt{n}}a_n+\frac{1}{2n})-n\theta[(\frac{1}{\sqrt{n}}a_n+\frac{1}{2n})^2]\longrightarrow-\infty$$

and hence

$$\sqrt{n}J_3=\sqrt{n}n^6(\sup_{|t|\ge\delta_n}\frac{1}{n}\sum_{j=1}^{n}|\nu_j(t)|+\frac{1}{2n})^n\longrightarrow 0,$$

that is $J_3=o(1/\sqrt{n})$. By Summarizing all of the results the conclusion of the theorem follows.

REMARK 3.2 To obtain the asymptotic expansions in Theorem 3.2, the weights $\varsigma_1, \ldots, \varsigma_n$ need only to statisfy $E(\varsigma_1-E\varsigma_1)^2=1$ and $E(\varsigma_1-E\varsigma_1)^3=1$.

By using of Theorem 3.2 we can get the asymptotic expansion for distributions in random weighting approach.

THEOREM 3.3 Suppose that T possesses a strong Frechet differential at F,

$$0<\sigma^2=\int\psi^2(x)dF(x)<\infty \text{ and } \int|\psi(x)|^3dF(x)<\infty.$$

Then

$$P^*\left\{W_n^*/[VAR^*(W_n^*)]^{1/2}\leq y\right\} = \Phi(y)-(1/6)\phi(y)(y^2-1)\sum_{j=1}^n r_{nj}^3+o(1/\sqrt{n}) \quad (3.9)$$

holds uniformly in y for almost all sample sequence X_1, X_2, ... , X_n, where

$$r_{n,j}=(T_{n,j}-T_n^*)/\left[\sum_{j=1}^n (T_{n,j}-T_n^*)^2\right]^{1/2}.$$

PROOF. Take $a_{n,j}=T_{n,j}-T_n^*$ in Theorem 3.2. Then given X_1, X_2, ... , X_n, $\{a_{n,j}\}_{j=1}^n$ is a triangle sequence of constants. Hence to prove this theorem we need only verify that conditions about a_{nj} in Theorem 3.2 are true for almost all sample sequence X_1, ... , X_n. First it is obvious that

$$\sum_{j=1}^n a_{n,j}=0$$

and from a theorem of Parr(1985),

$$\frac{1}{n}\sum_{j=1}^n (T_{n,j}-T_n^*)^2\longrightarrow\sigma^2<\infty \text{ a.s.},$$

hence what remain to prove are

$$(1)\ \overline{\lim_{n\to\infty}}\ \frac{1}{n}\sum_{j=1}^n |T_{n,j}-T_n^*|^3 \leq \int|\psi(x)|^3dF(x)<\infty \quad \text{a.s.} \quad (3.10)$$

and

$$(2)\ \lim_{n\to\infty} \frac{1}{\sqrt{n}} \max_{1\leq j\leq n} |T_{n,j}-T_n^*|\longrightarrow 0 \quad \text{a.s.} \quad (3.11)$$

Note that $\sup_{1\le i\le n} \|F_{n,i}-F_n\| \le 1/n$ and by the definition of differentiability we have

$$T(F_{n,i})=T(F_n)-\int\psi(x)d[F_{n,i}-F_n](x)+o(\|F_{n,i}-F_n\|)$$

$$=T(F_n)-\psi(x)/[n-1]+\{n(n-1)\}^{-1}\sum_{j=1}^{n}\psi(X_j)+o(1/n)$$

and hence

$$T_{n,j}-T_n^* = \frac{-\psi(X_j)}{n-1} - \frac{1}{n}\sum_{i=1}^{n}\psi(X_i)+o(1). \tag{3.12}$$

Therefore we have

$$\frac{1}{\sqrt{n}}\max_{1\le j\le n}|T_{n,j}-T_n^*|\le \frac{1}{\sqrt{n}}\max_{1\le j\le n}|\psi(X_j)|$$

$$+(1/\sqrt{n})\left[\frac{1}{n}\sum_{j=1}^{n}\psi(X_j)\right]+o(1/\sqrt{n})\longrightarrow 0 \text{ a.s.}$$

for the reason $E\psi(X_1)^2<\infty$ and also

$$\frac{1}{n}\sum_{j=1}^{n}|T_{n,j}-T_n^*|^3 \le \frac{3}{n}\sum_{j=1}^{n}|\psi(X_j)|^3 + \frac{3}{n}\left|\sum_{j=1}^{n}\psi(X_j)\right|^3+o(1)$$

$$\longrightarrow 3\int|\psi(x)|^3dF(x)<\infty \quad \text{a.s.} \tag{3.13}$$

Thus (3.10) and (3.11) are proved. The proof of the theorem is now completed.

REMARK 3.3 Comparing the asymptotic expansions of L_n and L_n^* given in Theorem 3.1 and 3.3 respectively, we can find that if we use the distribution of L_n^* to approximate that of L_n, the second order accuracy can not be obtained. The reason is that the second terms of corresponding expansions are not the same. The transformation given in (2.11) can help us to eliminate these differences. This is the conclusion of Theorem 2.1 which we will prove in the following.

For the proof of Theorem 2.1 we state a lemma.

LEMMA 3.2 The $h_n(y)$ defined above is a strictly increasing function with inverse function

$$u_n(x) = \frac{1}{\sqrt{n}}\beta_{0n} + x + \frac{1}{\sqrt{n}}\beta_{2n}x^2[1+o(x/\sqrt{n})] \qquad \text{a.s.}$$

The proof of the lemma can be obtained by elemetary mathematical analysis (see also Yao(1988)).

PROOF OF THEOREM 2.1 Since

$$F_n^*(y)=P^*\{h_n(W_n^*/[VAR^*(W_n^*)]^{1/2}\le y\}=P^*\{W_n^*/[VAR^*W_n^*]^{1/2}\le u_n(y)\}$$

we have from Theorem 3.3

$$\lim_{n\to\infty} \sqrt{n}\sup_y |F_n^*(y)-\Phi(u_n(y))+(1/6)\phi(u_n(y))(u_n^2(y)-1)\sum_{j=1}^n r_{nj}^3|=0. \quad (3.15)$$

Thus

$$\lim_{n\to\infty} \sqrt{n}\sup_y |F_n^*(y)-F_n(y)|$$

$$=\lim_{n\to\infty} \sqrt{n}\sup_y |\Phi(u_n(y))-(1/6)\phi(u_n(y))(u_n^2(y)-1)\sum_{j=1}^n r_{nj}^3 -\Phi(y)+\frac{1}{\sqrt{n}}\phi(y)[k_{31}/2-k_{32}(1-y^2)/6]|$$

$$\le\lim_{n\to\infty} \sqrt{n}\sup_y |\Phi(u_n(y))-(1/6)\phi(u_n(y))(u_n^2(y)-1)\sum_{j=1}^n r_{nj}^3 -\Phi(y)+\frac{\phi(y)}{\sqrt{n}}[\hat{k}_{31}/2-\hat{k}_{32}(1-y^2)/6]|$$

$$+\lim_{n\to\infty} \sup_y \phi(y)|(\hat{k}_{31}-k_{31})/2+(\hat{k}_{32}-k_{32})(1-y^2)/6|$$

$$\triangleq \lim_{n\to\infty} \sqrt{n}\sup_x |J_1(x)|+\lim_{n\to\infty} \sup_x |J_2(x)| \qquad (3.16)$$

From the definitions of $\hat{k}_{31}$ and $\hat{k}_{32}$, by the same arguments as in the proof of Theorem 3.3 of this paper and Theorem 2 of Parr(1985), we can prove that they are strongly consistent estimates of k_{31} and k_{32}. Hence we have

$$\lim_{n\to\infty} \sup_x |J_2(x)|=0 \quad \text{a.s.} \tag{3.17}$$

Now we need only to consider the asymptotic properties of $J_1(x)$. Note that

$$\begin{aligned}\sqrt{n}\sup_x|J_1(x)| &\le \sqrt{n}\sup_{|x|\le n^{1/12}}|J_1(x)|+\sqrt{n}\sup_{x\ge n^{1/12}}|J_1(x)| \\ &+\sqrt{n}\sup_{x\le n^{-1/12}}|J_1(x)| \\ &\triangleq J_{11}+J_{12}+J_{13}.\end{aligned} \tag{3.18}$$

For J_{11}, since

$$u_n(x)-x=\left(\frac{1}{\sqrt{n}}\hat{\alpha}_{0n} - \frac{1}{6}\sum_{j=1}^n r_{nj}^3\right)+ \frac{1}{\sqrt{n}}\left(\hat{\alpha}_{2n}+ \frac{1}{6}\sum_{j=1}^n r_{nj}^3\right)x^2[1+o(x/\sqrt{n})]$$

and

$$\Phi(u_n(x))-\Phi(x)=\phi(x)(u_n(x)-x)+r_n$$

where $|r_n|=(1/2)|\xi\phi(\xi)|(u_n(x)-x)^2$, ξ is between x and $u_n(x)$. We first prove that $\lim_{n\to\infty}\sqrt{n}\sup_{|x|\le n^{1/12}} r_n=0$. In fact, when $|x|\le n^{1/12}$, we have $|x/\sqrt{n}|\le n^{-5/12}$, hence $\lim_{n\to\infty}\sup_{|x|\le n^{1/12}}|o(x/\sqrt{n})|=0$. Observe that

$$\begin{aligned}\lim_{n\to\infty} n^{1/4}\sup_{|x|\le n^{1/12}}|u_n(x)-x| &=\lim_{n\to\infty}\Bigg[n^{-1/12}\left|\hat{a}_{0n}- \frac{\sqrt{n}}{6}\sum_{j=1}^n r_{nj}^3\right| \\ &+n^{-1/12}\left|\hat{\alpha}_{2n}+ \frac{1}{6}\sqrt{n}\sum_{j=1}^n r_{nj}^3\right|\left(1+\sup_{|x|\le n^{-1/12}}|o(x/\sqrt{n})|\right)\Bigg]=0 \quad \text{a.s.}\end{aligned}$$

Therefore

$$\begin{aligned}\sqrt{n}\sup_{|x|\le n^{1/12}}|r_n| &\le \frac{K}{2}\lim_{n\to\infty}\sqrt{n}\sup_{|x|\le n^{1/12}}(u_n(x)-x)^2 \\ &\le \frac{K}{2}\lim_{n\to\infty}[n^{1/4}\sup_{|x|\le n^{1/12}}|u_n(x)-x|]^2\longrightarrow 0 \text{ a.s.}\end{aligned} \tag{3.19}$$

However

$$J_{11}=\lim_{n\to\infty}\sqrt{n}\sup_{|x|\le n^{1/12}}\Big|\phi(x)(u_n(x)-x)-\frac{1}{6}\phi(u_n(x))(u_n^2(x)-1)\sum_{j=1}^{n}r_{nj}^3 - \frac{1}{\sqrt{n}}\phi(x)(\hat{\alpha}_{0n}+\hat{\alpha}_{2n}x^2)\Big|$$

$$=\lim_{n\to\infty}\sqrt{n}\sup_{|x|\le n^{1/12}}\Big|\frac{1}{6}[\phi(x)(x^2-1)-\phi(u_n(x))(u_n^2(x)-1)]\sum_{j=1}^{n}r_{nj}^3 +\phi(x)\Big(\frac{1}{\sqrt{n}}\hat{\alpha}_{2n}+\frac{1}{6}\sum_{j=1}^{n}r_{nj}^3\Big)x^2 o(x/\sqrt{n})\Big| \quad \text{a.s.}$$

$$=\lim_{n\to\infty}\sqrt{n}\sup_{|x|\le n^{1/12}}\Big|\frac{1}{6}[\phi(x)(x^2-1)-\phi(u_n(x))(u_n^2(x)-1)]\sum_{j=1}^{n}r_{nj}^3\Big|$$

and as in the proof of (3.13), we have

$$\phi(u_n(x))(u_n^2(x)-1)-\phi(x)(x^2-1)=\phi(x)(3x-x^3)(u_n(x)-x)+o(1/\sqrt{n}) \quad \text{a.s.}$$

Hence

$$J_{11}=\lim_{n\to\infty}\sup_{|x|\le n^{1/12}}\Big|\frac{1}{6}\phi(x)(3x-x^3)(u_n(x)-x)\sqrt{n}\sum_{j=1}^{n}r_{nj}^3\Big|$$

$$\le K\lim_{n\to\infty}\sup_{|x|\le n^{1/12}}|u_n(x)-x|\sqrt{n}\sum_{j=1}^{n}r_{nj}^3=0 \quad \text{a.s.} \tag{3.20}$$

For J_2, note that $\lim_{x\to-\infty}\phi(x)|x|^k=0\,(k=0,\ 1,\ 2)$, and hence

$$J_{12}\le \lim_{n\to\infty}\sqrt{n}\Phi(u_n(-n^{1/12}))+\lim_{n\to\infty}\Phi(-n^{1/12})$$

$$+\lim_{n\to\infty}\frac{1}{6}\phi(u_n(-n^{1/12}))(u_n^2(-n^{1/12})+1)\sqrt{n}\sum_{j=1}^{n}r_{nj}^3$$

$$+\lim_{n\to\infty}\phi(-n^{1/12})(|\hat{\alpha}_{0n}|+|\hat{\alpha}_{2n}|n^{1/6})$$

$$=\lim_{n\to\infty}\sqrt{n}\Phi(u_n(-n^{1/12})) \quad \text{a.s.}$$

However

$$\lim_{n\to\infty}|u_n(-n^{1/12})+n^{1/12}|$$

$$=\lim_{n\to\infty}\left[n^{-1/2}(\hat{\alpha}_{0n}-\frac{1}{6}\sqrt{n}\sum_{j=1}^{n}r_{nj}^{3})-n^{-5/12}\hat{\alpha}_{1n}\right.$$

$$\left.+n^{-1/3}(\hat{\alpha}_{2n}+\frac{1}{6}\sqrt{n}\sum_{j=1}^{n}r_{nj}^{3})\right]=0 \quad \text{a.s.}$$

so $J_{12}\le \lim_{n\to\infty}\sqrt{n}\Phi(-n^{1/12}+1)=0$. In the same way we can prove that $J_{13}\le 0$ a.s. Hence

$$\lim_{n\to\infty}\sqrt{n}\sup_{x}|J_1(x)| \le J_{11}+J_{12}+J_{13} \le 0 \quad \text{a.s.}$$

that is, $\sqrt{n}\sup_{x}|J_1(x)|\longrightarrow 0$ a.s. Summarizing the above results we find that the conclusion of the theorem is valid.

REMARK 3.4 From the proof of Theorem 3.2 we known that the conditions of Theorem 3.1 were only used to obtain the asymptotic expansion $P\{T_n\le x\}$. If for some special statistics T_n we can obtain the expansion for $P\{T_n\le x\}$ under some other(simpler) conditions, then Theorem 2.1 will be valid under those conditions and the conditions ensuring the consistency for the jackknife estimates defined in (2.13), such as

$$0<\sigma^2=\int\psi(x)^2dF(x)<\infty,\ \int|\psi(x)|^3dF(x)<\infty,\ E|\varphi(X_1,X_2)|<\infty$$

and

$$E|\psi(X_1)\psi(X_2)\varphi(X_1,X_2)|<\infty.$$

In the next section we will use this remark to give results for U-statistics and M-estimates.

4. TWO EXAMPLES

In this section we will give two examples. One is for the U-statistics and another for the M-estimates.

(1) U-statistics

Let $X_1, X_2, \ldots, X_n$ be n i.i.d. observations coming from an unknown population F. Then the U-statistics based on $X_1, X_2, \ldots,$, X_n can be written as

$$U_n=\binom{n}{2}^{-1}\sum_{i<j} h(X_i, X_j) \tag{4.1}$$

where h is a symmetric kernal function. In fact U_n is not a functional statistic but it has a strong relationship with the von-Mises statistic, a functional statistic defined by $V_n=n^{-2}\sum_{i=1}^{n}\sum_{j=1}^{n} h(X_i,X_j)$; so here we take it as an example. The U-statistics can have a decomposition

$$U_n = \frac{2}{n}\sum_{i=1}^{n} g(X_i)+\binom{n}{2}^{-1}\sum_{i<j}\varphi(X_i,X_j) \tag{4.2}$$

with

$$g(X_1)=E\{h(X_1,X_2)-\theta|X_1\},\ \varphi(X_1, X_2)=h(X_1,X_2)-\theta-g(X_1)-g(X_2)$$

and

$$\theta=Eh(X_1,X_2).$$

Based on this decomposition, Bickel et al.(1986) gave an Edgeworth expansion for U_n.

THEOREM 4.1 Suppose that there exist constants δ and C and functions $\chi_1(t)$ and $\chi_2(t)$ suth that

(1) $\lim_{t\to\infty}\chi_1(t)=0$;

(2) $r>2+\delta>2$, $E|\varphi(X_1,X_2)|^r\le C$;

(3) $Eg^3(X_1)I_{[t,\infty)}(|g(X_1)|)\le \chi_1(t)$ and $|E\exp(itg(X_1))|\le 1-\chi_2(t)<1(t>0)$, then there exist $\epsilon_n\longrightarrow 0$ dependending only on δ, C, χ_1 and χ_2 such that

$$\sup_x |P(\sigma_n^{-1}(U_n-\theta)\leq x)-\tilde{F}_n(x)| \leq \epsilon_n n^{-1/2} \tag{4.3}$$

where

$$\tilde{F}_n(x)=\Phi(x)-(1/6)\kappa_3 n^{-1/2}\phi(x)(x^2-1),$$

$$\sigma_n^2=n(n-1)^2 Eg^2(X_1)+\frac{n(n-1)}{2}E\varphi^2(X_1,X_2)$$

and

$$\kappa_3=\sigma_g^{-3}\{Eg^3(X_1)+3Eg(X_1)g(X_2)\varphi(X_1,X_2)\}$$

with $\sigma_g^2=Eg^2(X_1)>0$.

Let U_{ni} be the jackknife pseudo values of U_n. The jackknifed statistic of U_n is U_n itself. Hence the random weighting statistic of $U_n-\theta$ can be definied

$$W_n^*=\sum_{i=1}^n \varsigma_i(U_{n,i}-U_n) \tag{4.4}$$

where $\varsigma_1, \varsigma_2, \cdots, \varsigma_n$~iid with the probability density as in (2.8). By the same discussion as that in Section 3 we can obtain following results.

THEOREM 4.2 Suppose that $E|h(X_1,X_2)|^3<\infty$ and the other conditions of Theorems 4.1 are satisfied. Then for

$$\omega(y)=(y-\beta_n/\sqrt{n})^2 + \frac{\beta_n}{\sqrt{n}}(y-\beta_n/\sqrt{n})^2 + \frac{\beta_n^2}{\sqrt{n}}(y-\beta_n/\sqrt{n})^3 \tag{4.5}$$

where

$$\beta_n=\kappa_{3,n}-\frac{\sqrt{n}}{6}\sum_{j=1}^n(U_{n,j}-U_n)^3/[\sum_{j=1}^n(U_{n,j}-U_n)^2]^{3/2}$$

and

$$\kappa_{3,n}=\left[\frac{1}{n}\sum_{j=1}^n(U_{nj}-U_n)^2\right]^{-3/2}\left\{\frac{1}{n}\sum_{i=1}^n[U_{nj}-U_n]^3\right.$$

$$+\frac{6}{n}\sum_{i<j}[U_{nj}-U_n][U_{ni}-U_n][nU_n-(n-1)(U_{n-1}(i)+U_{n-1}(j))$$

$$+(n-2)U_{n-2}(k,j)]\Big\}$$

with

$$U_{n-1}(k)=\binom{n-1}{2}\sum_{\substack{i<j\\ i,j\neq k}}h(X_i,X_j),\quad U_{n-2}(k,l)=\binom{n-2}{2}\sum_{\substack{i<j\\ i,j\neq k,l}}h(X_i,X_j) \tag{4.6}$$

we have

$$\sqrt{n}\ \sup_x|P^*\{\omega(W_n^*/[VAR^*W_n^*]^{1/2})\leq x\}-P\{\sigma_n^{-1}(U_n-\theta)\leq x\}|\longrightarrow 0 \quad \text{a.s.} \tag{4.7}$$

A weaker form of this theorem can also be found in Tu and Shi (1988).

(2) M-estimates

Let Θ be an open interval in real line R^1 and $\mathfrak{F}$ be a set of distributions on the R^1 and $\theta(F)$ be the functional defined by the equation

$$\int\psi(x,\theta(F))dF(x)=0 \qquad F\in\mathfrak{F}. \tag{4.8}$$

The M-estimate of $\theta(F)$ is defined by the following empirical equation

$$\sum_{i=1}^{n}\psi(X_i,\hat{\theta}_n)=0. \tag{4.9}$$

It is a functional statistic. A series of M-estimate is said to be $\sqrt{n}$-consistent if for every $\epsilon>0$ we have $P\{|\hat{\theta}_n-\theta|\geq\epsilon\}=O(1/\sqrt{n})$. By following the asymptotic expansion for minimum constrast estimates of Pfanzagl(1973), Zheng(1988) gave the corresponding result for M-estimates.

THEOREM 4.3 Suppose that

(i) $\mathfrak{F}_\theta=\{F:\ F\in\mathfrak{F},\ \theta(F)=0\}$ is not an empty set.

(ii) $\psi(x,\theta)$ posesses second order derivative on Θ except for finite points and at these points $\psi''(x,\theta)$ have right and left limits.

(iii) For each θ, there exist U_θ and $M_1(x,\theta)$ such that

$$|\psi_\theta^{(i)}(x,\ \tau_2)-\psi_\theta^{(i)}(x,\ \tau_1)|\le|\tau_2-\tau_1|M_1(x,\theta)\qquad i=0,\ 1,\ 2,\ \tau_1,\ \tau_2\in U_\theta$$

with $\int M_1^2(x,\theta)dF(x)<\infty$.

(iv) For each θ and $F\in\mathfrak{F}_\theta$, $\psi(x,t)$ is two times differentiable at $t=\theta$.

(v) For each θ, there exist U_θ and $M_0(x,\theta)$ such that

$$\sup_{\tau\in U_\theta}|\psi(x,\tau)|\le M_0(x,\theta)$$

with

$$\int M_0^3(x,\theta)dF(x)<\infty.$$

(vi) For each θ, there exists a U_θ, such that for any $\delta>0$

$$\sup_{|t|\ge\delta}\ \sup_{\tau\in U_\theta}\left|\int\exp[it\psi(x,\tau)]dF(x)\right|<1$$

(vii) For each θ, $F\in\mathfrak{F}_\theta$,

$$V_\theta=\int\psi(x,\theta)^2dF(x)/[\int\psi_\theta'(x,\theta)dF(x)]^2\in(0,\ \infty).$$

Then uniformly in y we have

$$P_\theta\{\sqrt{n}(\hat{\theta}_n-\theta)/\sqrt{V(\theta)}\le y\}=\Phi(y)-\frac{\phi(y)}{\sqrt{n}}A(y,\theta)+o(1/\sqrt{n})$$

where

$$A(y,\theta)\ =\frac{1}{6}\left\{\frac{\int\psi^3(x,\theta)dF(x)}{[\int\psi^2(x,\theta)dF(x)]^{3/2}}(y^2-1)\right\}$$
$$-\frac{y^2}{2}\left\{\frac{\int\psi_\theta'(x,\theta)\psi(x,\theta)dF(x)}{(\int\psi^2(x,\theta)dF(x))^{1/2}(\int\psi_\theta'(x,\theta)dF(x))}\right.$$
$$\left.-\frac{\int\psi_\theta''(x,\theta)dF(x)}{(\int\psi^2(x,\theta)dF(x)dF(x))^{1/2}}V(\theta)\right\}.$$

The jackknife pseudo values of $\hat{\theta}_n$ can be definied as

$$\hat{\theta}_{n,i}=n\hat{\theta}_n-(n-1)\hat{\theta}^{(i)}_{n-1}$$

where $\hat{\theta}^{(j)}_{n-1}$ (j=1, 2, ..., n) satisfy

$$\int\psi(x,\hat{\theta}^{(j)}_{n-1})dF^{(j)}_{n-1}(x)=0$$

with

$$F^{(j)}_{n-1} = \frac{1}{n-1} \sum_{i\neq j} I\{X_i\leq x\}.$$

Let

$$\hat{\theta}_{J,n}=\frac{1}{n}\sum_{i=1}^{n}\hat{\theta}_{n,i},$$

$$E_n = \frac{1}{n}\sum_{i=1}^{n}(\hat{\theta}_{n,i}-\hat{\theta}_n)\zeta_i,$$

where the definition of ξ_i(i=1, 2, ... , n) is given above, and

$$\omega(y) =(y-b_n) - \frac{1}{\sqrt{n}} a_n(y-b_n)^2 + \frac{1}{3n} a_n(y-b_n)^3$$

with

$$a_n=\frac{c_n}{6}-\frac{2\sum_{i=1}^{n}[\hat{\theta}^{(i)}_{n-1}-\hat{\theta}_n]}{3s_n},\qquad b_n=-\frac{c_n}{6}+\frac{\sum_{i=1}^{n}[\hat{\theta}_{n,j}-\hat{\theta}_{J,n}]}{6s_n}$$

$$c_n=\frac{1}{s_n^3}\Big\{\frac{(n-1)^3}{n}\sum_{i=1}^{n}[\hat{\theta}^{(i)}_{n-1}-\hat{\theta}_n]$$

$$+\frac{3(n-1)^2}{n}\sum_{j\neq k}[\hat{\theta}^{(j)}_{n-1}-\hat{\theta}_n][\hat{\theta}^{(k)}_{n-1}-\hat{\theta}_n][n\hat{\theta}_n-(n-1)(\hat{\theta}^{(k)}_{n-1}+\hat{\theta}^{(j)}_{n-1})$$

$$+(n-2)\hat{\theta}^{(j,k)}_{n-2}]\Big\}$$

$$s_n=\frac{1}{n}\sum_{j=1}^{n}(\hat{\theta}_{n,j}-\hat{\theta}_{J,n})^2$$

and $\hat{\theta}^{(j,k)}_{n-2}$ satisfying

$$\int \psi(x, \hat{\theta}_{n-2}^{(j,k)}) dF_{n-2}^{(j,k)}(x)=0$$

$$F_{n-2}^{(j,k)}(x)=\frac{1}{n-2}\sum_{i\neq j,k} I\{X_i\leq x\}.$$

Then the same disscusion as that in Section 3 gives

Theorem 4.4 Under the conditions of Theorem 4.3 we have

$$\sqrt{n}\sup_y|P^*\{\omega(E_n/[VAR^*E_n]^{1/2})\leq y\}-P\{\sqrt{n}(\hat{\theta}_n-\theta)/\sqrt{V(\theta)}\leq y\}|\longrightarrow 0 \quad a.s.$$

Zheng(1988) gave another but more complicate random weighting approximation for

$$P\left\{\frac{\sqrt{n}(\hat{\theta}_n-\theta)}{\sqrt{V(\theta)}} \leq y\right\}$$

under the same conditions.

REMARK 4.1 In Theorem 4.2 the conditions are only put on $h(X_1,X_2)$, $g(X_1)$ and $\varphi(X_1,X_2)$. We need no conditions on $h(X_1,X_2)$. However for the consistency of bootstrap approximation we have to put some conditions on $h(X_1,X_1)$(refer to the counterexample given by Bickel and Freedman(1981)). Because U-statistics are not involved in the terms like $h(X_i,X_i)(i=1, 2, \ldots, n)$, the conditions for $h(X_1,X_1)$ are not the natural ones.

ACKNOWLEDGEMENTS

The author would like to thank Dr. C. R. Rao for helpful suggestions and and discussions and the encouragement he has given.

REFERENCES

Abramovitch, L. and Singh, K. (1985). Edgeworth corrected pivotal statistics and the bootstrap, *Ann. Statist.*, 13, 116-132.

Bai, Z. D. and Zhao, L. C. (1986). Asymptotic expansions of sums of independent random varibles, *Scientia Sinica*, Ser.A, 29, 1-22.

Beran, R. (1984), Jackknife approximations to bootstrap estimates, *Ann. Statist.*, 12, 101-118.

Bhattacharya, R. and Qumsiyeh, M. (1989). Second order and L^p-comparisons between the bootstarp and empirical Edgeworth expansion methodologies, *Ann. Statist.*, 17(1989), 160-169.

Bickel, P. J. and Freedman, D. (1981). Some asymptotic theory for the bootstrap, *Ann. Statist.*, 9, 1196-1217.

Bickel, P. J., Gotze, F. and van Zwet, W. R. (1986). The Edgeworth expansion for U-statistics of degree two, *Ann. Statist.*, 14, 1463-1484.

Efron, B. (1979). Bootstrap methods: another look at the jackknife. *Ann. Statist.*, 7, 1-26.

Efron, B. (1987). Better bootstrap confidence intervals, *J. Amer. Statist. Assc.*, 82, 171-185.

Frangos, C. C. (1987). An updated bibliography on the jackknife method, *Commun. Statist.-Theory Meth.*, 16, 1543-1584.

Hall, P. (1983). Inverting an Edgeworth expansion, *Ann. Statist.*, 9, 569-576.

Hinkley, D. and Wei, B. C. (1984). Improvement of jackknife confidence limit methods, *Biometrika*, 71, 331-339.

Konishi, S. (1989). Normalizing transformations and nonparametric confidence intervals, *Technical Report 89-01*, Center for Multivariate Analysis, The Penn State University.

Miller, R. G., JR. (1974). The jackknife: a review, *Biometrika*, 61, 1-17.

Parr, W. C. (1985). Jackknifing differentiable statistical functional, *J. Roy. Statist. Soc., Ser. B*, 47, 56-66.

Pfanzagl, J. (1973). Asymptotic expansions related to minimum contrast estimators, *Ann. Statist.*, 1, 993-1026.

Pfanzgal, J. (1985). *Asymptotic Expansions for General Statistical Models*, Springer-Verlag, Berlin.

Reeds, J. A. (1976). *On the Definitions of von-Mises Functional*, Ph. D. thesis, Harvard Unversity.

Sen, P. K. (1988). Functional jackknifing: Rationality and general asymptotics, *Ann. Statist.*, 16, 450-460.

Shi, X. Q. (1987). The jackknife method in nonparametric statistics, *Chinese J. Appl. Prob. Statist.*, 3, 69-76.

Takahashi, H. (1988). A note on Edgeworth expansions for the von-Mises functional, *J. Multi. Analysis*, 24, 56-65.

Tu, D. S. and Shi, X. Q. (1988). Bootstrapping and randomly weighting the U-statistics with jackknifed pseudo values, *Math. Statist. Appl. Prob.*, 3, 205-212.

Tu, D. S. and Zheng, Z. G. (1991). Randomly Weighting: an alternative approach to approximate the unknown distribution of pivotal statistics, *J. Combinatorics, Information & Sys. Scis.*, R. P. Krishniah Memorial Issue, to appear.

Withers, C. S. (1983). Expansion for the distribution and quantiles of a regular functional of the empirical distribution with applications to non-parametric confidence intervals. *Ann. Statist.*, 11, 577–587.

Yao, Z. Q. (1988). Approximations to distributions admitting Edgeworth expansions, *J. Sys. Sci. Math. Scis.*, 8, 113–126.

Zhang, L. and Tu, D. S. (1990). A comparison of some jackknife and bootstrap procedures in estimating sampling distributions of studentized statistics and constructing confidence intervals, *Technical Report No. 90-28*, Center for Multivariate Analysis, The Pennsylvania State University.

Zheng, Z. G. (1988). M-estimates and the random weighting method, *Acta Sci. Nat. Uni. Pekinensis*, 24, 277–286.

Part 3

Applications of The Bootstrap

BOOTSTRAPPING FOR ORDER STATISTICS SANS RANDOM NUMBERS (OPERATIONAL BOOTSTRAPPING)

William A. Bailey[1]

Abstract: *In contrast to what has come during the past decade to be known as boostrapping (using random numbers to resample from an original sample), what I am tentatively calling "operational" boostrapping does not use random numbers. Rather, it uses "generalized" numerical convolutions of distributions to generate distributions of the statistics of interest. The term "generalized" refers here to the convolutions (1) being not only for sums, but for any function of the numbers involved in the two distributions being convolved; (2) involving not only dimensions of 1 (i.e., univariate), but also dimensions higher than 1 (i.e., bivariate, trivariate,4- and 5-dimensional).*

The statistic of interest may be a sample mean or any order statistic which can be generated by a sequence of generalized convolutions. (Continuous distributions would first be discretized.) For example, given an empirical distribution obtained by attaching equal probabilities of 1/n to original sample values in a sample of size n,

(1) a distribution of means from resamples of size n can be obtained by convolving the univariate empirical distribution n times[2] (call the result $\left[(x_i,p_i)\right]_{i=1}^{\infty}$) and transforming the resulting distribution into $\left[\frac{1}{n} \cdot x_i, p_i\right]_{i=1}^{\infty}$; See Examples #1 and #3.[3]

Example #4 generates the same distribution considering ordered samples on the way.

(2) a distribution of ranges for resamples of size n can be obtained from the univariate empirical

[1]Kemper National Insurance Companies, Long Grove, IL 60049

[2]"n times" involves n-1 convolutions.

[3]The Examples (not shown herein) can be obtained from the author.

Exploring the Limits of Bootstrap. Edited by Raoul LePage and Lynne Billard.
 ISBN: 0471-53631-8.

distribution $\left[x_i, \quad p_i\right]_{i=1}^{i=n}$ by convolving that distribution n times for minimums and maximums jointly (call the result $\left[(x_i, y_i), p_i\right]_{i=1}^{\infty}$) and transforming the resulting distribution in to the univariate distribution $\left[y_i - x_i, \ p_i \ \right]_{i=1}^{\infty}$; See Example #2.

(3) a distribution of an order statistic such as a 25% trimmed mean for resamples of size n can be obtained by transforming the univariate empirical distribution

$$\left[x_i, \ p_i\right]_{i=1}^{i=n} \text{ into } \left[(x_i, \ 1, \ 0), \ p_i\right]_{i=1}^{i=n},$$

convolving this distribution with the empirical distribution n times using formulas specified in the Example #5 (call the result $\left[(x_i, y_i, z_i), p_i\right]_{i=1}^{\infty}$), and transforming the result into the univariate distribution $\left[\frac{1}{2 \cdot n} \cdot z_i, p_i\right]_{i=1}^{\infty}$. As the convolutions are performed, the variable in the x position is the largest amount encountered so far in the process, the variable in the y position is the number of duplicates of that amount so far, and the variable in the z position is the trimmed sum so far. See Example #5.

A different approach (referred to here as a "binomial approach) to generate this same result involves one less dimension, with the variable in the x position being the trimmed sum so far and the variable in the y position being the number of items entering into the trimmed sum so far (including even those items which did not contribute to the trimmed sum). See Examples #7 and #8.

(4) A distribution of ratios of 25% trimmed means to interquartile ranges is generated in Example #6.

MESHING AND VON MISES THEORY

As the size of the original sample increases, the number of calculations required to obtain distributions of order statistics will become prohibitive, even using fast computers. However, imposing meshes on the coordinate axes can reduce the volume of calculations to a manageable number while at the same time usually permitting results which are sufficiently fine. Using Von Mises Theorem and algorithm (see page 269 of Reference [ ]), the first three moments of each variable (X and Y) can be retained accurately. Using something akin to linear programming we can aim at also keeping accurate the covariance σ_{XY}.

Even if the solutions are expressed in terms of multivariate distributions, lower-dimensional distributions may be used if appropriate partial distributions can be captured in a fairly limited number of separate disk files. Such a reduction in the number of dimensions will be practical if the number of values which a marginal random variable[4] can take on is limited to (say) 2500. The original samples should be restricted to those with sample size 10 to 99, whereas we illustrate the approach using samples of size less than 10.

Illustration A: The solutions of **Example #4 and #5** were described in terms of 3-dimensional distributions [(X,Y,Z) P], where X is the maximum attained so far in the recursive process, and Y is the number of duplicates in the most recent sequence of duplicates in the recursive process.

Since the number of values Y can take on does not exceed 4, we could choose to restrict attention to the partial bivariate distributions[5] [(X,Z) P] | Y.

These partial distributions would be captured in 4 disk files, and combined at the time of the final transformation of Z into a univariate (1-dimensional) distribution of resample means. The same statements could be made about **Example #5**, except that there the focus is on resample 25% trimmed means.

---

[4]Univariate or bivariate.

[5]Each of these distributions is conditioned on a particular value of Y; however, they are partial distributions, since the probabilities have not be normalized to unity.

On the other hand, since the number of values the bivariate random variable (X,Y) can take on does not exceed 4+3+2+1=10, we could choose to restrict attention to the partial univariate distributions[6]
[Z P] | (X,Y).

These partial distributions would be captured in 10 disk files, and combined only at the time of the transformation of Z into a univariate distribution of resample mean. The same statements could be made about **Example #5**, except that there the focus is on resample 25% trimmed mean.

Illustration B: The solution of **Example #6** was described in terms of 4-dimensional distributions [(X,Y,Z,T) P], where X is the maximum attained so far, and Y is the number of duplicates in the most recent sequence of duplicates so far,in the recursive process.

Since the number of values of the bivariate random variable (X,Y) can take on does not exceed 4+3+2+1=10, we could choose to restrict attention to the partial bivariate (2-dimensional) distributions[7]
[(Z,T) P]|(X,Y).

These partial distributions would be captured in 10 disk files, and combined at the time of the final transformation of (Z,T) into a univariate distributions of resample ratios of 25% trimmed mean to interquartile range.

Illustration C: The solution of **Example #7** was described in terms of bivariate distributions [(X,Y) P], where Y is the number of items entering the sample sum so far in the recursive process.

Since the number of values the random variable Y can take on is 5, we could choose to restrict attention to the partial univariate distributions[8] [X P] | Y.
These partial distributions would be captured in 5 disk

---

[6]Each of these distributions is conditioned on a particular value of (X,Y); however, they are partial distributions, since the probabilities have not be normalized to unity.

[7]Ibid

[8]Each of these distributions is conditioned on a particular value of Y; however, they are partial distributions, since the probabilities have not be normalized to unity.

files, and combined at the end of the process.

BIVARIATE GENERALIZED NUMERICAL CONVOLUTIONS

If f_{X_1,Y_1} and f_{X_2,Y_2} are independent distributions of the bivariate discrete finite random variables (X_1,Y_1) and (X_2,Y_2), respectively, then the distribution $f_{W_1,W_2} = f_{X_1+Y_1,X_2+Y_2}$ of the bivariate random variable (X_1+X_2,Y_1+Y_2) is the bivariate convolution $f_{X_1,Y_1}+f_{X_2,Y_2}$ of f_{X_1,Y_1} and f_{X_2,Y_2} for sums.[9]

Let $f_{X_1,Y_1} = \left[(x1_i, y1_i) \quad p1_i \right]_{i=1}^{i=n_1}$. Similarly, let f_{X_2,Y_2} be $f_{X_2,Y_2} = \left[(x2_j, y2_j) \quad p2_j \right]_{j=1}^{j=n_2}$.

Then $f_{W_1,W_2} = f_{X_1+X_2,Y_1+Y_2} = f_{X_1,Y_1} + f_{X_2,Y_2} =$

$$\left[\left[(x1_i+x2_j, y1_i+y2_j) \quad p1_i \cdot p2_j \right]_{j=1}^{j=n_2} \right]_{i=1}^{i=n_1}$$

If n_1 and n_2 are (say) 10^3, then generating this matrix would involve as 10^6 lines.[10] This would be practical if we do not intend to use f_{W_1,W_2} for further convolutions. On the other hand, if (for example) we want to generate the distribution $f_{X_1+X_2+X_3+X_4,Y_1+Y_2+Y_3+Y_4}$ of the bivariate random variable $(X_1+...+X_4,Y_1+...+Y_4)$, then we might be dealing with having to generate $10^{12}=10^6 \cdot 10^6$ lines;[11] and this would be impractical, because of the amount of both computer storage and computing time required. The following algorithm has been designed to overcome these

---

[9] We are using + instead of * to indicate convolution for sums; that is, $f_{X_1,Y_1}+ f_{X_2,Y_2}$ instead of the usual $f_{X_1,Y_1} * f_{X_2,Y_2}$.

[10] There may be some collapsing due to identical amount pairs on different lines. The number of lines produced is reduced by representing on a single line all lines with identical amount pairs; on that single line is the amount pair and the sum of the original probabilities.

[11] Ibid

problems.

The Bivariate Convolution Algorithm

Choose $\varepsilon>0$. Typically ε is chosen to be 10^{-10} or 10^{-15}.

Loop #1: Perform the calculations indicated in Matrix #1 above, discarding any lines for which the resulting probability is less than ε; that is, discard lines for which $p1_i \cdot p2_j < \varepsilon$. The purpose of this is to avoid underflow problems and to increase the fineness of the partitions (meshes) to be imposed.

Calculate $low_x = \min_{\substack{i=1,2,\ldots,n \\ j=1,2,\ldots,m}} \{ x1_i+x2_j \neq 0 \mid p1_i \cdot p2_j > \varepsilon \}$ and

$$high_x = \max_{\substack{i=1,2,\ldots,n \\ j=1,2,\ldots,m}} \{ x1_i+x2_j \neq 0 \mid p1_i \cdot p2_j > \varepsilon \}$$

Let *nax* be a positive integer selected for the purpose of creating a partition. Let $\Delta = \dfrac{high_x - low_x}{nax/2 - 2}$; partition the open interval $(low_x-\Delta_x, high_x+\Delta_x)$ into $nax/2+1$ subintervals:

r	Subinterval I(r)
1	$[0,0]$
2	$(low_x-\Delta, low_x]$
3	$(low_x, low_x+\Delta)$
4	$[low_x+\Delta, low_x+2\cdot\Delta)$
.	.
.	.
$nax/2$	$[low_x+(nax/2 - 3)\cdot\Delta, high_x)$
$nax/2 + 1$	$[high_x, high_x+\Delta)$

Subinterval I(r) is the degenerate interval consisting of 0 alone. If for some $r_0>1$ $0 \in I(r_0)$, then 0 is deleted from $I(r_0)$; that is, that particular subinterval has a hole at 0.

Loop #2: For each r ($r=1,2,\ldots,nax/2 + 1$) calculate $low_y(r)=\min_{\substack{i=1,2,\ldots,n \\ j=1,2,\ldots,m}}\{y1_i+y2_j \neq 0 \mid p1_i \cdot p2_j>\varepsilon \text{ and } x1_i+x2_j \in I(r)\}$ and

$$high_y(r)=\max_{\substack{i=1,2,\ldots,n \\ j=1,2,\ldots,m}}\{y1_i+y2_j \neq 0 \mid p1_i \cdot p2_j>\varepsilon \text{ and } x1_i+x2_j \in I(r)\}$$

Let *nay* be a positive integer selected for the purpose of creating the following partition:

calculate $\Delta_y = \dfrac{high_{y(r)} - low_{y(r)}}{nay/2 - 2}$; partition each of the open intervals $(low_{y(r)} - \Delta_{y(r)},\ high_{y(r)} + \Delta_{y(r)})$ into $nay/2+1$ subintervals:

s	Subinterval $J_r(s)$
1	$[0,0]$
2	$(low_{y(r)} - \Delta_{y(r)}, low_{y(r)}]$
3	$(low_{y(r)}, low_{y(r)} + \Delta_{y(r)})$
4	$[low_{y(r)} + \Delta_{y(r)}, low_{y(r)} + 2\cdot\Delta_{y(r)})$
.	.
.	.
nay/2	$[low_{y(r)} + (nay/2-3)\cdot\Delta_{y(r)}, high_{y(r)})$
nay/2 + 1	$[high_{y(r)}, high_{y(r)} + \Delta_{y(r)})$

Loop #3: Let $I(r)\times J_r(s)$ (called a mesh rectangle) denote the Cartesian product of $I(r)$ and $J_r(s)$.

For each ordered pair (r,s) (r=1,2,...,*nax*/2+1; s=1,2,...,*nay*/2+1) set to zero the initial value of each of the following moment accumulators for each combination of two pairs (r,s) and (i,j) (r=1,2,...,*nax*/2+1;s=1,2,...,*nay*/2+1) (i=1,2,...,n_1; j=1,2,...,n_2) perform the accumulations

$$m_{00}[I(r)\times J_r(s)] = m_{00}[I(r)\times J_r(s)] + p1_i\cdot p2_j$$

$$m_{10}[I(r)\times J_r(s)] = m_{10}[I(r)\times J_r(s)] + (x1_i+x2_j)^1\cdot p1_i\cdot p2_j$$

$$m_{20}[I(r)\times J_r(s)] = m_{20}[I(r)\times J_r(s)] + (x1_i+x2_j)^2\cdot p1_i\cdot p2_j$$

$$m_{30}[I(r)\times J_r(s)] = m_{30}[I(r)\times J_r(s)] + (x1_i+x2_j)^3\cdot p1_i\cdot p2_j$$

$$m_{01}[I(r)\times J_r(s)] = m_{01}[I(r)\times J_r(s)] + (y1_i+y2_j)^1\cdot p1_i\cdot p2_j$$

$$m_{02}[I(r)\times J_r(s)] = m_{02}[I(r)\times J_r(s)] + (y1_i+y2_j)^2\cdot p1_i\cdot p2_j$$

$$m_{03}[I(r)\times J_r(s)] = m_{03}[I(r)\times J_r(s)] + (y1_i+y2_j)^3\cdot p1_i\cdot p2_j$$

$$m_{11}[I(r)\times J_r(s)] = m_{11}[I(r)\times J_r(s)] + (x1_i+x2_j)\cdot(y1_i+y2_j)\cdot p1_i\cdot p2_j$$

That is, we generate the probability, the 0th through 3rd x-moments, the 0th through 3rd y-moments, and the joint first moment for each mesh rectangle $I(r)\times J_r(s)$ (r=1,2,...,*nax*/2 + 1; s=1,2,...,*nay*/2 + 1).

Loop #4: Von Mises Theorem and algorithm guarantee that

for each pair (r,s) (r=1,2,...,*nax*/2+1; s=1,2,...,*nay*/2+1) there exist and we can find four pairs of real numbers $(x_1(r,s), p_{1o}(r,s))$, $(x_2(r,s), p_{2o}(r,s))$, $(y_1(r,s), p_{o1}(r,s))$ and $(y_2(r,s), p_{o2}(r,s))$ such that $x_1(r,s) \in I(r)$, $x_2(r,s) \in I(r)$, $y_1(r,s) \in I_r(s)$, and $y_2(r,s) \in I_r(s)$ and such that the following relationships hold:

$$
\begin{array}{|ll|}
\hline
p_{1o}(r,s)+p_{2o}(r,s) = p_{o1}(r,s)+p_{o2}(r,s) & =m_{00}[I(r)\times J_r(s)] \\
x_1(r,s)\cdot p_{1o}(r,s)+x_2(r,s)\cdot p_{2o}(r,s) & =m_{1o}[I(r)\times J_r(s)] \\
x_1^2(r,s)\cdot p_{1o}(r,s)+x_2^2(r,s)\cdot p_{2o}(r,s) & =m_{2o}[I(r)\times J_r(s)] \\
x_1^3(r,s)\cdot P_{1o}(r,s)+x_2^3(r,s)\cdot P_{2o}(r,s) & =m_{30}[I(r)\times J_r(s)] \\
y_1(r,s).p_{o1}(r,s)+y_2(r,s).p_{o2}(r,s) & =m_{01}[I(r)\times J_r(s)] \\
y_1^2(r,s).p_{o1}(r,s)+y_2^2(r,s).p_{o2}(r,s) & =m_{02}[I(r)\times J_r(s)] \\
y_1^3(r,s).p_{o1}(r,s)+y_2^3(r,s).p_{o2}(r,s) & =m_{03}[I(r)\times J_r(s)] \\
\hline
\end{array}
$$

We note that $(x_1(r,s), y_1(r,s))$, $(x_1(r,s), y_2(r,s))$, $(x_2(r,s), y_1(r,s))$ & $(x_2(r,s), y_2(r,s))$ all $\in [I(r)\times J_r(s)]$.

The "C" program *VONMISES* in **APPENDIX - VONMISES**[12] accepts the 0th through 3rd moments and produces at most two points and associated probabilities, with the feature that the first three moments are accurately retained. This program is called twice for each pair (r,s), once to produce $x_1(r,s)$, $p_{1o}(r,s)$, $x_2(r,s)$ and $p_{2o}(r,s)$ and once to produce $y_1(r,s)$, $p_{o1}(r,s)$, $y_2(r,s)$ and $p_{o2}(r,s)$.

Von Mises Theorem and algorithm do not assure that the covariance $m_{11}[I(r)\times J_r(s)]$ will equal

$$\sum_{i=1}^{2}\sum_{j=1}^{2} x_i(r,s)\cdot y_j(r,s)\cdot p_{ij}(r,s)$$

We can represent the (partial) distribution of (X_1+X_2, Y_1+Y_2) restricted to $I(r)\times J_r(s)$ by the four points and associated probabilities

[12]The Appendices (not shown herein) can be obtained from the author.

$$\begin{bmatrix} (x_1(r,s), y_1(r,s)) & p'_{11}(r,s) \\ (x_1(r,s), y_2(r,s)) & p'_{12}(r,s) \\ (x_2(r,s), y_1(r,s)) & p'_{21}(r,s) \\ (x_2(r,s), y_2(r,s)) & p'_{22}(r,s) \end{bmatrix}$$

where $p'_{11}(r,s)$, $p'_{12}(r,s)$, $p'_{21}(r,s)$ and $p'_{22}(r,s)$ are determined as follows:

We have $p_{1o}(r,s)$, $p_{2o}(r,s)$, $p_{o1}(r,s)$ and $p_{o2}(r,s)$. We want to determine $p'_{11}(r,s)$, $p'_{12}(r,s)$, $p'_{21}(r,s)$ and $p'_{22}(r,s)$, each in the interval [0,1], so that

$$\begin{aligned} p'_{11}(r,s) + p'_{21}(r,s) &= p_{o1}(r,s) \\ p'_{12}(r,s) + p'_{22}(r,s) &= p_{o2}(r,s) \\ p'_{11}(r,s) + p'_{12}(r,s) &= p_{1o}(r,s) \\ p'_{21}(r,s) + p'_{22}(r,s) &= p_{2o}(r,s) \end{aligned}$$

and so as to minimize the absolute value of the error

$$\left| m_{11}[I(r) \times J_r(s)] - \sum_{i=1}^{2} \sum_{j=1}^{2} x_i(r,s) \cdot y_j(r,s) \cdot p'_{ij}(r,s) \right|$$

in the joint moment.

We note that choosing a value for $p'_{11}(r,s)$ will determine the values for $p'_{12}(r,s)$, $p'_{21}(r,s)$ and $p'_{22}(r,s)$ in these four equations. Our choice of $p'_{11}(r,s)$ is not completely arbitrary, since $p'_{11}(r,s)$, $p'_{12}(r,s)$, $p'_{21}(r,s)$ and $p'_{22}(r,s)$ must each $\in [0,1]$. But, restricting $p'_{11}(r,s)$ to this extent, we can proceed to choose $p'_{11}(r,s)$ so as to minimize the absolute error in the joint moment. Choosing $p'_{11}(r,s)$ to do this is straightforward, because the error in the joint moment can be expressed as a linear function of $p'_{11}(r,s)$.[13]

The "C" program *LINEAR* IN **APPENDIX - LINEAR**[14] embodies this logic, enabling us to represent the (partial) distribution in each mesh rectangle $I(r) \times J_r(s)$ by the

---

[13]The absolute value of the error in the joint moment is linear in two pieces, because of the absolute value being taken.

[14]The Appendices (not shown herein) can be obtained from the author.

original four points and the newly derived associated probabilities. In the process the 0th through 3rd moments of X_1+X_2 and Y_1+Y_2, respectively, are retained accurately; and an attempt is made to also retain accurately the joint moment.

Having kept accurately the first three moments of each of X_1+X_2 and Y_1+Y_2, respectively, within each mesh rectangle, we have automatically kept accurately the corresponding global moments.

For each mesh rectangle $I(r)\times J_r(s)$ ($r=1,2,\ldots,nax/2+1$; $s=1,2,\ldots,nay/2+1$) there correspond the vectors

$$[(x_1(r,s),y_1(r,s)) \quad p'_{11}(r,s)]$$
$$[(x_1(r,s),y_2(r,s)) \quad p'_{12}(r,s)]$$
$$[(x_2(r,s),y_1(r,s)) \quad p'_{21}(r,s)]$$
$$[(x_2(r,s),y_2(r,s)) \quad p'_{22}(r,s)]$$

We can the express the full distribution $f_{X_1+X_2,Y_1+Y_2}$ of the bivariate random variable (X_1+X_2,Y_1+Y_2) as

$$\left\{\left\{\begin{matrix} (x_1(r,s),y_1(r,s)) & p'_{11}(r,s) \\ (x_1(r,s),y_2(r,s)) & p'_{12}(r,s) \\ (x_2(r,s),y_1(r,s)) & p'_{21}(r,s) \\ (x_2(r,s),y_2(r,s)) & p'_{22}(r,s) \end{matrix}\right\}_{r=1}^{r=nax/2+1}\right\}_{s=1}^{s=nay/2+1} .$$

REFERENCES & ACKNOWLEDGMENTS

Springer, M.D., The Algebra of Random Variables, John Wiley & Sons, Inc., New York, 1979

I want to thank Larry Hickey of PolySystems, Inc. in Chicago for focusing my attention on Von Mises Theorem and for patiently tutoring me in the "C" programming language. He has been helpful to me over the years as programmer, mathematician and codesigner. Josh Zirin, formerly of Kemper Group, assisted in developing the "C" program "vonmises". Thanks also to Anne Hoban of Kemper National and the CHIWRITER scientfic/multifont word processor.

A GENERALIZED BOOTSTRAP

By Edward J. Bedrick[1] and Joe R. Hill[2]
University of New Mexico and EDS Research

Abstract

This article defines a generalized boostrap that is based on the general framework for model-based statistics described in Hill (1990). This bootstrap includes Efron's (1979) frequentist, Rubin's (1981) Bayesian, and Hinkley and Schechtman's (1987) conditional bootstraps as special cases. It also includes new bootstraps for empirical Bayesian (EB) models. In particular, we present a conditional EB bootstrap used for EB inference that is conditional on EB ancillary statistics.
Key words: Bayesian bootstrap; Conditional bootstrap; Empirical Bayesian bootstrap; Frequentist bootstrap; General framework.

1 Introduction

The generalized bootstrap has four inputs:

(i) a general statistical model, $P = \{p_\lambda(\theta, y), \lambda \in \Lambda\}$, which is a collection of joint probability distributions for the parameters θ and the data y, indexed by $\lambda \in \Lambda$;

(ii) an estimator of λ, $\hat{\lambda} = \hat{\lambda}(y)$;

(iii) a vector of fixed features of the experiment, $z = z(\theta, y)$; and

(iv) observed data, y_{obs}.

Given these four inputs, a bootstrap replication is conceptually defined in two steps:

1. generate z^* from the distribution of z given $y = y_{obs}$ assuming that $\lambda = \hat{\lambda}_{obs} = \hat{\lambda}(y_{obs})$; then

2. generate (θ^*, y^*) from the distribution of (θ, y) given $z = z^*$ assuming that $\lambda = \hat{\lambda}_{obs}$.

[1]Department of Mathematics and Statistics, University of New Mexico, Albuquerque, New Mexico, 87131.
[2]EDS Research, 5951 Jefferson St. NE, Albuquerque, New Mexico, 87110.

Exploring the Limits of Bootstrap. Edited by Raoul LePage and Lynne Billard.

REMARK A. The joint distributions in a general statistical model can be represented in two ways: for all $\lambda \in \Lambda$, $p_\lambda(\theta, y) = p_\lambda(y \mid \theta)\, p_\lambda(\theta) = p_\lambda(\theta \mid y)\, p_\lambda(y)$. From left to right these distributions form the joint, the sampling, the prior, the posterior, and the marginal families. This definition of a statistical model includes frequentist, Bayesian, and classical empirical Bayesian (EB) models as special cases (Hill, 1990).

For a frequentist model (Neyman, 1935),

$$p_\lambda(\theta, y) = p(y \mid \theta)\, \delta_\lambda(\theta) = \delta_\lambda(\theta)\, p_\lambda(y)$$

where $\delta_\lambda(\theta) = 1$ if $\theta = \lambda$, 0 otherwise and $p_\lambda(y) = p(y \mid \theta = \lambda)$. That is, the parameter θ is a "fixed but unknown constant" (Neyman, 1977). Consequently, the prior and posterior families are identical and the marginal family is the sampling family with θ replaced by λ.

A Bayesian model is a completely specified joint distribution for the parameter θ and the data y,

$$p(\theta, y) = p(y \mid \theta)p(\theta) = p(\theta \mid y)p(y).$$

The index λ takes a single value for Bayesian models, so $\hat{\lambda} = \lambda$ for all y.

Classical empirical Bayesian models (Robbins, 1956) are families of joint distributions for parameter $\theta = (\theta_1, ..., \theta_k)'$ and data $y = (y_1, ..., y_k)'$. Usually, (θ_i, y_i), $i = 1, ..., k$, are exchangeable. Section 3 discusses the empirical Bayesian model for a one-way layout.

REMARK B. A summary combines a probability model P with observed data y_{obs} to form an estimate of the underlying probability mechanism (Efron, 1982b). For a frequentist model, $P = \{p_\lambda(y), \lambda \in \Lambda\}$, Efron (1982b) defined the maximum likelihood summary (MLS), $\hat{p}_{obs}(y)$, to be the member of P that satisfies the property: $\hat{p}_{obs}(y_{obs}) \geq p_\lambda(y_{obs})$, for all $\lambda \in \Lambda$. Hence, the MLS is the maximum likelihood estimate (MLE) of $p_\lambda(\cdot)$; that is, $\hat{p}_{obs}(y) = p_{\lambda=\hat{\lambda}_{obs}}(y)$ where $\hat{\lambda}_{obs} = \hat{\lambda}(y_{obs})$ is the observed MLE of λ. Applying this concept to a general statistical model P and an estimator $\hat{\lambda}$, a summary is a single joint distribution for θ and y; that is, $\hat{p}_{obs}(\theta, y) = p_{\lambda=\hat{\lambda}_{obs}}(\theta, y)$.

The generalized bootstrap generates fixed features z^* from the posterior distribution of z given $y = y_{obs}$ determined by the summary $\hat{p}_{obs}(\theta, y)$, then it generates outcomes (θ^*, y^*) from the conditional distribution of (θ, y) given $z = z^*$ determined by the summary. As such, the generalized bootstrap is essentially an estimated Bayesian bootstrap.

REMARK C. Rubin (1984) defined the concept of "fixed features of an experiment." If the experiment were going to be repeated, then certain features like the sample size, the configuration of missing data, and the observational units might be kept the same as in the original experiment. Because bootstraps hypothetically repeat the experiment as though the summary were the actual underlying distribution, the bootstrap variability of the fixed features is given by their estimated posterior distribution. The bootstrap variability of the other features of the experiment is determined by their estimated distributions conditioned on the fixed features.

REMARK D. Let $Q = Q(\theta, y)$ be a random variable whose expectation $\bar{Q}$ with respect to repetitions defined by fixed features z is of interest. The bootstrap estimate of $\bar{Q}$ is (assuming that all variables are absolutely continuous)

$$\bar{Q}_{obs} = \left[\int \left\{ \int \int Q(\theta^*, y^*) p_\lambda(\theta^*, y^* \mid z^*) d\theta^* \, dy^* \right\} p_\lambda(z^* \mid y = y_{obs}) dz^* \right]_{\lambda = \hat{\lambda}_{obs}} .$$

The bootstrap takes advantage of the fact that $\hat{\lambda}_{obs}$ will be plugged-in for λ by directly calculating

$$\bar{Q}_{obs} = \int \left\{ \int \int Q(\theta^*, y^*) \hat{p}_{obs}(\theta^*, y^* \mid z^*) d\theta^* \, dy^* \right\} \hat{p}_{obs}(z^* \mid y = y_{obs}) dz^* .$$

That is, the summary $\hat{p}_{obs}(\theta, y)$ is the only member of P used by the bootstrap. Hence the bootstrap can avoid calculations involving all λ in Λ, which contrasts with most analytical methods.

REMARK E. The inputs P, $\hat{\lambda}$, and z make this bootstrap very general. Section 2 shows that for appropriate choices of these inputs, the generalized bootstrap reduces to specific well-known bootstraps including Efron's (1979) frequentist bootstrap, Rubin's (1981, 1984) Bayesian bootstraps, and Hinkley and Schechtman's (1987) conditional bootstrap. Section 3 defines a conditional EB bootstrap appropriate for inference (Hill, 1987, 1990).

2 Well-Known Examples

2.1 Efron's frequentist bootstrap

The frequentist bootstrap (Efron, 1979, 1982a, b) has inputs: (i) P is a frequentist model as described in Remark A of Section 1; (ii) usually, $\hat{\lambda}$ is taken to be the MLE, but this is not required; and (iii) z is null. A bootstrap replicate sets $\theta^* = \hat{\lambda}_{obs}$ and generates y^* from the MLS $\hat{p}_{obs}(y)$. Because the estimated prior and the posterior distributions are both the same point distribution $\theta = \hat{\lambda}_{obs}$ with probability one, this bootstrap also arises if the fixed features $z = \theta$.

EXAMPLE 1. Let $y_1, \ldots, y_n$ be independently and identically distributed with cumulative distribution function (CDF) $\theta = \lambda \in \Lambda$, where Λ is the collection of all continuous distributions on $\Re$. The vector of order statistics $t = (y_{(1)}, \ldots, y_{(n)})'$ is complete sufficient for λ and the empirical CDF, $\hat{\lambda}$, is the nonparametric MLE of λ. The bootstrap sets $\theta^* = \hat{\lambda}_{obs}$ and generates y^* by randomly sampling from $\hat{\lambda}_{obs}$, which amounts to sampling with replacement from y_{obs}. This is a nonparametric bootstrap. Parametric bootstraps are also possible as illustrated by the following two examples.

EXAMPLE 2. Let $x_1, ..., x_n$ be a random sample from the discrete sample space $\{1, ..., k\}$ with $\Pr[x_i = j \mid \theta = (\theta_1, ..., \theta_k)'] = \theta_j$, $j = 1, ..., k$. The vector of cell counts, $y = (y_1, ..., y_k)'$, $y_j = \#\{i : x_i = j\}$, is a multinomial random variable with

$$y \mid \theta \sim \text{Mult}_k(n, \theta).$$

Assume that $\theta = \lambda$ with probability one for some unknown vector of probabilities λ. Then the marginal distribution of y is

$$y \sim \text{Mult}_k(n, \lambda).$$

The MLE of λ is $\hat{\lambda} = y/n$. The bootstrap sets $\theta^* = \hat{\lambda}_{obs}$ and generates y^* from the MLS

$$y^* \sim \text{Mult}_k(n, \hat{\lambda}_{obs}).$$

EXAMPLE 3. Let

$$y_i \sim \text{N}(\mu, \sigma^2)$$

independently for $i = 1, ..., n$ where $\lambda = \theta = (\mu, \sigma^2)$ indexes the family. The MLE of λ is $(\bar{y}, \hat{\sigma}^2)$ where $\bar{y} = \sum y_i/n$ and $\hat{\sigma}^2 = \sum(y_i - \bar{y})^2/n$. The bootstrap sets $\theta^* = \hat{\lambda}_{obs}$ and generates y^* from the MLS

$$y_i^* \sim \text{N}(\bar{y}_{obs}, \hat{\sigma}^2_{obs})$$

independently for $i = 1, ..., n$. For example, the bootstrap estimate of the bias and variance of $\bar{y}$ are 0 and $\hat{\sigma}^2_{obs}/n$, respectively.

2.2 Rubin's Bayesian bootstraps

The basic Bayesian bootstrap (Rubin, 1981) has inputs: (i) P is a Bayesian model as defined in Remark A of Section 1; (ii) $\hat{\lambda} = \lambda$ because there is only one distribution in P; and (iii) $z = y$. A bootstrap replicate sets $y = y_{obs}$ and generates θ^* from the posterior distribution $p(\theta \mid y = y_{obs})$.

EXAMPLE 4. Following Example 2, assume that $y = (y_1, ..., y_k)'$ has a multinomial sampling distribution

$$y \mid \theta \sim \text{Mult}_k(n, \theta).$$

Furthermore, assume that $\theta = (\theta_1, ..., \theta_k)'$ has a Dirichlet prior distribution

$$\theta \sim \text{Dir}_k(m_1, ..., m_k)$$

where the m_j are known. Then the posterior distribution of θ is also Dirichlet; in particular,

$$\theta \mid y \sim \text{Dir}_k(y_1 + m_1, ..., y_k + m_k).$$

A sample from $p(\theta \mid y = y_{obs})$ can be generated using the gaps between ordered uniform random variables (Rubin, 1981).

If $m_j = 0$, $j = 1, ..., k$, and $\hat{\theta}_{obs} = y_{obs}/n$, then the posterior mean and variance-covariance matrix of $\theta - \hat{\theta}_{obs}$ are

$$\text{E}(\theta - \hat{\theta}_{obs} \mid y = y_{obs}) = 0$$

and

$$\text{Var}(\theta - \hat{\theta}_{obs} \mid y = y_{obs}) = [\text{Diag}(\hat{\theta}_{obs}) - \hat{\theta}_{obs}\hat{\theta}'_{obs}]/(n+1),$$

respectively.

These results are close to those for the frequentist bootstrap given in Example 2. For that situation, the frequentist bootstrap distribution of $\theta - \hat{\theta}$, which assumes that θ equals the observed MLE $\hat{\theta}_{obs} = y_{obs}/n$, has mean 0 and variance-covariance matrix

$$[\text{Diag}(\hat{\theta}_{obs}) - \hat{\theta}_{obs}\hat{\theta}'_{obs}]/n.$$

This similarity was noted by Rubin (1981) and by Efron (1982a).

Rubin (1984) defined two other Bayesian bootstraps. When the fixed features $z = \theta$ the bootstrap generates θ^* from the posterior $p(\theta \mid y = y_{obs})$ as before. But now it also generates y^* from the sampling distribution $p(y \mid \theta = \theta^*)$. Marginally, y^* is generated from the predictive distribution of a future observation given the original observed data. Rubin (1984) used this bootstrap to evaluate tests of model adequacy when the experimental units where considered fixed features of the experiment.

When z is null the bootstrap generates (θ^*, y^*) from the joint distribution $p(\theta, y)$. Rubin (1984) used this bootstrap to evaluate the unconditional calibration of procedures.

2.3 Hinkley and Schechtman's conditional bootstrap

The conditional bootstrap (Hinkley and Schechtman, 1987) has inputs: (i) P is a frequentist model; (ii) $\hat{\lambda}$ is taken to be the MLE; and (iii) $z = a$, where a is ancillary for λ (that is, $(\hat{\lambda}, a)$ is minimally sufficient for λ and $p_\lambda(a) = p(a)$ for all λ). A conditional bootstrap replication sets $\theta^* = \hat{\lambda}_{obs}$ and $a^* = a_{obs}$, then generates y^* from the estimated conditional distribution $p_{\lambda=\hat{\lambda}_{obs}}(y \mid a = a_{obs})$.

Efron (1982b) related this estimated conditional distribution to his discussion of maximum likelihood summarization. Fisher (1934, 1956) argued strongly that inference should be conditioned on any ancillary information in the minimal sufficient statistic in order to recapture information lost by the MLE. Hinkley (1980) argued that inference must be conditional in order to account for the use of the ancillary as a model checking statistic.

For a real-valued parameter, Efron and Hinkley (1978) showed that the conditional bootstrap estimate of the variance of the MLE equals, approximately, one over the observed Fisher information. This result contrasts with the unconditional bootstrap estimate which equals, approximately, one over the expected Fisher information.

3 A Conditional EB Bootstrap

We illustrate the conditional EB bootstrap using the following example.

EXAMPLE 5. Let $\theta = (\theta_1, ..., \theta_k)'$ and $y = (y_1, ..., y_k)'$ have joint distributions described by

$$y_i \mid \theta \sim^{ind} \mathrm{N}(\theta_i, 1), \quad \theta_i \sim^{ind} \mathrm{N}(0, \tau^2), \; i = 1, ..., k,$$

indexed by τ^2. This model is equivalent to the model

$$y_i \sim^{ind} \mathrm{N}(0, 1/B), \quad \theta_i \mid y \sim^{ind} \mathrm{N}\{(1-B)y_i, 1-B\}, \; i = 1, ..., k,$$

indexed by $B = 1/(1+\tau^2)$. Let $\eta = \theta_1$ be the parameter of interest. If $S = \sum y_i^2$, $a = y_1/(S/k)^{1/2}$, and $t = (S, a)$, then t is minimally sufficient and a is ancillary for η (Hill, 1990).

Let $\hat{B} = \hat{B}(S)$ be an estimator of B based on S and let $\hat{B}_{obs} = \hat{B}(S_{obs})$. A conditonal EB bootstrap replicate

1. sets $a^* = a_{obs}$,
2. generates $S^* \sim \chi_k^2/\hat{B}_{obs}$,
3. calculates $y_1^* = a^*(S^*/k)^{1/2}$, and
4. generates $\eta^* \sim \mathrm{N}\{(1-\hat{B}_{obs})y_1^*, 1-\hat{B}_{obs}\}$.

Note that this bootstrap differs from the one defined by Laird and Louis (1987).

To find the conditional mean squared error of an estimator $\hat{\eta}$ of η, in the notation of Remark D of Section 1, let $Q(\theta, y) = (\eta - \hat{\eta})^2$ and $z = a$. Hill (1990) proved that if $\hat{\eta}(t) = (1 - B(S))y_1$, then

$$\bar{Q} = \mathrm{MSE}_B(\hat{\eta} \mid a) = 1 - B + \mathrm{RSL}_B(\hat{\eta})Ba^2$$

where $\mathrm{RSL}_B(\hat{\eta})$ is the relative savings loss of $\hat{\eta}$ defined by Efron and Morris (1973).

For Stein's estimator $\hat{\eta} = (1 - \hat{B})y_1$, where $\hat{B} = (k-2)/S$ is an unbiased (in the marginal family) estimator of B,

$$\bar{Q} = \text{MSE}_B(\hat{\eta} \mid a) = 1 - B + (2/k)Ba^2.$$

So the bootstrap estimate of the conditional EB MSE of Stein's estimator is

$$\bar{Q}_{obs} = \text{MSE}_{B=\hat{B}_{obs}}(\hat{\eta} \mid a) = 1 - \hat{B}_{obs} + (2/k)\hat{B}_{obs}a^2.$$

This estimate equals the posterior MSE of η when $\tau^2 \sim \text{Unif}[-1, \infty)$, the hyperprior that gives Stein's rule.

Efron and Morris (1973) derived the relative savings loss for a class of truncated Bayes rules. We can use their results to provide analytic conditional EB bootstrap estimates of the conditional MSE for those rules.

Our results on EB conditional confidence bounds and more complicated EB models will be reported elsewhere.

References

EFRON, B. (1979). Bootstrap methods: Another look at the jackknife. *Ann. Statist.* **7**, 1-26.

EFRON, B. (1982a). *The Jackknife, Bootstrap, and Other Resampling Plans.* Society for Industrial and Applied Mathematics, Philadelphia.

EFRON, B. (1982b). Maximum likelihood and decision theory. *Ann. Statist.* **10**, 340-356.

EFRON, B., AND HINKLEY, D. V. (1978). Assessing the accuracy of the MLE: Observed versus expected Fisher information (with discussion). *Biometrika* **65**, 457-487.

EFRON, B., AND MORRIS, C. (1973). Stein's estimation rule and its competitors — An empirical Bayes approach. *J. Amer. Statist. Assoc.* **68**, 117-130.

FISHER, R. A. (1934). Two new properties of mathematical likelihood. *Proc. Roy. Soc. London Sect. A* **144**, 285-307.

FISHER, R. A. (1956). *Statistical Methods and Scientific Inference.* Oliver and Boyd, Edinburgh.

HILL, J. R. (1987). Comment on "Empirical Bayes confidence intervals based on bootstrap samples," by N. M. Laird and T. A. Louis, *J. Amer. Statist. Assoc.* **82**, 752-754.

HILL, J. R. (1990). A general framework for model-based statistics. *Biometrika* **77**, 115-126.

HINKLEY, D. V. (1980). Likelihood. *Can. J. Statist.* **8**, 151-163.

HINKLEY, D. V., AND SCHECHTMAN, E. (1987). Conditional bootstrap methods in the mean-shift model. *Biometrika* **74**, 85-93.

LAIRD, N. M., AND LOUIS, T. A. (1987). Empirical Bayes confidence intervals based on bootstrap samples. *J. Amer. Statist. Assoc.* **82**, 739-750.

NEYMAN, J. (1935). On the problem of confidence intervals. *Ann. Math. Statist.* **6**, 11-116.

NEYMAN, J. (1977). Frequentist probability and frequentist statistics. *Synthese* **36**, 97-131.

ROBBINS, H. (1956). An empirical Bayes approach to statistics. *Proc. Third Berkeley Symp. Math. Statist. Probab.* **1**, 157-164.

RUBIN, D. (1981). The Bayesian bootstrap. *Ann. Statist.* **9**, 130-134.

RUBIN, D. (1984). Bayesianly justifiable and relevant frequency calculations for the applied statistician. *Ann. Statist.* **12**, 1151-1172.

BOOTSTRAPPING ADMISSIBLE LINEAR MODEL SELECTION PROCEDURES

David Brownstone, University of California, Irvine*

Abstract

This paper investigates the use of the bootstrap for estimating sampling distributions of standard and limited–translation Stein–rule estimation procedures, based on Li (1985 and 1987) and Stein (1981). These estimators incorporate uncertainty in the model selection process and dominate both Ordinary Least Squares on the largest model and common pretest estimators. Since there are no asymptotic approximations available for these estimators, the bootstrap is necessary to obtain standard errors needed for their practical application. Monte Carlo studies are performed to verify the performance of the new procedures and the accuracy of the bootstrap approximations to the second moments of their sampling distributions. The bootstrap generates much more accurate standard errors than the delete–one jackknife for the designs considered here. Additional Monte Carlo studies are used to study the accuracy of various confidence bands calculated from the bootstrap distributions. Although reasonably accurate, there is room for improvement using double bootstrapping or prepivoting.

1. Introduction

This paper extends the work in Brownstone (1990 a and b) analyzing the bootstrap estimator for the sampling distributions of standard and limited–translation Stein–rule procedures for the linear model. These procedures, based on Li (1985 and 1987) and Stein (1981), are admissible alternatives to pretest estimators commonly used in the presence of model uncertainty. Since there are no asymptotic approximations of the sampling distribution of these procedures available, the bootstrap is necessary to obtain standard errors needed for their practical application.

Brownstone (1990b) describes a Monte Carlo study, summarized in Section 4 of this paper, which verifies the improved performance (relative to Least Squares) of the Stein–rule procedures and the accuracy of their bootstrapped variances. Section 5 of this paper describes similar Monte Carlo studies for models with more exogenous variables, and it also examines two alternative versions of the delete–one jackknife (see Wu,

* Department of Economics, University of California, Irvine, CA 92717.
Financial support from the UCI Institute of Transportation Studies and the UCI Research Unit in Mathematical Behavioral Sciences is gratefully acknowledged.

Exploring the Limits of Bootstrap. Edited by Raoul LePage and Lynne Billard.
 ISBN: 0471-53631-8.

1986) variance estimators. These jackknife estimators require less computation than the bootstrap, but they are badly biased for the cases considered here.

Sampling variances are frequently used to construct approximate confidence intervals for parameter estimates. The sixth section investigates the accuracy of these intervals and other simple confidence intervals calculated directly from the bootstrap distributions. The intervals based on the bootstrapped variances perform as well as the "percentile" and "bias–corrected percentile" methods given in Efron and Tibshirani (1986), but all of these intervals can be improved using prepivoting and/or "double bootstrapping" (see Beran, 1986, and Hall, 1988). These improved intervals are discussed in Section 7.

2. Bootstrapping and Jackknifing Multiple Regression Model Estimators

This section describes how Efron's (1979) bootstrap procedure can be applied to estimate sampling distributions of estimators for the multiple regression model. The bootstrap method can be applied to much more general situations (see Efron, 1982), but all of the essential elements of the method are clearly seen by concentrating on the familiar multiple regression model:

$$y = X\beta + \epsilon, \tag{1}$$

where X and β are fixed $(T{\times}K)$ and $(K{\times}1)$ matrices with full rank and $T \geq K$. The components of ϵ are independent identically distributed random variables with zero mean and common variance σ^2.

The nonparametric bootstrap estimate of the sampling distribution of an estimator β^* of β is generated by repeatedly drawing with replacement from the residual vector

$$\epsilon^* = y - X\beta^*. \tag{2}$$

If e_b is a $(T{\times}1)$ vector of T independent draws from ϵ^*, then the corresponding bootstrap dependent variable is given by

$$y_b = X\beta^* + e_b. \tag{3}$$

For each vector y_b the estimator is recomputed, and the sampling distribution of the estimator is estimated by the empirical distribution of these estimates computed over a large number of y_b. This nonparametric bootstrap has been applied to real data by Freedman and Peters (1984) and others (see Veall's, 1989, survey for further references).

The bootstrap method described above first approximates the true distribution of the residuals by the empirical cumulative distribution function of the estimated residuals from eq. 2. It then evaluates the exact sampling distribution of the estimator by Monte Carlo simulations assuming that this empirical cumulative distribution function is true. In contrast, the traditional (non–bootstrap) approach first derives an asymptotic approximation to the sampling distribution, and then estimates the unknown parameters of this approximation. Theoretical results by Bickel and Freedman (1981) and Beran (1986) show that the bootstrap distribution converges (as the original sample size, T, and the number of bootstrap repetitions gets large) to the correct sampling distribution. The key assumption, which is satisfied by all of the procedures considered in this paper, is that the empirical cumulative distribution function of the residuals in eq. 2 converges uniformly to the true distribution. If one is uncertain about the consistency of β^*, then β^* could be replaced by LS (or some other consistent estimator) to define ϵ^* in (2). Since the resulting residuals satisfy the convergence criterion, the bootstrap sampling distribution converges no matter how badly behaved β^* is.

Jackknifing is an older resampling method which has recently been reviewed by Wu (1986). I will only consider the delete–one jackknife here since it is the only one offering significant computational advantages over the bootstrap. If β^*_{-i} is the estimate of β computed with the ith observation deleted from the sample, then define the pseudovalues $p_i = T\beta^* - (T-1)\beta^*_{-i}$. The ordinary jackknife variance estimator for β^* is given by :

$$\nu_J = \frac{1}{T(T-1)} \sum_i (p_i-\beta^*)(p_i-\beta^*)' \;. \tag{4}$$

Hinkley (1977) proposed an improved version, called the balanced jackknife, which uses the weighted pseudovalues $Q_i = \beta^* + T(1-w_i)(\beta^*-\beta^*_{-i})$, where $w_i = x_i'(X'X)^{-1}x_i$ and x_i is the ith row of X. The balanced jackknife variance estimator is given by:

$$\nu_H = \frac{1}{T(T-K)} \sum_i (Q_i-\beta^*)(Q_i-\beta^*)' \;. \tag{5}$$

Both of these jackknife variance estimators are consistent but biased under assumptions on the X matrix and the smoothness of the mapping from the least squares estimator to β^*. These jackknife estimators are much faster to compute than the bootstrap since the β^*_{-i} can be computed without recomputing regressions for each reduced data set.

Of course, good asymptotic properties are only useful if the asymptotic approximation holds for realistic models and sample sizes. Monte Carlo studies by Brownstone (1990 a and b) and Freedman, Navidi, and Peters (1988) indicate that, with sufficient degrees of freedom, the bootstrap can generate accurate standard error estimates for estimated parameters in linear models with reasonable degrees of freedom. Section 5 of this paper shows that the jackknife estimators may have poor small sample properties in similar situations.

3. Stein–Rule Estimators and Model Selection

This section briefly reviews the results in Li (1985) and Li and Hwang (1984) and shows how they can be applied to the models considered in the rest of this paper. Stein–rule or "shrinkage" estimators were first investigated by James and Stein (1961), who showed that these estimators dominate LS under a squared error loss function. There has been a large amount of work extending these estimators to different models and situations. Judge and Bock (1978) and Judge and Yancey (1986) provide good reviews of the application of these developments to multiple regression and other econometric models, and Stigler (1990) gives an excellent intuitive proof of Stein's (1956) result. These authors have also proposed Stein–rule estimators as an alternative to commonly used pretest procedures which are known to have poor sampling properties (see Brownstone, 1990a, and Hill and Judge, 1987). Li and Hwang (1984) further point out that appropriately defined Stein–rule estimators are "robust" with respect to misspecified models. The development in this section, following Li (1985), assumes that there is a true base model, and all approximate models are derived by imposing linear restrictions on this base model. Although it is relatively straightforward to consider nonlinear restrictions, the assumption that the base model is correct is critical.

I will only consider the orthonormal linear model where $X'X = I$. The general linear model in equation (1) can be transformed to this case using the singular value decomposition. The LS estimator of β, β^*, is just $X'y$. If we temporarily assume that the ϵ are Normally distributed, then :

$$\beta^*_i = \beta_i + \delta_i, \quad i = 1,\ldots K, \tag{6}$$

where δ_i are independent normally distributed with mean zero and variance σ^2. The estimation of β from β^* in eq. 6 is the classic Multivariate Normal mean problem studied by Stein (1956), where he showed that LS is inadmissible under squared error loss ($\|\beta^*-\beta\|^2$).

Consider a class of linear estimators, $\Gamma^*(h)$, of the unknown mean vector β. Associated with each h in a finite index set H there is a K×K matrix, M(h), such that $\Gamma^*(h) = M(h)\beta^*$. The model selection problem is how to choose h. For example, if the columns of X are derived from a singular value decomposition of a collinear design matrix, then some of the characteristic values and the associated β_i will be close to zero. This suggests considering a class of restricted models with the last (K–h) β^*_i set to zero (i.e. M(h) is a diagonal matrix with the first h elements equal to one and the rest zero)[1].

The next step is to choose h to minimize the risk function, but for typical linear estimators (including LS) this risk function is unbounded. Li (1985) suggests replacing the class of linear estimators, $\Gamma^*(h)$, with their Stein–rule counterparts:

$$\beta^s(h) = \beta^* - \frac{\sigma^2}{\beta^{*\prime} B(h)\beta^*} A(h)\beta^*, \tag{7}$$

[1]Li(1987) shows that this general framework can also be applied to ridge regression and nearest–neighbor nonparametric regression.

where $A(h) = I-M(h)$ and, if $M(h)$ is symmetric[2],

$$B(h) = \{(\text{trace}A(h)I - 2A(h)\}^{-1} A(h)^2. \tag{8}$$

Note that eq. (8) requires that the largest characteristic root of A(h) be less than half the trace of A(h). This condition, which can be restrictive in practice, will be discussed in later sections. The asymptotic properties of β^s are unaffected if σ^2 is replaced by a consistent estimator such as s^2.

Li (1985) shows that these Stein–rule estimators have the same asymptotic properties as the associated linear estimators, but for any finite sample size they have bounded risk. These Stein–rule estimators shrink the restricted estimator, $\Gamma^*(h)$, toward the unrestricted estimator (β^*). In the linear model selection example discussed above, if the model indexed by h is appropriate (i.e., the deleted coefficients are small) then the shrinkage factor $\sigma^2/(\beta^{*\prime}B\beta^*)$ is close to one and the Stein–rule estimates are close to the restricted estimate. If the model indexed by h is inappropriate, then the Stein–rule estimates are closer to the unrestricted LS estimates.

Finding an optimal h among the class of Stein–rule estimators, $\beta^s(h)$, requires an estimator for their risks. Stein (1981) shows that, under regularity conditions,

$$S(h) = \sigma^2 - \frac{\sigma^4 \|A(h)\beta^*\|^2}{K\ (\beta^{*\prime}B(h)\beta^*)^2} \tag{9}$$

is an unbiased estimator for the risk of $\beta^s(h)$. Li(1985) further shows that even when S(h) is not a consistent estimator of the risk, it is still a consistent estimator of the true loss ($\|\beta^s(h) - \beta\|^2$) divided by K. Note that choosing h to minimize S(h) is equivalent to minimizing

$$\frac{K\ (\beta^{*\prime}B(h)\beta^*)^2}{\|A(h)\beta^*\|^2} \tag{10}$$

which is independent of σ^2. Li (1987) shows that minimizing eq. (10) is identical to the Generalized Cross Validation model selection criterion (see Craven and Wahba, 1979) and is asymptotically optimal in the sense that

$$\frac{L(\beta^s(h^*))}{\min_{h\in H} L(\beta^s(h))} \to 1 \text{ in probability (as } T\to\infty), \tag{11}$$

where $L(\cdot)$ is the loss function and h^* denotes the model chosen to minimize eq. (10).

Li's methodology outlined above can be applied to general linear model selection problems, but the remainder of this paper will only consider the problem of choosing how many of the β^*_i to restrict to zero. In this case

[2]Li (1987) gives an alternative formula for general M(h). All the cases considered in this paper have symmetric M(h).

the trace condition on A(h) implies that there are at least three restrictions. An alternative approach for improving LS in this situation is Stein's (1981) truncated limited–translation estimator, which uses different shrinkage values for each component and therefore limits the bias. Stein's estimator is defined for each component by:

$$\beta_i^n = \left[1 - \frac{(q-2)\,\sigma^2 \min(1, Z_q/|\beta^*_i|)}{\sum_{j=1}^{k} \min(\beta^{*2}_i, Z_q^2)}\right] \beta^*_i, \tag{12}$$

where q is a large fraction of K, and $Z_1 \leq \cdots \leq Z_k$ are the order statistics of the $|\beta^*_i|$. Note that if q=K , eq. 12 simplifies to the James–Stein (1961) estimator. If q<K, then the estimator defined by eq. 12 shrinks the β^*_is more if they are closer to zero, but some method for determining q is needed. Dey and Berger (1983) propose choosing $q \geq 3$ to maximize

$$\frac{(q-2)^2}{\sum_{j=1}^{K} \min(\beta^{*2}_i, Z_q^2)}, \tag{13}$$

which maximizes the improvement in Bayes risk[3] over LS. Stein (1981) showed that the estimator defined by eq. (12) is minimax for fixed q. If the data–dependent q maximizing eq. (13) is used, then it is very difficult to analyze the resulting estimator, but a limited Monte Carlo study reported in Dey and Berger suggests that the resulting estimator is also minimax and compares favorably with other Stein–Rule estimators.

In actual applications, the shrinkage factors for both $\beta^s(h^*)$ and β^n should be truncated so that the resulting "positive part" estimates lie between the restricted and unrestricted estimates. σ^2 also needs to be replaced by an estimator such as s^2. The main difference between β^n and β^s is that, relative to β^s, the limited–translation estimator limits the bias at the cost of somewhat less efficiency. The Monte Carlo studies in the following sections give some indications of the size of the efficiency losses incurred using β^n relative to β^s.

For fixed q and A(h) with σ^2 known, Stein (1981) shows that the two estimators discussed above are minimax and admissible. Stein's (1981) unbiased risk estimators also provide a basis for generating confidence bands for these estimators without bootstrapping or jackknifing. Unfortunately, these methods do not appear to yield estimators for the sampling distributions of the estimators used here where q and h depend on the data in a complex fashion. The remainder of this paper reports results from Monte Carlo studies of the bootstrap and jackknife sampling

[3]Efron and Morris (1973) show that most Stein–rule estimators can also be derived as Empirical Bayes procedures, which justifies considering the Bayes risk here.

distributions for these Stein–rule procedures.

4. Estimator Performance

This section describes a Monte Carlo study of the positive part versions of Li's estimator ($\beta^s(h^*)$), hereafter called LISTEIN, and Stein's limited translation estimator (β^n), hereafter called NEWSTEIN. The columns of X are generated as orthonormalized independent draws from a Uniform distribution on the unit interval, and they are held fixed throughout the experiments. There are T=100 observations (rows of X), and the first set of experiments have K=10 regressors. The dependent variables (y) are generated from equation 1 with ϵ generated as independent draws from a standard Normal distribution. This design has the risk of LS = K for all of the experiments. The GAUSS (1989) programming system was used for all calculations on a 386 personal computer, and the pseudo–random numbers were all generated using GAUSS' normal and uniform random number generators.

The parameter vector β was chosen according to $\beta = L\lambda$, where L is a scalar. As L increases, the signal to noise ratio, as measured by the population $R^2 = \beta'\beta \,/\, (\beta'\beta + T)$, varies between 0 and .988. This definition of R^2 is appropriate since the regression plane passes through the origin, and σ^2 is always equal 1. Table 1, abridged from Brownstone (1990b), shows the results of a Monte Carlo study for two choices of the direction vector, λ. "Case 1" has all components of $\lambda = 1$, and "Case 2" has the first 5 components = 1.9999 and the remaining 5 = .0001.

Each entry in Table 1 is based on 400 Monte Carlo replications. In addition to population R^2, these tables give

$$\sum_{i=1}^{400} (\beta^*_i-\beta)'(\beta^*_i-\beta)/400\,, \tag{14}$$

which is the empirical (mean) risk or expected prediction error for each of the two Stein–rule estimators. By construction, LS has theoretical risk 10 for all cases, and its empirical risk was very close to 10 for all cases, so these empirical risks are omitted from Tables 1 and 2. The tables also provide the average value of h^* (the number of β^*_i set to zero), denoted "hopt," for LISTEIN and the average value of q for NEWSTEIN. The standard errors of the empirical risk measures are not reported because they are always less than .30.

Unlike the pretest and Stein–rule pretest estimators considered in Brownstone (1990a), both LISTEIN and NEWSTEIN dominate LS in these experiments. For "Case 1", the risk of LISTEIN and NEWSTEIN rises smoothly toward 10 (the risk of LS in all of these experiments). This is to be expected since as R^2 increases all parameters are estimated more precisely and there is less gain from imposing approximate restrictions. As LISTEIN and NEWSTEIN approach LS, both hopt and q (which are restricted to be no less than three) approach their maximum values (10). When hopt and q are at their maxima, both LISTEIN and NEWSTEIN are just the standard positive–part James–Stein estimator. Since the shrinkage value (i.e. $\beta^{*\prime}B(h)\beta^{*}$ in eq. 8) is large, the resulting estimators are essentially LS. In actual applications with larger models there will usually be at least three valid restrictions.

Some of the restricted models (i.e. at least three $\beta_i = 0$) considered in the model selection phase are valid for "Case 2," which is why the risks of the Stein–rule procedures remain below 10. Except for the smallest R^2 values, LISTEIN has lower risk than NEWSTEIN since it does not attempt to limit the bias in the resulting estimates. As R^2 increases, the average values of hopt and q approach 5, which is the number of approximately correct zero parameter restrictions for this case. The hump in the risk function for the Stein–rule procedures around $R^2 = 0.26$ is caused by noise in the model selection criteria (equations 10 and 13 respectively) used to identify the best approximate model.

Table 1: Empirical Risks and Truncation Values for 10 Regressors

Case	Estimators	R^2 0.00	0.04	0.26	0.50	0.80	0.90	0.96
1	LISTEIN	2.23	4.79	8.48	9.37	9.91	9.80	10.1
	hopt	9.47	9.50	9.86	9.98	10.0	10.0	10.0
	NEWSTEIN	1.30	4.22	8.27	9.34	9.91	9.80	10.1
	q	8.50	8.68	9.90	10.0	10.0	10.0	10.0
2	LISTEIN	2.64	4.83	8.11	6.85	6.72	7.45	7.15
	hopt	9.41	8.98	6.06	4.95	4.87	4.82	4.87
	NEWSTEIN	1.62	4.12	9.17	9.64	9.09	9.66	9.60
	q	8.39	8.68	8.06	6.55	4.45	4.47	4.52

Table 2: Empirical Risks and Truncation Values for 20 Regressors

Estimators	R^2 0.00	0.04	0.26	0.50	0.80	0.90	0.96
LSR	11.1	10.8	10.9	11.0	11.1	10.6	11.0
LISTEIN	3.64	6.88	14.1	12.5	12.9	12.1	12.8
hopt	18.9	18.2	13.0	9.57	9.33	9.47	9.33
NEWSTEIN	1.87	5.01	15.2	17.8	18.9	17.7	18.1
q	15.5	14.8	15.7	12.2	7.35	7.23	7.22

Table 2 gives the results of a similar Monte Carlo study where 10 additional independent regressors are added to the design matrix used in Table 1. The true parameter vector was set to have the last 10 values equal zero and the first equal to 1. Table 2 gives the empirical risks and truncation values for the Stein–rule estimators. It also gives the risk for a pretest estimator (LSR), which is defined to be LS if the F–test for the null hypothesis that the last 10 parameter values are zero in principal component space is significant at the 5 percent level. Otherwise it is the restricted least squares estimator with these last principal components set to zero.

The risk for LS is always equal to 20 for these designs, and all of these estimators clearly dominate LS. The LSR estimator performs so well here because it is performing only one pretest of a clearly true restriction. In actual applications, investigators do not know which set of parameters to restrict to zero, so multiple pretests must be performed. The risk of LSR is very close to 10, which is the theoretical value of least squares on the first 10 regressors, so there was very little uncertainty in the pretest in this design. Considering that it is searching over all possible zero restrictions, the LISTEIN procedure performs very well here. NEWSTEIN performs worse than LISTEIN here because it is limiting the bias relative to LS, which is a poor estimator for these designs.

5. Bootstrap and Jackknife Variance Estimation

Minimizing risk is not the only criterion for choosing an estimator. Econometricians are frequently interested in estimating, and making inferences about, components of β. Two feasible approaches for estimating the sampling variances of the LISTEIN and NEWSTEIN estimators are the bootstrap and jackknife procedures described in Section 2. Since these are only asymptotically valid, Monte Carlo studies are needed to establish their finite sample properties.

Table 3 gives the results of a small study of the bootstrap for three of the designs in Table 2. There are 300 Monte Carlo repetitions and 300 bootstrap repetitions for each Monte Carlo repetition. The first two rows for each estimator give the empirical means and standard errors of the estimators of the first component of β over these Monte Carlo repetitions. For each Monte Carlo sample the empirical standard errors of the estimators of the first component of β across the bootstrap samples were computed. The italicized rows give the mean ("*Bootstrap SE*") and standard deviation ("*SD Boot*") of these bootstrap standard error estimates over the Monte Carlo repetitions for each estimator.

Comparisons between the Monte Carlo ("Std. Error") and Bootstrap ("*Bootstrap SE*") rows in Table 3 show that the bootstrap gives good estimates with no noticeable biases. The theoretical standard error of the LS estimator is 1.8 for all these designs, so all these estimators perform much better than LS. There is some evidence of bias in the Stein–rule estimators for the $R^2 = .26$ case, but these biases are not significant. The results in Table 3 are similar to those obtained in Brownstone (1990b) which examined the bootstrap for the designs in Table 1.

The generally good results from this study are in contrast to one case in Freedman, Navidi, and Peters (1988) which showed the poor performance of the bootstrap for LS in an orthonormal design with 100 observations and 75 regressors. Two possible causes for the poor performance in that case are the lack of degrees of freedom and the presence of 50 extraneous regressors (i.e., those with true parameter values set to zero). When there are extraneous regressors there is excess variation in the LS residuals, so it is not surprising that the bootstrap based on these residuals will be biased. The two Stein–rule procedures considered here should be less subject to this "extraneous regressor" bias since they shrink the coefficients of these extraneous regressors towards zero before bootstrap residuals are calculated.

Table 4 gives the results of additional Monte Carlo studies of the bootstrap and the two jackknife estimators described in Section 2 for the same designs as Table 3. There were only 100 bootstrap repetitions performed for each Monte Carlo repetition in Table 4 so that the bootstrap would be more computationally comparable to the jackknife estimators. The jackknife estimators are still much faster to compute since they are based on the delete–one LS estimates which only require one regression to compute. In addition to the three estimators investigated in Table 3, Table 4 includes results for the LS estimator on the full model, which has risk = 20 for all the designs. The "Std. Error" rows in Table 4 give the standard errors of the estimates of the first component of β over the 300 Monte Carlo repetitions. These entries differ from Table 3 and across estimators because, for computational reasons, different random numbers were drawn for each estimator. The "Est." and "MSE" rows give the mean and mean squared error of the jackknife and bootstrap estimators over the Monte Carlo repetitions.

Table 3: Bootstrap Estimates for First Component of β
(300 Monte Carlo and 300 Bootstrap Repetitions)

Estimators		True Parameter Value (R^2) 0 (0)	.64 (.26)	3.2 (.90)
	Mean	0.026	0.64	3.2
LSR	Std. Error	1.2	1.2	1.1
	Bootstrap SE	*1.1*	*1.1*	*1.1*
	SD Boot.	*0.10*	*0.093*	*0.098*
	Mean	–0.048	0.40	3.2
LISTEIN	Std. Error	0.64	1.2	1.1
	Bootstrap SE	*0.75*	*1.1*	*1.2*
	SD Boot.	*0.14*	*0.12*	*0.10*
	Mean	–0.023	0.44	3.1
NEW–	Std. Error	0.41	0.98	1.2
STEIN	*Bootstrap SE*	*0.48*	*0.89*	*1.2*
	SD Boot.	*0.083*	*0.13*	*0.11*

Table 4: Comparison of Bootstrap and Jackknife Standard Errors

True Parameter Value 0.00 ($R^2 = 0$)

Estimator		LS	LSR	LISTEIN	NEWSTEIN
Jackknife	Std. Error	1.39	1.17	0.674	0.462
	Est.	*1.57*	*1.37*	*1.25*	*0.959*
	MSE	*0.0579*	*0.540*	*2.38*	*0.636*
Balanced Jackknife	Std. Error	1.38	1.17	0.665	0.471
	Est.	*1.40*	*1.26*	*1.16*	*0.859*
	MSE	*0.0193*	*0.516*	*2.07*	*0.469*
Bootstrap (100 Rep.)	Std. Error	1.28	1.07	0.597	0.379
	Est.	*1.25*	*1.12*	*0.729*	*0.477*
	MSE	*0.0195*	*.0167*	*0.0376*	*0.0176*

True Parameter Value 0.636 ($R^2 = .26$)

Estimator		LS	LSR	LISTEIN	NEWSTEIN
Jackknife	Std. Error	1.41	1.16	1.23	1.02
	Est.	*1.58*	*1.42*	*1.86*	*1.48*
	MSE	*0.0552*	*0.921*	*1.20*	*0.527*
Balanced Jackknife	Std. Error	1.46	1.19	1.26	1.06
	Est.	*1.39*	*1.23*	*1.63*	*1.31*
	MSE	*0.0253*	*0.674*	*0.728*	*0.293*
Bootstrap (100 Rep.)	Std. Error	1.47	1.17	1.28	1.08
	Est.	*1.26*	*1.13*	*1.11*	*0.881*
	MSE	*0.0616*	*.0179*	*0.0525*	*0.0553*

True Parameter Value 3.18 ($R^2 = .90$)

Estimator		LS	LSR	LISTEIN	NEWSTEIN
Jackknife	Std. Error	1.41	1.17	1.21	1.30
	Est.	*1.58*	*1.40*	*1.55*	*1.60*
	MSE	*0.0548*	*0.658*	*0.281*	*0.171*
Balanced Jackknife	Std. Error	1.31	1.12	1.13	1.19
	Est.	*1.40*	*1.23*	*1.40*	*1.41*
	MSE	*0.0279*	*0.335*	*0.243*	*0.0990*
Bootstrap (100 Rep.)	Std. Error	1.45	1.19	1.24	1.33
	Est.	*1.26*	*1.12*	*1.15*	*1.22*
	MSE	*0.0549*	*.0179*	*0.0240*	*0.0289*

Except for LS, the jackknife estimators perform much worse than the bootstrap, and the balanced jackknife performs slightly better than the regular jackknife. Since the mapping from LS to LSR and the Stein–rule procedures is not continuous, the jackknife estimators are only known to be consistent for LS. The balanced jackknife was designed to improve the jackknife for LS, so its better performance here is not surprising. Even though the balanced jackknife is an inferior variance estimator (relative to the bootstrap) for these designs, it may still be good enough to be used to prepivot the bootstrap sampling distributions.

6. Bootstrap Confidence Bands

One important use of sampling variances is the construction of confidence bands for unknown parameters. Since the Stein–rule procedures considered in this paper may not have Normal sampling distributions in finite samples, it is also important to consider the performance of different confidence bands computed directly from the bootstrap sampling distributions. This section describes Monte Carlo studies of a number of nominal 90 percent confidence procedures using the same design as in Table 3. Estimating tail probabilities of the bootstrap distributions requires more bootstrap repetitions than variance estimation, so all of the results in this section are based on 1500 bootstrap repetitions for each Monte Carlo repetition.

Table 5: Level 2α Confidence Band Definitions

Name	Definition
Boot Std.	$[\beta^* + \sigma_b t(\alpha)\ ,\ \beta^* + \sigma_b t(1-\alpha)]$
Percentile	$[G^{*-1}(\alpha),\ G^{*-1}(1-\alpha)]$
BC Percentile	$[G^{*-1}(\phi\{2z_0+z(\alpha)\}),\ G^{*-1}(\phi\{2z_0+z(1-\alpha)\})]$
Std.	$[\beta^* + st(\alpha)\ ,\ \beta^* + st(1-\alpha)]$
T Percentile	$[\beta^* + sH^{*-1}(\alpha)\ ,\ \beta^* + sH^{*-1}(1-\alpha)]$

Notes: σ_b is the bootstrap standard error for β^* and $t(\alpha)$ is the level α cutoff value from a student T distribution with 80 degrees of freedom. G^* is the bootstrap distribution of β^*, ϕ is the standard normal cumulative distribution function, $z(\alpha)$ is the level α cutoff value from this distribution, and $z_0 \equiv \phi^{-1}\{G^*(\beta^*)\}$. s, which is only calculated for LS, is the square root of the usual LS variance estimate. H^* is the bootstrap distribution of $(\beta^b{}_j - \beta^*)/s_j$, where $\beta^b{}_j$ and s_j are the values of the estimators at the jth bootstrap repetition.

Table 5 gives the definitions of the intervals, but the "Std." and "T Percentile" are only calculated for the LS estimator since they require a reliable estimate of σ. Efron and Tibshirani (1986) show that the "Boot Std.," "Percentile," and "BC Percentile" intervals are correct under increasingly general conditions. Note that if the bootstrap bias is zero ($G^*(\beta^*)=.5$), then the Percentile and BC Percentile bands are identical. Since the "Boot Std." band only requires bootstrap standard errors, it can be computed with fewer bootstrap repetitions.

Table 6 shows the performance of 3 different 90 percent confidence bands for the first coefficient of β. Except for the Stein–rule estimators at true coefficient = 0, the bootstrap standard errors ("Boot SE" columns in Table 6) underestimate the true values ("Std. Error" columns in Table 6). This negative bias is related to the negative bias in the nonparametric bootstrap standard error estimates for LS[4]. The coverage probability errors generally follow the biases in the bootstrap standard errors. The "Boot Std." and "Percentile" bands behaved similarly in these runs. The bias–corrected intervals ("BC Percentile") behaved similarly for the LSR estimator, but they displayed much more variability and had poor coverage for the Stein–rule estimators. The culprit in the bad performance of the bias–corrected intervals is high variation in the estimates of z_0.

One difficulty with interpreting Table 6 is that there are no well–defined "correct" confidence bands for these estimators to compare against. Table 7 considers the confidence bands for the LS estimator where the usual t interval ("Std." column in Table 7) provides a natural basis for comparison. The bootstrap intervals studied in Table 6 all have some undercoverage due to the negative bias in the bootstrap standard errors ("Boot SE" column in Table 7). The similarity of these three intervals is expected since the LS estimator is normally distributed in this example. Even though it is based on the biased bootstrap residuals ("Boot s" column in Table 7), the "T Percentile" intervals are almost identical to the correct "Std." intervals. The exact correctness of the "T Percentile" interval in this example is not surprising since $(\beta^b{}_i-\beta^*)/s_i$ is an exact pivot for the LS estimator with normally distributed residuals. In more general settings, Hall (1988) shows that the "T Percentile" intervals are asymptotically more accurate than the other bootstrap intervals in Tables 6 and 7 since they capture additional terms in the Cornish–Fisher expansions of the estimators' sampling distributions. Direct application of the "T Percentile" intervals to the LISTEIN or NEWSTEIN procedure requires a bootstrap or jackknife variance estimate for each bootstrap repetition. This "double bootstrapping" is computationally beyond the scope of the current study.

Tables 6 and 7 also give values of the "Bootstrap Risk," which is defined as

---

[4] The negative bias for LS is expected, since the sampling variability of the LS residuals underestimates the variability of the true residuals. In particular, $E(e_i^2)=\sigma^2(1-h_i)$, where e_i is the LS residual for the ith observation and h_i is the ith element of the diagonal of $X(X'X)^{-1}X'$. One solution is to divide each LS residual by the square root of $(1-h_i)$, but this is only valid for LS. This bias disappears in large samples.

$$\sum_{i=1}^{N} (\beta^{b}_{i}-\beta^{*})'(\beta^{b}_{i}-\beta^{*})/N, \tag{15}$$

where N is the number of bootstrap repetitions. This is a consistent, but biased, estimator for the risk of these estimators (see Bunke and Droge, 1984). The results in Tables 6 and 7 show that this bias can be substantial in these designs, but these "Bootstrap Risks" generally give the correct ordering between the estimators.

7. Further Refinements

The Monte Carlo studies presented here and in Brownstone (1990a and b) show that bootstrapping estimation strategies for linear regression models yields reasonably accurate estimates of their sampling distributions. Although the bootstrap standard errors and confidence intervals are adequate for most applications with sufficient degrees of freedom, the biases in Tables 6 and 7 suggest the need for more accurate estimates. One approach is to use exact finite sample theory and/or higher–order asymptotic expansions. These theoretical results are available for simple pretest estimators, such as LSR (Judge and Yancey, 1986), and for simple Stein–rule estimators (Phillips, 1984), but not for the more complex Stein–rule procedures (LISTEIN and NEWSTEIN) considered here. Even when exact finite sample theory is available, the results are typically quite sensitive to the functional form of the residual distribution.

Another approach which preserves the distributional robustness of the bootstrap is to use improved bootstrap techniques (see Beran, 1986, Efron, 1987, Hall, 1988, and Loh, 1987). These work by improving the rate of convergence of the bootstrap sampling distribution, but they generally require an order of magnitude more computations than the simple bootstrap. The most general approach (Beran, 1986 and Hall, 1988) is to use the bootstrap distribution as an approximate pivot, which is a generalization of the "T Percentile" interval in Table 7. Efron's (1987) accelerated bias–corrected bootstrap intervals have the same second–order asymptotic properties as the T Percentile, but they are easier to compute in many cases. Nevertheless these accelerated intervals require the equivalent of a jackknife variance computation for each bootstrap repetition for the LISTEIN and NEWSTEIN procedures. Monte Carlo studies of these improved techniques require supercomputer resources, but their application to even large applied problems is feasible on current high–end workstations.

Even though the techniques described above will improve the accuracy of the methods studied in this paper, the Stein–rule procedures with the simple bootstrap confidence intervals are a substantial improvement relative to current popular techniques. For the setting in Tables 6 and 7, most applied econometricians would use extensive pretesting to find the "best" set of regressors, and then perform inference conditional on the final model chosen being correct. Brownstone (1990a) and Hill and Judge (1987) show that these procedures may have much higher risk than LS applied to the full model, and the conditional inferences

Table 6: 90% Bootstrap Confidence Intervals for First Component of β (275 Monte Carlo Repetitions and 1500 Bootstrap Repetitions)

True coefficient = 0.00

Estimator	Mean	Std. Error	Risk	Boot SE	Boot Risk
LSR	0.0452	1.18	11.0	*1.13*	*10.3*
LISTEIN	0.0153	0.683	4.32	*0.762*	*5.55*
NEWSTEIN	0.0496	0.495	2.90	*0.516*	*3.78*

Mean confidence bands and actual coverage probabilities

Estimator	Boot Std.	Percentile	BC Percentile
LSR	[−1.83, 1.92]	[−1.85, 1.84]	[−1.77, 1.88]
Cov. Pr.	0.892	0.900	0.852
LISTEIN	[−1.25, 1.28]	[−1.23, 1.22]	[−1.48, 1.23]
Cov. Pr.	0.956	0.980	0.836
NEWSTEIN	[−0.809, 0.908]	[−0.783, 0.807]	[−0.996, 0.977]
Cov. Pr.	0.940	1.00	0.720

True coefficient = 0.636

Estimator	Mean	Std. Error	Risk	Boot SE	Boot Risk
LSR	0.652	1.20	11.1	*1.12*	*9.98*
LISTEIN	0.474	1.28	14.5	*1.10*	*11.6*
NEWSTEIN	0.422	1.11	15.0	*0.892*	*11.1*

Mean confidence bands and actual coverage probabilities

Estimator	Boot Std.	Percentile	BC Percentile
LSR	[−1.21, 2.51]	[−1.18, 2.48]	[−1.18, 2.48]
Cov. Pr.	0.887	0.865	0.887
LISTEIN	[−1.35, 2.30]	[−1.51, 2.10]	[−1.16, 2.50]
Cov. Pr.	0.858	0.855	0.805
NEWSTEIN	[−1.06, 1.91]	[−1.33, 1.59]	[0.771, 2.31]
Cov. Pr.	0.816	0.876	0.674

(Table 6 continued)

True coefficient = 3.18

Estimator	Mean	Std. Error	Risk	Boot SE	Boot Risk
LSR	3.16	1.24	10.2	*1.13*	*10.2*
LISTEIN	3.11	1.31	11.9	*1.16*	*11.7*
NEWSTEIN	2.93	1.39	17.2	*1.23*	*15.8*

Mean confidence bands and actual coverage probabilities

Estimator	Boot Std.	Percentile	BC Percentile
LSR	[1.28, 5.04]	[1.31, 5.01]	[1.30, 5.01]
Cov. Pr.	0.844	0.844	0.829
LISTEIN	[1.19, 5.04]	[1.23, 5.03]	[1.19, 5.00]
Cov. Pr.	0.829	0.848	0.829
NEWSTEIN	[0.886, 4.97]	[0.670, 4.70]	[1.16, 5.20]
Cov. Pr.	0.856	0.848	0.844

Table 7: Bootstrap Confidence Intervals for LS for First Component of β (275 Monte Carlo Repetitions and 1500 Bootstrap Repetitions)

True Coef.	Mean	Std. Error	Risk	Boot SE	Boot s	Boot Risk
0.00	0.0655	1.40	20.1	*1.26*	*1.27*	*16.0*
0.636	0.608	1.37	19.4	*1.26*	*1.26*	*16.0*
3.18	3.37	1.53	20.6	*1.26*	*1.26*	*15.9*

Mean confidence bands and actual coverage probabilities

Coeff.	Std.	Boot Std.	Percentile	BC Percentile	T Percentile
	[–2.28, 2.41]	[–2.03, 2.16]	[–2.02, 2.13]	[–2.02, 2.13]	[–2.29, 2.39]
0	0.896	0.860	0.864	0.860	0.888
	[–1.73, 2.95]	[–1.48, 2.70]	[–1.46, 2.67]	[–1.47, 2.66]	[–1.73, 2.94]
0.636	0.922	0.896	0.887	0.878	0.926
	[1.03, 5.71]	[1.28, 5.46]	[1.30, 5.43]	[1.30, 5.43]	[1.03, 5.70]
3.18	0.872	0.830	0.819	0.823	0.872

ignoring model selection can be badly biased. Relative to this type of pretesting procedure, the Stein–rule procedures in Table 6 perform very well.

8. References

Beran, R. (1986), "Simulated Power Functions," *The Annals of Statistics*, V. 14, Pp. 151–173.

Bickel, P. J. and D. A. Freedman (1981), "Some Asymptotic Theory for the Bootstrap," *The Annals of Statistics*, V. 9, Pp. 1196–1217.

Brownstone, D. (1990a), "Bootstrapping Improved Estimators for Linear Regression Models," *Journal of Econometrics*, V. 44, Pp. 171–188.

Brownstone, D. (1990b), "How to 'Data Mine' If You Must: Bootstrapping Pretest and Stein–Rule Estimators," UC Irvine Research Unit in Mathematical Behavioral Science Technical Report MBS 90–08, June, 1990.

Bunke, O. and B. Droge (1984), "Bootstrap and Cross–Validation Estimates of the Prediction Error for Linear Regression Model," *The Annals of Statistics*, V. 12, Pp. 1400–1424.

Craven, P. and G. Wahba (1979), "Smoothing Noisy Data with Spline Functions: Estimating the Correct Degrees of Smoothing by the Method of Generalized Cross–Validation," *Numer. Math.*, V. 31, Pp. 377–403.

Dey, D. K. and J. O. Berger (1983), "On Truncation of Shrinkage Estimators in Simultaneous Estimation of Normal Means," *Journal of the American Statistical Association*, V. 78, Pp. 865–869.

Efron, B. (1979), "Bootstrap Methods: Another Look at the Jackknife," *The Annals of Statistics*, V. 7, Pp. 1–26.

Efron, B. (1982), *The Jackknife, Bootstrap and Other Resampling Plans*, SIAM, Philadelphia.

Efron, B. (1987), "Better Bootstrap Confidence Intervals (with discussion)," *Journal of the American Statistical Association*, V. 82, Pp. 171–200.

Efron, B. and C. Morris (1973), "Stein's Estimation Rule and its Competitors – An Empirical Bayes Approach," *Journal of the American Statistical Association*, V. 68, Pp. 117–130.

Efron, B. and R. Tibshirani (1986), "Bootstrap Methods for Standard Errors, Confidence Intervals, and Other Measures of Statistical Accuracy," *Statistical Science*, V. 1, Pp. 54–74.

Freedman, D.A. and S.C. Peters (1984), "Bootstrapping a Regression Equation: Some Empirical Results," *Journal of the American Statistical Association*, V. 79, Pp. 97 – 106.

Freedman, D.A., W. Navidi, and S.C. Peters (1988), "On the Impact of Variable Selection in Fitting Regression Equations," in T. K. Dijkstra (ed.), *On Model Uncertainty and its Statistical Implications*, Springer–Verlag, Berlin, Pp. 1–16.

GAUSS (1989), Version 2.0 of the Gauss Programming Language, Aptech Systems, Kent, Washington.

Hall, P. (1988), "Theoretical Comparison of Bootstrap Confidence Intervals," *Annals of Statistics,* V. 16, Pp. 927–953.

Hill, R.C. and G.G. Judge (1987), "Improved Prediction in the Presence of Multicollinearity," *Journal of Econometrics*, V. 35, Pp. 83 –100.

Hinkley, D.V. (1977), "Jackknifing in Unbalanced Situations," Technometrics, V. 19, Pp. 285–292.

James, W. and C. Stein (1961), "Estimation with Quadratic Loss," *Proceedings of the Fourth Berkeley Symposium on Mathematical Statistics and Probability*, University of California Press, Berkeley, CA, Pp. 316 – 379.

Judge, G.G. and M.E. Bock (1978), *The Statistical Implications of Pre–Test and Stein–Rule Estimators in Econometrics*, North–Holland, Amsterdam.

Judge, G.G. and T.A. Yancey (1986), *Improved Methods of Inference in Econometrics*, North–Holland, Amsterdam.

Li, K–C. (1985), "From Stein's Unbiased Risk Estimates to the Method of Generalized Cross Validation," *Annals of Statistics*, V.13, Pp. 1352–1377.

Li, K–C. (1987), "Asymptotic Optimality of Cp, Cl, Cross–Validation and Generalized Cross Validation: Discrete Index Set," *Annals of Statistics*, V.15, Pp. 958–975.

Li, K–C. and J. T. Hwang (1984), "The Data–Smoothing Aspect of Stein Estimates," *Annals of Statistics*, V.12, Pp. 887–897.

Loh, W–Y. (1987), "Calibrating Confidence Coefficients," *Journal of the American Statistical Association*, V. 82, Pp. 155–162.

Phillips, P. (1984), "The Exact Distribution of the Stein–Rule Estimator," *Journal of Econometrics*, V. 25, Pp. 123–132.

Stein, C. (1956), "Inadmissibility of the Usual Estimator for the Mean of a Multivariate Normal Distribution," *Proceedings of the Third Berkeley Symposium on Mathematical Statistics and Probability*, University of California Press, Berkeley, Pp. 197–206.

Stein, C. (1981), "Estimation of the Parameters of a Multivariate Normal Distribution," *Annals of Statistics*, V. 9, Pp. 1135–1151.

Stigler, S. M. (1990), "A Galtonian Perspective on Shrinkage Estimators," *Statistical Science*, V. 5, Pp. 147–155.

Veall, M.R. (1989), "Applications of Computationally–Intensive Methods to Econometrics," Working Paper No. 89–05, Dept. of Economics, McMaster University, Hamilton, Ontario.

Wu, C. F. J. (1986), "Jackknife, Bootstrap and Other Resampling Methods in Regression Analysis," *Annals of Statistics*, V. 14, Pp. 1261–1295.

A HAZARD PROCESS FOR SURVIVAL ANALYSIS

John J. Hsieh, University of Toronto

Abstract This article develops a hazard process and applies it to non-parametric survival analysis. The approach is a direct generalization of the filtered counting process martingale techniques and can easily accommmodate right censoring, competing risks and time-varying covariates. The hazard processes arising from the point process, marked point process and marked process with covariate path, respectively, are derived from the corresponding deterministic hazard functions. The derived processes are employed to obtain non-parametric estimates of various deterministic hazard and other functions as well as asymptotic distributions of the estimators and test statistics , using martingale methods. In addition to the usual functions describing the distribution of the failure time, we also introduce the occurrence/exposure rates and show their applications in significant tests comparing the hazard level of a cohort with that of the general population.

1. Introduction

Let T be a non-negative random variable representing the length of the interval from some initial zero point to the occurrence of an absorbing or non-recurrent event of interest such as waiting time, failure time, recurrence time. For a given sample of n individuals, $T(\omega_i)$ represents the duration time for the member ω_i , i=1,2,...,n. Since the probability measure induced by T is equivalent to the original one defined on the σ -algebra **F** generated by subsets of the sample space **S** , we will henceforth deal directly with the former.

The (cumulative) distribution function $F(t) \equiv P[T \le t]$ of T is right continuous, monotone non-decreasing on $[0, \infty)$ with F(0)=0 and $\lim_{t\uparrow\infty} F(t)=1$. The complementary probability S(t)=1-F(t) is the survival function. If T is further assumed to be absolutely continuous, then the density function

$$f(t) \equiv \lim_{\Delta\downarrow 0} P[t \le T < t+\Delta]/\Delta$$

exists for $t \ge 0$, almost everywhere, and is given by f(t)=F'(t)=-S'(t). The hazard rate function is defined for S(t)>0 by

$$h(t) \equiv \lim_{\Delta\downarrow 0} P[T < t+\Delta \mid T \ge t]/\Delta = f(t)/S(t), \tag{1}$$

with the cumulative hazard given by

$$H(t) \equiv \int_{(0,t]} h(u)du = -\ln S(t),$$

upon using the initial condition S(0)=1, so that

$$S(t) = exp[-H(t)] = exp[-\int_{(0,t]} h(u)du]. \tag{2}$$

Exploring the Limits of Bootstrap. Edited by Raoul LePage and Lynne Billard.
 ISBN: 0471-53631-8.

Note that if the distribution function of T is not absolutely continuous, then the hazard rate function cannot be defined. If T has a finite mean, then the life expectancy or mean residual life function is defined by

$$e(t) \equiv E[T-t \mid T \geq t] = \int_{(t,\infty)} u dF(u)/S(t^-) = \int_{(t,\infty)} S(u) d(u)/S(t^-), \tag{3}$$

for $S(t^-)>0$ and e(t) ≡ 0 whenever S(t)=0, where $S(t^-) \equiv \lim_{u \uparrow t} S(u) = P[T \geq t]$. Note that $S(t^-)=S(t)$ if T is continuous. This fact was used in (1). The five functions defined above are equivalent.

Finally, we define an important rate function -- the duration-specific occurrence/exposure rate. The $[t_1, t_2]$-specific O/E rate is defined for $t_2 > t_1 \geq 0$ by

$$\frac{S(t_1^-)-S(t_2)}{\int_{(t_1,t_2]} S(t)dt}. \tag{4}$$

Expression (4) represents the ratio of the number of events in $[t_1,t_2]$ to the number of individual-time units of exposure in $(t_1,t_2]$. If T is (absolutely) continuous, then $S(t_1^-)=S(t_1)$ and an application of the mean value theorem of integral calculus yields

$$\lim_{t_2 \downarrow t_1} \frac{S(t_1)-S(t_2)}{\int_{(t_1,t_2]} S(t)dt} = -\frac{S'(t_1)}{S(t_1)} = h(t_1), \tag{4a}$$

and an instantaneous O/E rate reduces to a hazard rate function. Thus, an O/E rate is an average rate and as such is a set function while a hazard rate is an instantaneous rate and as such is a point function. Death rates and incidence rates are examples of the former while force of mortality and force of morbidity are examples of the latter. For more functions useful in describing the distribution of the duration time, see Hsieh (1990).

2. Deterministic Hazards

To cover both discrete and continuous ditributions, we begin by defining the differential of F as follows:

$$dF(t) \equiv \begin{cases} F(t)-F(t^-)=P[T=t], & \text{if } F \text{ is discrete,} \\ \dfrac{dF(t)}{dt}dt = f(t)dt, & \text{if } F \text{ is continuous,} \end{cases} \tag{5}$$

where $F(t^-)=\lim_{u \uparrow t} F(u)=P[T<t]$. Note that in the discrete case dF(t) becomes the probability mass function representing the size of jump of F at t which is zero for all t except at a countable number of jump points $T(\omega_i)$. The differential of H is then defined for $S(t^-)>0$ as

$$dH(t) \equiv \frac{dF(t)}{S(t^-)} = \begin{cases} P[T=t \mid T \geq t], & \text{if } F \text{ is discrete,} \\ f(t)dt/S(t)=h(t)dt, & \text{if } F \text{ is continuous,} \end{cases} \tag{6}$$

the second equality following directly from (5).

We shall call *dH(t), the hazard (probability) of T at time t.* The quantity dH(t) is the fundamental function on which the developments in this paper will be based.In the remainder of this section we shall derive other functions of use in terms of dH(t) for the discrete case which is most relevant to estimation and hypothesis testing to be developed in Sections 4 and 5.(The expressions for the continuous distribution are as given in Section 1).

Clearly the cumulative hazard function H(t) is simply a Stieltjes integral over (0,t], which is

$$H(t)=\int_{(0,t]} dH(t)=\sum_{u\le t} dH(u). \tag{7}$$

Like F(t), H(0)=0 and H(t) is nondecreasing and right-continuous. Furthermore, H(t) is a function of bounded variation and has the same set of discontinuity points as F(t) which is countable and contains no limit (cluster) points. Since we have from (6),

$$1-dH(t)=1-P[T=t \mid T\ge t]=P[T>t \mid T\ge t],$$

this, together with the left-continuity of $S(t^-)=P[T\ge t]$ (implied by the right-continuity of F(t)=P[T ≤ t]) implies that the survival function can be expressed in terms of dH(t) as follows:

$$S(t)=P(T>t)=\prod_{u\le t}[P(T>u)/P(T\ge u)]=\prod_{u\le t}[1-dH(u)]. \tag{8}$$

The distribution function F(t) is obtained from (5) and (6) as

$$F(t)=\int_{(0,t]} dF(u)=\int_{(0,t]} S(u^-)dH(u)=\sum_{u\le t} S(u^-)dH(u), \tag{9}$$

where $S(u^-)=\prod_{s<u}[1-dH(s)]$ (10).

The mean residual life function e(t) and the $[t_1,t_2]$ -specific O/E rate for discrete distribution are as given in (3) and (4), respectively, with S(t) and $S(t^-)$ replaced by (8) and (10), respectively.

3. The Hazard Process

To make statistical inference on dH(t), it is necessary to relate the hazard defined in (6) to the observed information, Let $\mathbf{F}_{t^-}$ represent all the accumulated information available up to just before time t. $\mathbf{F}_{t^-}$ is called the *filtration on* **F** or *pre-t information* or *history*. We assume from now on that t takes values in some bounded interval $[0,\tau]$, for some $\tau>0$. $\mathbf{F}_t$ is a right-continuous nested σ -algebra such that $\mathbf{F}_s\subset \mathbf{F}_t$ ($\mathbf{F}_t$ is finer than $\mathbf{F}_s$) for all $0\le s\le t$. A typical subcollection of $\mathbf{F}_t$ is the *internal history*

$$\mathbf{F}_t^T=\sigma(I(T(\omega_i)\le s), 0\le s\le t, i=1,..,n), \tag{11}$$

the σ -algebra generated by the observable indicator process { $I(T(\omega_i)\le s)$ } up to time s ≤ t. Clearly, the accumulated information represented by $\mathbf{F}_t^T$ increases as the observation time increases and so $\mathbf{F}_t^T$ is an increasing family of σ -algebra. In Section 3 through 5 we shall take $\mathbf{F}_{t^-}^T$ as the natural filtration.

Now, for each member ω_i of the sample space Ω, equation (6) may be rewritten as

$$dH_i(t)=dF_i(t)/S_i(t^-)=E[dI(\omega_i)\le t) \mid T(\omega_i)\ge t]. \tag{12}$$

This follows because

$dF_i(t)=dP\,[T(\omega_i)\le t]=dE\,[I(T(\omega_i)\le t)]=E\,[I\,(T\omega_i)\le t)]-E\,[I\,(T\,(\omega_i)<t)]=E\,[dI\,(T\omega_i)\le t)]$.

The indicator $I(T(\omega_i)\le t)$ is a random measure, so that $\{\ I(T(\omega_i)\le t, t\ge 0\ \}$ can be seen as a *point process or counting process* which counts whether or not the event (of interest) has occurred to ω_i in (0,t]. We assume that $I(T(\omega_i)\le 0)=0$.

We are now in a position to randomize the deterministic hazard $dH_i(t)$ given in (12) by conditioning on the observed pre-t information $\mathbf{F}_{t}-$ to obtain the *stochastic hazard* for the member ω_i defined by

$$d\lambda_i(t)=E\,[dI\,(T\,(\omega_i)\le t)\mid \mathbf{F}_{t}-]. \tag{13}$$

If we let the filtration $\mathbf{F}_t=\mathbf{F}_t^T$, the sample path of $I(T(\omega_i)\le s)$, for $s\le t$, defined in (11), then since the accumulated information represented by $\mathbf{F}_t^T-$ changes only at the stopping times $T(\omega_i)$, equation (13) reduces to a dichotomy solely determined by whether or not the event occurred before t:

$$d\lambda_i(t)=\begin{cases} E\,[dI\,(T\omega_i)\le t)\mid T(\omega_i)\ge t], & \text{if } I\,(T\,(\omega_i)\ge t)=1,\\ 0, \ \text{if } I\,(T\,(\omega_i)\ge t)=0, & \end{cases} \tag{14}$$

Using (12), eqation (14) may be more compactly written as

$$d\lambda_i(t)=y_i(t)dH_i(t), \tag{15}$$

where $y_i(t)\equiv I\,(T\,(\omega_i)\ge t)$, indicating (by the value 1 or 0) whether or not ω_i is still at risk at time t^-. The observable process $\{\ y_i(t), t\ge 0\ \}$ is *predictable* , i.e., *fixed given* $\mathbf{F}_t^T-$. We see from (15) that, unlike $dH_i(t)$, $d\lambda_i(t)$ is random because $y_i(t)$ is random. Hence, we call $d\lambda_i(t)$ stochastic hazard and $dH_i(t)$ deterministic hazard.

If the distribution of $T(\omega_i)$ has a density, then (15) may be written as

$$\frac{d\lambda_i(t)}{dt}=y_i(t)h_i(t). \tag{15'}$$

We shall call (15') the *stochastic hazard rate of* T (while $h_i(t)$ is a deterministic hazard rate). Integrating (13) and (15) over (0,t] we obtain (noting that $\lambda_i(0)=0$)

$$\lambda_i(t)=\int_{(0,t]} y_i(u)dH_i(u)=H_i[min\,(t,T\,(\omega_i))] = E\,[I\,(T\,(\omega_i)\le t\mid \mathbf{F}_{t-}]. \tag{16}$$

We shall call $\lambda_i(t)=\int_{(0,t]} y_i(u)dH_i(u)$ the *hazard process* associated with the counting process $\{\ I(T(\omega_i)\le t)\}$ for member ω_i. Thus a hazard process is nothing but an integrated stochastic hazard. It is a random process, though considerably less random than its associated counting process if the process admits a stochastic hazard rate.

If we define the stochastic differential $dM_i(t)$ by

$$dM_i(t)=dI\,(T\,(\omega_i)\le t)-y_i(t)dH_i(t), \tag{17}$$

which represents the non-predictable portion (i.e., the innovation) of the observable jump process $dI\,(T\omega_i)\le t)$, then, since $y_i(t)$ is predictable, it follows from (17), (13) and (15) that

$$E\,[dM_i(t)\mid \mathbf{F}_t-]=0, \tag{18}$$

so that the integral of (17) (with $M_i(0)=0$)

$$M_i(t)=I\,(T\,(\omega_i)\le t)-\int_{(0,t]} y_i(u)dH_i(u) \tag{19}$$

is an $\mathbf{F}_t$- *-martingale* with zero mean

$$E[M_i(t)]=0. \tag{20}$$

In other words, the difference between the observable counting process $I(T(\omega_i)\le t)$ and its smooth, the hazard process $\lambda_i(t)=\int_{[0,t]} y_i(u)dH_i(u)$, represents a random noise (i.e., zero- mean martingale $M_i(t)$). In general, the conditional expectation of a counting process given the history $\mathbf{F}_t$- is a predictable measure called the *compensator* of the counting process with respect to $\mathbf{F}_t$- (see, e.g., Liptser and Shiryayev 1978 and Karr 1986). Thus a hazard process is a compensator. But a compensator (of a counting process) need not be a hazard process. For example, if in (13) we let $\mathbf{F}_t=\{\Phi,\Omega\}$, the smallest possible σ -algebra, then $d\lambda_i(t)=dF_i(t)$ and the compensator $E[I(T(\omega_i)\le t)\mid \mathbf{F}_t\text{-}]=F_i(t)$ is deterministic and hence not a hazard process. Similarly, both Poisson and Wiener processes have deterministic compensators as a result of the independent increment property.

Now, if the ω_i's in Ω are all copies of ω (i.e., if the population is homogeneous), then, for each i=1,2,...,n,

$$dH_i(t)=dH(t), \tag{21}$$

(i.e., same hazard for every individual in the population). In this case, we can superimpose all the counting processes $\{ I(T(\omega_i)\le t) \}$ for individual members ω_i, i=1,2,...,n, to obtain the counting process

$$N(t)=\sum_{i=1}^{n} I(T(\omega_i)\le t) \tag{22}$$

for the whole sample space Ω which counts the total number of occurrencies of the event of interest in (0,t]. The stochastic hazard $d\Lambda(t)$ and the hazard process $\Lambda(t)$ of (22) are respectively obtained by summing both sides of (15) and (16) over i and using (13) and (21):

$$d\Lambda(t)=\sum_{i=1}^{n} d\lambda_i(t)=E[dN(t)\mid \mathbf{F}_t\text{-}], \tag{23}$$

where $d\Lambda(t)=\Lambda(t)-\Lambda(t^-)$, and

$$\Lambda(t)\equiv\sum_{i=1}^{n}\lambda_i(t)=\int_{[0,t]} Y(u)dH(u)=E[N(t)\mid \mathbf{F}_t\text{-}], \tag{24}$$

where

$$Y(t)\equiv\sum_{i=1}^{n} y_i(t)=\sum_{i=0}^{n} I(T(\omega_i)\ge t) \tag{25}$$

is the *number at risk just before t*. If there are no censoring , as is the case here, then Y(t)= n - $N(t^-)$. Clearly, both N(t) and Y(t) are bounded by n. That $\Lambda(t)$ given by (24) is the *hazard process* of N(t) follows from the " no simultaneous occurrences " assumption which says that with probability one, each jump of N(t) is of unit magnitude (i.e., no two individual counting processes jump at the same time.)

By summing each of the formulas (17)-(20) for the individual counting processes over i and using the relations (21), (22) and (25), we obtain the corresponding formulas for the space counting process (22), with $M(t)\equiv\sum_{i=1}^{n} M_i(t)$:

The innovation gain

$$dM(t) = dN(t) - Y(t)dH(t) \tag{17'}$$

with zero conditional mean

$$E[dM(t) \mid \mathbf{F}_{t^-}] = 0 \tag{18'}$$

and the F_{t^-} - martingale

$$M(t) = N(t) - \int_{(0,t]} Y(u)dH(u), \tag{19'}$$

with zero mean

$$E[M(t) \mid \mathbf{F}_0] = 0. \tag{20'}$$

If F is absolutely continuous, then so is the hazard process. In this case the hazard rate exists so that the $dH_i(t)$ in (15),(16),(17),(19) and (21) may be replaced by $h_i(t)dt$ and the dH(t) in (21),(23),(24),(17') and (19') replaced by h(t)dt. The existence of the hazard rate function implies the continuity of the hazard process and hence simplifies results in statistical inference (see (27) below and Section 5). It is noted that the counting processes $I(T(\omega_i) \le t)$ and N(t) are right-continuous functions of t with left-hand limits, while the predictable (stochastic) processes $y_i(t)$ and Y(t) are left continuous with right-hand limits.The relationship among the hazard process, its associated counting process and the martingale is akin, in that order, to the relationship among the random effects $\mu(\mathbf{X})$, the response data $\mathbf{Y}$ and the error $\mathbf{e}$ in the random-effects (linear statistical) model:

$$\mathbf{Y} = \mu(\mathbf{X}) + \mathbf{e}, \tag{26}$$

with $\mathbf{X}$ playing the role of the pre-t information $\mathbf{F}_{t^-}$.

While the conditional mean of a martingale M(t) given $\mathbf{F}_{t^-}$ is $M(t^-)$ (see (20')), the conditional variance is not. The conditional variance of M(t) given $\mathbf{F}_{t^-}$ is equal to the *predictable variation process* given by

$$\langle M \rangle(t) = \int_{(0,t]} Y(u)dH(u), \tag{27}$$

as a result of Theorem 2.21 of Karr (1986) and equation (24), provided that the hazard process is a continuous function of t. Note that the variance of M(t) given by (27) is a random variable because Y(t) is. Note also that the hazard process given on the right side of (27) equals both conditional mean and variance of its associated counting process given $\mathbf{F}_{t^-}$. This implies that a counting process associated with a continuous hazard process is conditionally Poisson given F_{t^-}. However, a noncontinuous hazard process does not yield (27) for <M>(t) and hence does not have a conditional Poisson counting process. Under the "no simultaneous occurrences" assumption, it can similarly be shown that the predictable covariation process of any two $\mathbf{F}_{t^-}$ -martingales defined by (19) or (19') is zero, provided that the hazard process is a continuous function of t. That is to say, the martingales defined by (19) or (19') for continuous hazard processes are orthogonal.

4. Estimation of Deterministic Functions

In this and the next sections we shall use the martingale method to calculate the nonparametric estimators of the deterministic functions and to derive various test statistics. The martingale method is employed in preference to other methods (such as the likelihood method) mainly because of its direct link to the counting and hazard processes (see (19) and (19')) and its ease in making such calculations (see, e.g., Aalen 1978, Andersen et.al. 1982 and Karr 1986).

To estimate the deterministic functions, we shall first obtain the estimate for the unobserved dH(t). The martingale estimate of dH(t) is obtained by solving (17') for dH(t) and using the assumption of asymptotic negligibility (dM(t)/Y(t) $\rightarrow 0$, as $n\rightarrow\infty$) to yield

$$d\hat{H}(t)=I\,(Y(t)>0)\frac{dN(t)}{Y(t)}, \tag{28}$$

where $dN(t)=N(t)-N(t^-)$ is the number of events at time t or in $(t^-,t]$, which is zero for all t except at the jump points $t=T(\omega_i)$.The indicator function inserted in (28) as multiplier is to cover the possibility that Y(t) may take zero values. In view of the definition of dH(t) given in (6), formula (28) is the natural estimate of dH(t). Integrating (28) over (0,t] one obtains the martingale estimate of H(t):

$$\hat{H}(t)=\int_{(0,t]} I\,(Y(s)>0)\frac{dN(s)}{Y(s)}=\sum_{s_i\le t}\frac{1}{Y(s_i)} \tag{29}$$

where $s_i, i=1,...,n$, are the sequence of ordered occurrence times. Formula (29) is the natural estimate of the cumulative hazard H(t) (see also Section 6) and, as expected, is essentially identical to the ones obtained by Nelson (1969), Aalen (1975) and Altshuler (1980). The estimator (28) is then substituted into (8) and (10) to obtain the estimates of S(t) and $S(t^-)$, respectively, which in turn are substituted into (3) and (4) to obtain the estimates of the mean residual life e(t) and the O/E rate, respectively.

To obtain the (asymptotic) distribution of the estimator we will need to use two main results from the martingale theory (see, e.g., Aalen 1978, Gill 1984, Kopp 1984, Liptser and Shiryayer 1978, and Rebolledo 1980). The first is the *martingale transform theorem* which states that stochastic integrals of predictable processes with respect to martingales are martingales. The proof is very simple. The second is the *central limit theorem* for a sequence of martingales, which states that for a sequence of (mean zero) square-integrable martingale $M_1, M_2, \ldots, M_n, \ldots$, if there exists a non-negative continsuous non-decreasing deterministic function V(t) with V(0)=0 such that (a) for each t $\in(0,\tau]$, $<M_n>(t)$ converges to V(t) in distribution and (b) there exist constants $c_n \downarrow 0$ such that as $n\rightarrow\infty$, $P\,[\sup_t |dM_n(t)|\le c_n]\rightarrow 1$, then, for $t\in(0,\tau]$, $M_n(t)$ converges to M(t) in distribution, where M(t) is normally distributed with mean zero and variance equal to <M>(t)=V(t). The proof of this theorem is more technical and makes use of the fact that a martingale has orthogonal increments to reach a weak convergence to a time-transformed Wiener process as n increases (For detaills of the proof with different degrees of generality, see Rebolledo 1980, Liptser and Shiryayev 1980, Hall and Heyde 1980, and Helland 1982.)

[Note: Equation (29) is to be followed by equation (33).]

5. Asymptotic Distributions of Estimates and Hypothesis Testing

In this section we assume that F is absolutely continuous. Our next task is to obtain the asymptotic variance of the estimator $H(t)$ and to derive test statistics for testing hypotheses about the unknown H(t). To do so, we shall investigate the asymptotic distribution of the process $M_1(t)$ defined by

$$M_1(t) \equiv \sqrt{n}\,[\hat{H}(t) - H(t)]. \tag{33}$$

The first step is to show that $M_1(t)$ is asymptotically a martingale: Using (29) and (17'), we obtain

$$\begin{aligned}\sqrt{n}\,[\hat{H}(t)-H(t)] &= \sqrt{n}\,[\int_{[0,t]} I(Y(s)>0)\frac{dN(s)}{Y(s)} - dH(s)] \\ &= \sqrt{n}\int_{[0,t]} I(Y(s)>0)\frac{dM(s)}{Y(s)} + \sqrt{n}\int_{[0,t]} [I(Y(s)>0)-1]dH(s) \\ &\xrightarrow{P} \sqrt{n}\int_{[0,t]} I(Y(s)>0)\frac{dM(s)}{Y(s)}, \quad as \ n\to\infty,\end{aligned} \tag{34}$$

(where $\xrightarrow{P}$ signifies convergence in probability) which by the martingale transformation theorem, is a martingale because Y is $\mathbf{F}_t$-predictable. Note that $E[dM_1(t) \mid \mathbf{F}_t-]=0$ as the size of n increases. This follows upon taking expectation of (33) and using (29) and (23).

The second step is to obtain the asymptotic variation process of $M_1(t)$: Using (34),(27) and the $\mathbf{F}_t$- - predictability of Y, we obtain for large n,

$$d\langle M_1\rangle(t) = nI(Y(t)>0)Var[dM(t) \mid \mathbf{F}_t-]/[Y(t)]^2 = \frac{nI(Y(t)>0}{Y(t)}dH(t). \tag{35}$$

Hence, we have the large sample variation process

$$\langle M_1\rangle(t) = \int_{[0,t]} d\langle M_1\rangle(s) = n\int_{[0,t]} \frac{I(Y(s)>0)}{Y(s)}dH(s). \tag{36}$$

The third step is to find the asymptotic distribution of $M_1(t)$. We will use the so-called *asymptotic stability condition (the Glivenko-Cantelli theorem)* which states that Y(t)/n converges uniformly (see Pollard 1984) with probability one to the deterministic mean function $\rho(t) \equiv E[Y(t)/n]$, as $n\to\infty$. (If there are no censoring, as is the case here, $\rho(t)=S(t^-)$ .This, by the way, establishes the asymptotic negligibility condition of the martingale estimate (28)). Now we are ready to show that the two conditions of the martingale central limit theorem are satisfied by the martingale $M_1(t)$, so that it has (continuous) Gaussian distribution with mean zero and variance estimated by (36) for large n. We have, as $n\to\infty$,

$$\langle M_1\rangle(t) = \int_{[0,t]} \frac{I(Y(s)>0)}{[Y(s)/n]}dH(s) \xrightarrow{P} \int_{[0,t]} \frac{dH(s)}{\rho(s)},$$

which is deterministic and hence satisfies the first condition (a). To show that condition (b) holds, we note that jumps in the martingale $\sqrt{n}[H(t)-H(t)]$ arise only from $H(t)$ whose jumps are of size 1/Y(t)(see (28)), which is of order 1/n and hence the Lindeberg condition holds for $c_n = n^{-1/4}$ so that $\sqrt{n}[H(t)-H(t)]$ tends to zero , as $n\to\infty$. Note that conditions weaker than (b) such as those of Lindeberg, Lyapounov and Rebolledo suffice to yield the same results(see Andersen and Gill 1982 and Karr 1986). We therefore conclude that the estimator $H(t)$ given by (29) is consistent and asymptotically normal.

Now, from (36) we immediately obtain the conditional variance of $\hat{H}(t)$ as

$$Var[\hat{H}(t) \mid \mathbf{F}_{t^-}] = \int_{(0,t]} \frac{I(Y(s)>0)}{Y(s)} dH(s) \tag{37}$$

and, upon substituting (28) in (37), the sample variance

$$\hat{Var}[\hat{H}(t) \mid \mathbf{F}_{t^-}] = \int_{(0,t]} \frac{I(Y(s)>0)}{[Y(s)]^2} dN(s). \tag{38}$$

We now derive test statistics for various hypotheses on h(t), $t \in (0,\tau]$. In general, two types of hypotheses are of interest. One type is to compare the unknown hazard h(t) in the target population with some known hazard $h^o(t)$ (such as that in the general population). The second type is to compare the unknown hazards among different target populations.

In the first type, the hypothesis of interest may be either

$$Hypo\ \ 1: h(t) = h^o(t) \tag{39}$$

or

$$Hypo\ \ 2: h_j(t) = h_j^o(t),\ \ j=1,\ldots,k, \tag{40}$$

depending on whether we are dealing with one target population or k target populations. In (39) we might be interested in comparing the hazard level of a cohort with that of a general population of which the cohort is a subset. In (40) we might be interested in comparing the hazard levels for various occupational groups with those for the nation as a whole.

We shall derive two test statistics for (39). Using the asymptotic distribution of $M_1(t)$, namely, the Gaussian distribution with mean zero and variance estimated by (36), one immediately obtains the asymptotic unit normal test statistic for Hypo 1 as follows:

$$Z = \frac{M_1(\tau)}{[<M_1>(\tau)]^{1/2}} = \frac{\hat{H}(\tau) - H^o(\tau)}{[<\hat{H} - H^o>(\tau)]^{1/2}}$$

$$= \frac{\int_{(0,\tau]} I(Y(s)>0) \frac{dN(s)}{Y(s)} - \int_{(0,\tau]} h^o(s)ds}{[\int_{(0,\tau]} \frac{I(Y(s)>0)}{Y(s)} h^o(s)ds]^{1/2}}, \tag{41}$$

where we have used (29) and (38) to obtain the first term in the numerator and the denominator, respectively. In actual applications, the hazard rate function $h^o(t)$ is normally given in terms of the duration-specific occurrence/exposure rates $\{h_j\}_{j=1}^r$, with $(0,\tau]$ partitioned into r subintervals $I_j \equiv (t_j, t_{j+1}]$, where $t_1 = 0$ and $t_{r+1} = \tau$, the cutoff date of the study. Accordingly, the two terms involving h^o on the right side of (41) are written in terms of the h_j^o's as follows:

$$\int_{(0,\tau]} h^o(s)ds = \sum_{j=1}^{r} h_j^o (t_{j+1} - t_j), \tag{41a}$$

where h_j^o is defined by (4) and

$$\int_{(0,\tau]} \frac{I(Y(s)>0)}{Y(s)} h^o(s)ds = \sum_{j=1}^{r} h_j^o \int_{I_j} \frac{I(Y(s)>0)}{Y(s)} ds. \tag{41b}$$

An alternative test statistic can be derived by utilizing the result obtained at the end of Section 3 that for the continuous hazard process

$$\Lambda(t)=\int_{(0,t]} Y(s)h(s)ds, \tag{24'}$$

the associated counting process N(t) is conditionally Poisson with mean and variance both equal to (24'). This implies an asymptotic unit normal test statistic for Hypo 1 given by

$$Z'=\frac{N(\tau)-\int_{(0,\tau]} Y(s)h^o(s)ds}{[\int_{(0,\tau]} Y(s)h^o(s)ds]^{1/2}}. \tag{42}$$

When, as above, $h^o(t)=h_j^o(t)$ for $t\in I_j$, thenthe terms involving h^o in (42) become

$$\int_{(0,\tau]} Y(s)h^o(s)ds=\sum_{j=1}^{r} h_j^o \int_{I_j} Y(s)ds. \tag{42a}$$

The test statistic Z' given in (42) could also be derived by defining the transformed martingale (see Andersen et al. 1982, and Andersen and Borgan 1985)

$$M'_1(t)=\int_{(0,t]} Y(s)dM_1(s), \tag{43}$$

with $M_1(t)$ as defined in (33). $M'_1(t)$ as defined in (43), being a sum of linear functions of zero-mean Gaussian processes $M_1(t)$, is a zero-mean Gaussian process with variance estimated by

$$<M'_1>(t)=n\int_{(0,t]} Y(s)dH(s), \tag{44}$$

since we have from (35) and (43),

$$d<M'_1>(t)=[Y(t)]^2 d<M_1>(t)=nY(t)dH(t).$$

Finally, substitution of (33) into (43), followed by a normalization by the square root of (44), yields the test statistic (42) for Hypo 1.

In many applied disciplines, the hazard process (24') is referred to as the "expected" counts. (The use of the word "expected" is inappropriate because as noted earlier, the hazard process is a random measure). Thus, the test statistic (42) simply represents a standardized difference between observed and "expected" counts (O-E)/$\sqrt{E}$. The ratio O/E of the observed to the "expected" counts is known in demography as the *standardized mortality ratio (SMR)* and in economics as the *Paasche index.*

Now, we come to the second hypothesis (40). Since the martingales defined by (33) are orthogonal, we may form the sum of squares of k test statistics of (42) to obtain

$$\chi_k^2=\sum_{j=1}^{k}\frac{(O_j-E_j)^2}{E_j}, \tag{45}$$

where O_j and E_j represent, respectively, the two terms in the numerator of (42) for each sample j=1,...,k, as a test statistic for (40) which has an asymptotic Chi-square distribution with k degrees of freedom.

Hypotheses without specification of hazard values are also of interest. For example, an investigator might be interested in comparing the hazard levels among various population groups classified by certain fixed external discrete or categorical covariates of interest. Suppose the intersection of these discrete covariates produces a total of k categories. Then we are here interested in testing the second type of hypothesis given by

$$Hypo:\ h_1(t)=h_2(t)=...=h_k(t). \tag{46}$$

To derive the k-sample test statistic for the hypothesis (46), we make use of the fact that under the null hypothesis, the k samples are homogeneous with common hazard rate h(t)= $h_j(t)$, j=1,2,...,k. Hence, both the counting processes $N_j(t)$ and the $\mathbf{F}_{t^-}$-predictable processes $Y_j(t)$ can be superimposed over j to obtain the martingale:

$$M(t)=N.(t)-\int_{(0,t]} Y.(s)h(s)ds, \tag{47}$$

where $N.(t)=\sum_{j=1}^{k} N_j(t)$, with $n_j=Y_j(0)$, so that n.=Y.(0)=$\sum_{j=1}^{k} n_j$ and $Y.(t)=\sum_{j=1}^{k} Y_j(t)$. The unknown h(t) or equivalenty H(t) in this case has to be estimated in order to construct a test statistic for (46).

Corresponding to (29), we obtain the martingale estimate of H(t) from (47) as follows:

$$\hat{H(t)}=\int_{(0,t]} I(Y.(s)>0)\frac{dN.(s)}{Y.(s)}=\sum_{s_i\le t}\frac{1}{Y.(s_i)}, i=1,...,n.. \tag{48}$$

Corresponding to (42), we obtain the test statistic for the hypothesis (46) as

$$\chi^2_{k-1}=\sum_{j=1}^{k}(O_j-E_j)^2/E_j, \tag{49}$$

where $O_j \equiv N_j(\tau)$ and

$$E_j \equiv \int_{(0,\tau]} Y_j(s)d\hat{H(s)}=\int_{(0,\tau]} I(y.(s)>0)Y_j(s)\frac{dN.(s)}{Y.(s)}=\sum_{s_i\le t}\frac{Y_j(s_i)}{Y.(s_i)}, i=1,...,n.. \tag{49a}$$

The test statistic (49) is known as the *Chi-square test of homogeneity*. It has an asymptotic Chi-square distribution with k-1 degrees of freedom. The loss of one degree of freedom from the total k degrees of freedom is due to the requirement of estimating the unknown H(t).

6. Censoring, Competing Risks and Time-Varying Covariates

With no loss of generality, we shall from now on take "failure" as the absorbing or nonrecurrent event of interest. Suppose there are q modes (causes) of failure and, for any member ω_i of the sample space Ω, let $T_\delta(\omega_i)$ be ω_i 's failure time due to cause δ in the absence of all other causes, δ=1,2,...,q. T_δ is known as the *latent life* in competing risks theory. With $F_\delta(t)\equiv P[T_\delta \le t]$, the hazard rate h_δ, hazard probability dH_δ, cumulative hazard H_δ, density f_δ, survival S_δ and mean residual life function e_δ of T_δ are all as defined in Sections 1 and 2 with T there replaced by T_δ. These functions are not identifiable because the T_δ's are *not observable*.

Let

$$T^*(\omega_i) \equiv \min(T_1(\omega_i),...,T_q(\omega_i)). \tag{50}$$

Then $T^*(\omega_i)$ is the failure time of ω_i in the presence of all causes. Note that right censoring is considered and included in (50) as a mode of failure. To be specific, let $T_q(\omega_i)$ be the censoring time for ω_i. Just as in the last paragraph, with $F^*(t) \equiv P[T^* \le t]$, the functions describing the distribution of T^*, namely, h^*, dH^*, H^*, f^*, S^* and e^*, are similarly defined. These functions are identifiable because $T^*(\omega_i)$ is observable.

Let $X(\omega_i)$ index the mode (cause) of failure and $Z_t(\omega_i))$ be a vector of *internal* covariates ω_i assumes at time t. Clearly, $X(\omega_i)$ takes values from the finite index set $\Delta \equiv \{1,2,...,q\}$ which lists the causes of failure. Let $\Delta' \equiv \{1,2,...,q-1\}$ which is the subset of Δ with censoring excluded. Suppose for simplicity that $Z_t(\omega_i)$ takes values from a set of discrete vector values $\Gamma \equiv \{z_1, \ldots, z_p\}$. Note that when $Z_t(\omega_i)$ is continuous we may categorize it and use $z_\gamma, \gamma=1,2,...,p$, to represent the mean value for the γ th category. Clearly, $Z_t(\omega_i)$ is a bounded predictable vector covariate process and can change the values it takes, from z_1 to z_2, say, as t progresses from 0 to $T^*(\omega_i)$.

As there are several possible ways for the covariate to vary with time, different classifications of the time-varying covariates prevail in the literature. The covariates $Z_t(\omega_i)$ or Z(t) introduced here corresponds to the class of *internal* covariates in Kalbfleisch and Prentice's (1980) classification. Most of work on statistical inference involving time-varying covariates done so far employed either parametric approach or semiparametric regression models and used likelihood or least-squares methods which require (1) a given functional form for the whole or part of the hazard rate function with covariates written as explicit functions of time and (2) integration of the hazard rate function to obtain the cumulative hazard and survival functions (see Petersen 1986, Blossfeld et al 1989 and references therein). These procedures are mathematically unwieldy and computaionally lengthy. Besides, it is not always possible to write the time path of covariate as an explicit function of time. Our approach to the treatment of the time-varying covariate is nonparametric and is akin to those of Leung and Wong (1990) and McKeague and Utikal (1990). It requires a Markovian assumption on the covariates and provides a straight forward approach to the estimation and hypothesis testing problem.

In this section we formulate the competing-risks problem as a *marked point process* $\{T^*(\omega_i), X(\omega_i)\}$ with the internal history $\mathbf{F}_{t^-}^{T^* \times X}$ defined by

$$\mathbf{F}_t^{T^* \times X} = \sigma\{I(T^*(\omega_i) \le s, X(\omega_i) = \delta);\ 0 \le s \le t, \delta \in \Delta, i=1,...,n\} \tag{51}$$

and employ the sub σ-algebra

$$\mathbf{F}_t^{T^*} = \sigma\{I(T^*(\omega_i) \le s), 0 \le s \le t, i=1,...,n\} \tag{52}$$

and the covariate history

$$\mathbf{F}_t^{Z} = \sigma\{Z_s(\omega_i) = z; 0 \le s \le \min(T^*(\omega_i), t), z \in \Gamma, i=1,...,n\} \tag{53}$$

as natural filtration to accommodate time-varying covariates. Note that the two random variables $T^*(\omega_i)$ and $X(\omega_i)$ are realized simultaneously. We cannot observe one without the other. Thus, when failure occurs to ω_i, in addition to the observed failure time $T^*(\omega_i)$, we also observe the mode of failure $X(\omega_i)$ and the covariate history $Z_t(\omega_i)$ up to the time of failure $T^*(\omega_i)$. This is a considerable enlargement from the previous sections where we only dealt with the point process $\{ T(\omega_i) \}$ and employed the internal history $\mathbf{F}_{t^-}^{T}$ defined in (11) as natural filtration.

An appropriate extension of the definition of the deterministic hazard rate h(t) given in (1) to incorporate censoring/competing risks and time-varying covariates yields the two *net hazard rate functions* of the latent life T_δ given by

$$h_\delta(t) \equiv \lim_{\Delta \downarrow 0} P[T_\delta < t+\Delta \mid T_\delta \geq t]/\Delta \tag{54a}$$

and

$$h_\delta(t;z) \equiv \lim_{\Delta \downarrow 0} P[T_\delta < t+\Delta \mid T_\delta \geq t, Z(t)=z]/\Delta, \tag{54b}$$

and the two *crude hazard rate functions* of the marked point $\{T^*, X\}$ given by

$$h^*(t,\delta) \equiv \lim_{\Delta \downarrow 0} P[T^* < t+\Delta, X=\delta \mid T^* \geq t]/\Delta \tag{55a}$$

and

$$h^*(t,\delta;z) \equiv \lim_{\Delta \downarrow 0} P[T^* < t+\Delta, X=\delta \mid T^* \geq t, Z(t)=z]/\Delta. \tag{55b}$$

Similarly, extensions for the deterministic hazard probability (6) yields the two *net hazard probabilities:*

$$dH_\delta(t) = \begin{cases} P[T_\delta = t \mid T_\delta \geq t], & \text{if } F_\delta \text{ is dicrete,} \\ h_\delta(t)dt, & \text{if } F_\delta \text{ is continuous,} \end{cases} \tag{56a}$$

and

$$dH_\delta(t;z) = \begin{cases} P[T_\delta = t \mid T_\delta \geq t, Z(t)=z], & \text{if } F_\delta \text{ is discrete,} \\ h_\delta(t;z)dt, & \text{if } F_\delta \text{ is continuous;} \end{cases} \tag{56b}$$

and the two *crude hazard probabilities:*

$$dH^*(t,\delta) = \begin{cases} P[T^* = t, X=\delta \mid T^* \geq t], & \text{if } F^* \text{ is discrete,} \\ h^*(t,\delta)dt, & \text{if } F^* \text{ is continuous,} \end{cases} \tag{57a}$$

$$dH^*(t,\delta;z) = \begin{cases} P[T^* = t, X=\delta \mid T^* \geq t, Z(t)=z], & \text{if } F^* \text{ is discrete,} \\ h^*(t,\delta;z)dt, & \text{if } F^* \text{ is continuous.} \end{cases} \tag{57b}$$

In (54b), (55b), (56b) and (57b) above we have replaced the conditioning events $\{ T_\delta \geq t, Z(s), 0 \leq s \leq t \}$ and $\{ T^* \geq t, Z(s), 0 \leq s \leq t \}$ in the general definitions of conditional hazard functions by $\{ T_\delta \geq t, Z(t)=z \}$ and $\{ T^* \geq t, Z(t)=z \}$ because we assume the Markovian Property for Z(t) (see Heckman and Singer (1984) for illustrations) which implies that $h_\delta(t;z)$ $h^*(t,\delta;z)$, $dH_\delta(t;z)$ and $dH^*(t,\delta;z)$ will remain unchanged with the condition Z(t)=z in these formulas replaced by $Z(s)=z, 0 \leq s \leq t$. That is , the hazard at duration time t depends on the covariates only through their values at the duration time t and the above four hazard formulas are identical to those with the covariate values remaining unchanged at z over the entire course of the spell and hence are interpretable as such.

Just as in Sections 4 and 5, our interest here lies in making inference on $dH_\delta(t)$ or $h_\delta(t)$ and $dH_\delta(t;z)$ or $h_\delta(t;z)$. However, these net hazard functions are not identifiable, i.e., cannot be estimated from the observed information (51)-(53) without further assumptions. If we make the common assumption that the latent lives T_δ , $\delta \in \Delta$, are mutually independent or conditionally independent given $\mathbf{F}_t^Z$, it follows that (54a) equals (55a), and (54b) equals (55b), (56a) equals (57a) and (56b) equals (57b) (see, e.g., Hsieh 1989 for the proof of these and related results). Since the crude hazards are identifiable, the net hazards can thus be estimated through their relationships to the crude hazards under the assumption of independent or conditional independent latent lives.

As in Section 3 we shall employ the homogeneity assumption that the ω_i's in Ω are all copies of the same ω (homogeneous population) and the "no simultaneous occurrences" assumption which in the present context means that with probability 1, $T_{\delta_1}(\omega_i) \neq T_{\delta_2}(\omega_i)$ for $\delta_1 \neq \delta_2$, $i=1,...,n$, and $T_\delta(\omega_{i_1}) \neq T_\delta(\omega_{i_2})$, for $i_1 \neq i_2$, $\delta=1,...,q$. The counting process N(t) of Section 3 (see (22)) is directly extended as follows: For the problem of competing risks accounting for right censoring, we have

$$N(t,\delta)=\sum_{i=1}^{n} I(T^*(\omega_i)\leq t, X(\omega_i)=\delta), \quad t\geq 0, \quad \delta \in \Delta', \tag{58a}$$

which counts the number of failures in (0,t] due to cause $\delta \in \Delta'$. For the problem of competing risks accounting for right censoring and time-varying covariates, we have

$$N(t,\delta;z)=\sum_{i=1}^{n} I(T^*(\omega_i)\leq t, X(\omega_i)=\delta, Z_{T^*}(\omega_i)=z), \quad t\geq 0, \quad \delta \in \Delta', \quad z \in \Gamma, \tag{58b}$$

which counts the number of failures in (0,t] due to cause $\delta \in \Delta'$ with covariates z at the time of failure.

The $\mathbf{F}_{t^-}$ -predictable process for the counting process $N(t,\delta)$ defined in (58a) is

$$Y^*(t)=\sum_{i=1}^{n} y_i^*(t)=\sum_{i=1}^{n} I(T^*(\omega_i)\geq t), \quad t\geq 0, \tag{59a}$$

which represents the number at risk at time t of failure from any cause (including censoring), and the $\mathbf{F}_{t^-}$ -predictable process for $N(t,\delta;z)$ defined in (58b) is

$$Y^*(t;z)=\sum_{i=1}^{n} y_i^*(t;z)=\sum_{i=1}^{n} I(T^*(\omega_i)\geq t, Z_t(\omega_i)=z), \quad t\geq 0, \quad z\in\Gamma, \tag{59b}$$

which represents the number at risk at time t with covariate z at time t. Note that $y_i^*(t)$ and $Y^*(t)$ are $\mathbf{F}_{t^-}^T$ -predictable while $y_i^*(t;z)$ and $Y^*(t;z)$ are $\mathbf{F}_{t^-}^T$ $\mathbf{F}_{t^-}^Z$ -predictable. Note also that Y(t) defined in (25) and $Y^*(t)$ defined in (59a) are *decrement-life-table functions* and hence are monotone nonincreasing function of t, while for fixed covariate z, $Y^*(t,z)$ defined in (59b) is an *increment-decrement-life-table* function and hence may either increase or decrease with t.

The hazard process $\Lambda(t)$ of Section 3 (see (24) and (24')) is extended to correspond to (58a) and (58b), respectively, as follows:

$$\Lambda(t;\delta)=\int_{(0,t]} Y^*(s)dH^*(s,\delta)=\int_{(0,t]} Y^*(s)dH_\delta(s), \tag{60a}$$

and

$$\Lambda(t,\delta;z)=\int_{(0,t]} Y^*(s;z)dH^*(s,\delta;z)=\int_{(0,t]} Y^*(s;z)dH_\delta(s;z). \tag{60b}$$

The second equalities in (60a) and (60b) follow from the assumption of independent and conditonal independent latent lives.

The assumption of independent latent lives implies preservation of martingale property. The martingale equations (19') and (20') of Section 3 are extended to correspond to the counting process (58a) and (58b) and their respective compensators (60a) and (60b) as follows:

$$M(t,\delta)=N(t,\delta)-\int_{(0,t]}Y^*(s)dH_\delta(s),\ \ \delta\in\Delta', \tag{61a}$$

with $E[M(t,\delta)\,|\,\mathbf{F}_{t^-}]=0$, where $\mathbf{F}_t=\mathbf{F}_t^{T^*\times X}$, and

$$M(t,\delta{:}z)=N(t,\delta;z)-\int_{(0,t]}Y^*(s,z)dH_\delta(s;z),\ \ \delta\in\Delta',\ z\in\Gamma, \tag{61b}$$

with $E[M(t,\delta;z)\,|\,\mathbf{F}_{t^-}]=0$, where $\mathbf{F}_t=\mathbf{F}_t^{T^*\times X}\ \mathbf{F}_t^Z$.

It follows immediately from (61a) and (61b) that

$$E[N(t,\delta)\,|\,\mathbf{F}_{t^-}]=\int_{(0,t]}Y^*(s)dH_\delta(s)$$

and

$$E[N(t,\delta;z)\,|\,\mathbf{F}_t]=\int_{(0,t]}Y^*(s;z)dH_\delta(s;z),$$

which are the corresponding extensions of (24) of Section 3.

Thus, martingale estimation formulas (28) and (29) of Section 4 for the deterministic hazard dH(t) are extended to those for $dH_\delta(t)$ as follows: For each $\delta\in\Delta'$, we have

$$d\hat{H}_\delta(t)=I(Y^*(t)>0)\frac{dN(t,\delta)}{Y^*(t)},\ \ t\ge 0, \tag{62a}$$

and

$$\hat{H}_\delta(t)=\int_{(0,t]}I(Y^*(s)>0)\frac{dN(s,\delta)}{Y^*(s)}=\sum_{s_i\le t}\frac{dN(s_i^*,\delta)}{Y^*(s_i^*)},\ \ t\ge 0, \tag{62b}$$

where s_i^*, i=1,...,n, is the sequence of ordered failure times irrespective of cause and $dN(t,\delta)=N(t,\delta)-N(t^-,\delta)$ is the number of failures due to cause $\delta\in\Delta'$ at time t, which is zero except at the mark points $(T^*(\omega_i)=t, X(\omega_i)=\delta)$. Similar extensions for estimation of $dH_\delta(t;z)$ and $H_\delta(t;z)$ yield for each $(\delta,z)\in\Delta\times\Gamma$:

$$d\hat{H}_\delta(t;z)=I(Y^*(t;z)>0)\frac{dN(t,\delta{:}z)}{Y^*(t;z)},\ \ t\ge 0, \tag{63a}$$

and

$$\hat{H}_\delta(t;z)=\int_{(0,t]}I(Y^*(s;z)>0)\frac{dN(s,\delta;z)}{Y^*(s;z)}=\sum_{s_i\le t}\frac{dN(s_i^*,\delta;z)}{Y^*(s_i^*;z)},\ \ t\ge 0, \tag{63b}$$

where $dN(t,\delta;z)=N(t,\delta;z)-N(t^-,\delta;z)$ is the number of failures due to cause $\delta\in\Delta'$ at time t with covariate z at the time of failure, which is zero except at the points $(T^*(\omega_i)=t,\ X(\omega_i)=\delta, Z_{T^*}(\omega_i)=z)$. Note that just as in Section 4, we have used the assumption of asymptotic negligibility in (62) and (63). The validity of this assumption will be shown in the next paragraph.

The hypotheses and test statistics of Section 5 on the h(t)'s are to be extended to the corresponding hypotheses and test statistics on the $h_\delta(t)$'s and $h_\delta(t;z)$'s. In order for the corresponding test statistics for these hypotheses to hold, one has to show that the two martingale extensions of $M_1(t)$ defined by (33), namely,

$$M_2(t)=\sqrt{n}[\hat{H}_\delta(t)-H_\delta(t)] \tag{64a}$$

and

$$M_3(t)=\sqrt{n}[\hat{H}_\delta(t;z)-H_\delta(t;z)] \tag{64b}$$

satisfy the two conditions of the martingale central limit theorem stated in Section 4. In the case of (64a), these two conditions are satisfied because just as Y(t) in Section 5, $Y^*(t)$ in this section is monotone non-increasing, for which almost surely uniform convergence of $Y^*(t)/n$ holds and Glivenko-Cantelli theorem applies, yielding the deterministic function $E(Y^*(t)/n)=S^*(t^-)=\prod_{\delta=1}^{q}S_\delta(t^-)$. In the case of (64b), the two conditions of the martingale central limit theorem are also satisfied but, because of the non-monotonicity of $Y^*(t;z)$ in this case, one arrives at convergence in probability rather than almost surely uniform convergence (see Pollard 1984 and McKeague and Utikal 1990) and the asymptotic stability condition implies that $Y^*(t;z)/n$ converges in probability to a deterministic function $\rho(t,z)$ as $n\to\infty$. (The exact expression for $\rho(t,z)$, which is rather complicated, is not required to obtain the asymptotic Gaussian distribution of $M_3(t)$). These results also imply the validity of the asymptotic negligibility, namely, $1/Y^*(t)\to 0$ and $1/Y^*(t;z)\to 0$, as $n\to\infty$, used in the martingale estimation of $dH_\delta(t)$ in (62a) and $dH_\delta(t;z)$ in (63a) above. We thus conclude that both $M_2(t)$ and $M_3(t)$ have asymptotic Gaussian distributions with mean zero and respective variances $<M_2(t)>(t)$ and $<M_3>(t)$ estimated by (36) with Y(s), H(s) replaced by $Y^*(s),H_\delta(s)$ and by $Y^*(s;z),H_\delta(s;z)$, respectively, and that the estimators $\hat{H}_\delta(t)$ given by (62b) and $\hat{H}_\delta(t;z)$ given by (63b) are consistent and asymptotically normal.

Corresponding to Hypo 1 (Equation (39)) of Section 5 we have two extensions in this section. For the competing risks problem we have:

$$Hypo\,1'\!: \; h_\delta(t0=h_\delta^o(t), \;\; \delta\in\Delta'. \tag{39'}$$

It follows from the results of the last paragraph that the first test statistic for (39') is given by (41) with Y(s), N(s), $h^o(s)$ replaced by $Y^*(s)$, $N(s,\delta)$, $h_\delta^o(s)$, respectively, together with similar replacements for (41a) and (41b), and that the second test for (39') is given by (42) with $N(\tau)$, Y(s), $H^o(s)$ replaced by $N(\tau,\delta)$, $Y^*(s)$, $h_\delta^o(s)$, respectively, together with similar replacements for (42a). For the problem of competing risks with time-varying covariates we have the extension:

$$Hypo\,1''\!: \; h_\delta(t;z)=h_\delta^o(t;z), \;\; \delta\in\Delta'; \;\; z\in\Gamma. \tag{39''}$$

The two test statistics for (39'') are given by (41) with Y(s), N(s), $h^o(s)$ replaced by $Y^*(s;z)$, $N(s,\delta;z)$, $h_\delta^o(s;z)$, and given by (42) with $N(\tau)$, Y(s),$h^o(s)$replaced by $N(\tau,\delta;z)$, $Y^*(s;z)$, $h_\delta^o(s;z)$, respectively, together with similar replacements for (42a).

Corresponding to Hypo 2 (equation (40)) we also have two extensions:

$$Hypo\;2'\!: \; H_{\delta j}(t)=h_{\delta j}^o(t), \;\; j=1,\ldots,k, \;\; \delta\in\Delta', \tag{40'}$$

and

$$Hypo\,2''\!: \; h_{\delta j}(t;z)=h_{\delta j}^o(t;z), \;\; j=1,\ldots,k, \;\; \delta\in\Delta', \;\; z\in\Gamma. \tag{40''}$$

The test statistic for (40') and that for (40'') are given by (45) wherein O_j and E_j represent the two terms in the numerator of (42) for each sample j=1,...,k, with appropriate replacements of $N_j(\tau),Y_j(s)$ and $h_j^o(s)$ as described in the last paragraph for the competing risks problems with and without time-varying covariates,respectively.

Finally, for the second type of hypothesis, we have in correspondence to the hypothesis (46) the following two extensions:

$$Hypo': \; h_{\delta 1}(t)=h_{\delta 2}(t)=...=h_{\delta k}(t), \;\; \delta\in\Delta', \tag{46'}$$

and

$$Hypo'': \; h_{\delta 1}(t;z)=...=h_{\delta k}(t;z), \;\; \delta\in\Delta', \;\; z\in\Gamma. \tag{46''}$$

The k-sample test statistics for (46') and (46'') are given by (49) and (49a) with $N_j(\tau)$, $Y_j(s)$, Y.(s) and N.(s) replaced by $N_j(\tau,\delta)$, $Y_j(s,\delta)$, $Y.(s,\delta)$ and $N.(s,\delta)$, respectively, in the case of (46'), and by $N_j(\tau,\delta;z)$, $Y_j(s,\delta;z)$, $Y.(s,\delta;z)$ and $N.(s,\delta;z)$, repectively, in the case of (46'').

REFERENCES

Aalen, O. O. (1978): " Non-parametric Inference for a Family of Counting Processes," *Ann. Statist.* **6,** 701-726.

Altshuler, B. (1970): " Theory for the Measurement of Competing Risk in Animal Experiments," *Math. Biosci.* **6,** 1-11.

Andersen, P. K. and Borgan, O. (1985): ' Counting Process Model for Life History Data: a review, " *Scand J. Statist.* **12,** 97-158.

Andersen, P. K., Borgan, O., Gill, R. D. and Keiding, N. (1982): " Linear Nonparametric Tests for Comparison of Counting Processes, with Applications to Censored Survival Data (with discussion)," *Int. Statist. Rev.* **50,** 219-258. Correction **52,** 225.

Andersen, P. K., and Gill, R. D. (1982): " Cox's Regression Model for Counting Processes: a Large Sample Study," *Ann. Statist.* **10,** 1100-1120

Barlow, R. and Proschan, F. (1975): *Statistical Theory of Reliability and Life Testing,* New York: Holt, Rinehart and Winston.

Blossfeld, H. P., A. Hamerle, and K. U. Mayer (1989): *Event History Analysis: Statistical Theory and Application in Social Sciences.* Hillsdale, NJ: Lawrence Erlbaum Associates.

Cox, D. R., and D. Oakes (1984): *Analysis of Survival Data.* London: Chapman and Hall.

Gill, R. D. (1984): Understanding Cox's Regression Model: a Martingale Approach, *J. Am. Statisti. Ass.* 441-447

Hall, P., and C. Heyde (1980): *Martingale Limit Theory and Its Applications.* New York. NY.: Academic Press.

Heckman, J. J., and B. Singer (1984): " Econometric Duration Analysis, " *Journal of Econometrics,* **24,** 63-132.

Helland, I. S. (1982): " Central Limit Theorem for Martingales with Continuous Time," *Scand J. Statist.* **9,** 79-94

Hsieh, J. J. (1989): "A Probability Approach to the Construction of Competing-Risk Life Tables, " *Biometrical Journal,* **31,** 339-357

------------ (1990): " Construction of Expanded Continuous Life Tables --- A Generalization of Complete and Abridged Life Tables," *Mathematical Biosciences,* (In Press).

Kalbfleisch, J. D., and Prentice, R. L. (1980): *The Statistical Analysis of Failure Time Data.* New York: Wiley.

Karr, A. F. (1986): *Point Processes and their Statistical Inference.* New York:

Dekker.

Kopp, P. E. (1984): *Martingales and Stochastic Integrals.* Cambridge: Cambridge University Press.

Leung, S. F., and W. H. Wong (1990): " Nonparametric Hazard Estimation with Time-Varying Discrete Covariates," *Journal of Econometrics,* **45,** 309-330

Lipster, R. S. and Shiryayev, A. N. (1978): *Statistics of Random Processes II: Applications,* New York: Springer-Verlag.

Mckeague, I. W., and K. J. Utikal(1990): " Inference for a Nonlinear Counting Process Regression Model," *Annals of Statistics* **18,** 1172-1187.

Nelson, W. (1969): " Hazard Plotting for Incomplete Failure Data, " *J. Qual. Technol.* **1,** 27-52

Petersen, T (1986): " Fitting Parametric survival Models with Time-Dependent Covariates," *Journal of Royal Statistical Society,series C,* **35,** 281-288

Pollard, D. (1984): *Convergence of Stochastic Process.* New York, NY.: Springer-Verlag.

Rebolledo, R. (1980): " Central Limit Theorems for Local Martingales, " Z. *Wahrsch. verw.Gebiete* **51, 269-286.**

Bootstrap Assessment of Prediction in Exploratory Regression Analysis

Victor Kipnis
University of Southern California*

Abstract

The paper concentrates on the evaluation of the impact of exploratory analysis for selecting the 'best' predictor with respect to the mean squared error of prediction (MSPE). The very selection process affects the distribution of the conventional MSPE estimators and, in particular, leads to their substantial overoptimism. To allow for the selection effect, it is suggested to construct a pseudomodel for generating bootstrap pseudodata and to apply to them the selection procedure used for the original data. The simulation study demonstrates that the suggested approach can be used not only for assessing existing procedures, but it also may be helpful in providing a criterion for a stopping rule in multistep model building.

1. Introduction.

As opposed to the traditional inference, which is based on an *a priori* specified model, a major feature of modern regression analysis is model building within the framework of exploratory data analysis. This process may include checking model assumptions , applying regression diagnostics , assessment of influence, selection of regressors, choosing functional form of the equation, along with parameter estimation, hypotheses testing, *etc.* As was pointed out by Box (1983), not one but at least two kinds of inference are involved within this exploratory process: fitting the data according to a tentative model and data and model criticism. The latter includes assessment of the fit with regard to a specified criterion and replacement, if possible, of the current model with a new one. It is used iteratively and in alternation with fitting. This makes contemporary regression analysis a multistep, iterative process in which the regression model is not fixed in advance but rather is continually evolving. In practice, when exploratory activity is usually carried out using and reusing the same data, the conventional means of inference both at the intermediate steps and for the finally chosen equation could be misleading, sometimes badly so (e.g., Freedman, 1983; Lovell, 1983; Miller, 1984; Pinsker et al., 1985, 1987).

When the main goal of regression analysis is prediction of new observations, model building is usually reduced to selection of a regression equation that will give the 'best' predictions. There is a good deal of literature on different selection procedures with regard to their computational (logical) scheme (e.g., see a review in Hocking, 1976 and Thompson, 1978), but much less has been done concerning their impact on statistical properties of a selected predictor. This paper concentrates on estimating the mean squared error of prediction

*University Park, Los Angeles, California 90089-0641.

Exploring the Limits of Bootstrap. Edited by Raoul LePage and Lynne Billard.
 ISBN: 0471-53631-8.

(MSEP) of a regression equation selected by an exploratory procedure. For illustrative purposes, regression procedures based on subset selection methods are considered. This choice has been motivated by the fact that subset selection methods have a relatively simple logic and can be easily formalized and, on the other hand, they well demonstrate the vast conceptual and statistical problems associated with model selection. Besides, they have been very popular in applied work and have received much attention in the literature. It should be emphasized, though, that the general concepts of the present study apply to any other exploratory procedure for model building.

After formulating a problem of MSEP estimation in exploratory regression analysis in section 2, three bootstrap estimators are introduced and discussed in section 3. The results of the experimental comparison of the bootstrap and conventional estimators are analyzed in section 4. The final section 5 is a conclusion.

2. Problem Formulation.

Consider the linear regression model

$$Y = Y^o + \epsilon = X\beta + \epsilon, \tag{1}$$

where Y is a n–vector of observations on the response variable, $Y^o = X\beta$ is a n-vector of unobservable mean components of the model, X is a $n \times k$ full rank matrix of observations on each of the k predictor variables, β is a k–vector of unknown parameters, and ϵ is a n–vector of unobservable independent and identically distributed disturbances with mean 0 and variance σ^2. The X's are assumed fixed. The given $n > k$ observations on the response and the predictor variables constitute, we suppose, a construction set $V = (X_V, Y_V)$. Let $W = (X_W, Y_W)$ be a new set of n_W observations based on the same model (1), where Y_W is independent from Y_V. Given the construction set V and the matrix X_W, the regression goal is prediction of hypothetical new observations Y_W.

Consider an exploratory procedure

$$g : (V; X_W) \to \hat{Y}_W$$

that selects a p-subset, $p \leq k$, of the predictor variables and yields the n_W-vector $\hat{Y}_W$ of the predicted values of Y_W based on the OLS fitting of the selected subset. To be more specific, let $S_p = [i_1, i_2, \ldots, i_p]$ be a set of indices of the predictor variables to be included in the selected p-subset. Let D_p be a corresponding $k \times p$ 'indicator' matrix consisting of zeros and ones such that $X_{Vp} = X_V D_p$ contains only those p columns of X_V with indices from S_p. Then the procedure g consists of (i) choosing subset S_p; (ii) estimating vector β of the regression coefficients by $\hat{\beta} = D_p \hat{\beta}_p$, where $\hat{\beta}_p = (D_p' X_V' X_V D_p)^{-1} D_p' X_V' Y_V$; (iii) predicting Y_W by $\hat{Y}_W = X_W \hat{\beta}$. Note that although $\hat{\beta}_p$ is the OLS estimator of $\beta_p = D_p' \beta$, the resulting vector $\hat{\beta}$ is not the OLS estimator of β. Its distribution, as well as the distribution of the predictor $\hat{Y}_W = g(V; X_W)$, depends on the procedure g. It is important to emphasize here that the chosen predictor does not pretend to represent the 'real' model and, actually, is based on a random subset S_p of predictor variables. Its justification should rest on a justification of the prediction procedure g by which the data are transformed into actual prediction values.

For the quadratic loss function

$$L(Y_W; \hat{Y}_W) = \frac{1}{n_W}(Y_W - \hat{Y}_W)'(Y_W - \hat{Y}_W),$$

two distinct risk functions are important measures of efficiency of the procedure g. The first one is the conditional MSEP at X_W, given the construction set V,

$$\begin{aligned} MSEP_W(g; V) &= E[L_W(g; V) \mid V] \\ &= \sigma^2 + \frac{1}{n_W}(\beta - \hat{\beta})' X_W' X_W (\beta - \hat{\beta}). \end{aligned} \tag{2}$$

$MSEP_W(g; V)$ is a random variable, together with $\hat{\beta}$, and could be considered as measuring both the predictive ability of a selected model and the efficiency of the selection procedure g for a given construction set V. To be able to analyze existing procedures and to invent new ones, one needs to know their statistical properties for different realizations of V. A summarizing characteristic is the unconditional MSEP, *i.e.*, the average risk over all possible construction sets:

$$\begin{aligned} MSEP_W(g) &= E[MSEP_W(g; V)] \\ &= \sigma^2 + \frac{1}{n_W}(\beta - E\hat{\beta})' X_W' X_W (\beta - E\hat{\beta}) \\ &\quad + \frac{1}{n_W} tr(X_W' X_W \ \mathrm{VAR}\ [\hat{\beta}]), \end{aligned} \tag{3}$$

where $\mathrm{VAR}\ [\hat{\beta}] = E[(\hat{\beta} - E\hat{\beta})(\hat{\beta} - E\hat{\beta})']$ is the variance-covariance matrix of $\hat{\beta}$. As opposed to the conditional risk (2), the unconditional MSEP measures the average efficiency of the selection procedure g, not of a selected model that may vary for different realizations of V.

The simplest empirical estimator of MSEP is the autoloss function, or 'apparent error' in terminology of Efron (1982),

$$\begin{aligned} AL(g; V) &= L(Y_V; g(V; X_V)) \\ &= \frac{1}{n}(Y_V - \hat{Y}_V)'(Y_V - \hat{Y}_V), \end{aligned} \tag{4}$$

that measures goodness-of-fit of the procedure g over the construction set V. Usually AL tends to underestimate MSEP, because the same data have been used for both construction and evaluation. This is a familiar fact that could be easily demonstrated for a very simple procedure consisting of OLS–fitting of an *a priori* specified subset $X_{Vp} = X_V D_p$, with fixed D_p, and $X_W = X_V$, i.e., when a new response vector Y_W is predicted for the same set X_V of observations on the explanatory variables. For future reference we will denote this procedure by g_p. We have

$$\hat{Y}_W = g_p(V) = P_p Y_V , \tag{5}$$

where $P_p = X_{Vp}(X'_{Vp}X_{Vp})^{-1}X'_{Vp}$ is the projection matrix onto the fixed linear space spanned by the column-vectors of the matrix X_{Vp}. In this case, $X_V E\hat{\beta} = P_p Y_V^o$, and it follows from (3)-(4) that

$$E\big[AL(g_p;V)\big] = \frac{\sigma^2}{n}(n-p) + \frac{1}{n}Y_V^{o\prime}(I_n - P_p)Y_V^o \tag{6}$$

and

$$MSEP(g_p) = \frac{\sigma^2}{n}(n+p) + \frac{1}{n}Y_V^{o\prime}(I_n - P_p)Y_V^o, \tag{7}$$

so that AL underestimates $MSEP$ with the negative bias $(-2p\sigma^2/n)$.

In the literature different 'adjusted' forms of AL have been suggested (e.g. J_p, Rothman, 1968; Amemiya, 1980; C_p, Mallows, 1973; AIC, Akaike, 1973; BIC, Sawa, 1978, and others). These statistics, or their simple transformations in the case of information criteria, are unbiased estimators of MSEP under certain conditions. The major assumption requires that g be based on OLS-fitting a subset S_p specified independently from the construction data set V. Since model selection violates this assumption, it may seriously affect the distribution of the conditional $MSEP(g,V)$ in a way that is not reflected by any of the adjusted statistics. The fact that the conventional estimators become quite unsatisfactory in exploratory analysis has been repeatedly pointed out (e.g., Berk, 1978; Hjorth, 1982; Miller, 1984), but up to now no theory has been put forward to account for it. In our view, such a theory could be developed only after an explicit introduction of the concept of exploratory procedure into regression analysis and making it an object of statistical study. For instance, the 'procedural approach' allows one to clearly understand and interpret the fact that the conventional adjusted estimators become substantially biased under selection process. These estimators have been derived for one particular procedure, namely g_p, that does not involve selection, and cannot be used with other procedures without further adjustment. Many misunderstandings in the literature on model selection have resulted from underestimation of this fact. Without the concept of regression procedure, it is very tempting to 'forget' how S_p has been arrived at and to apply the theory developed for an *a priori* fixed S_p in the case when S_p is a random subset selected by some procedure g. As a result, conventional estimators, such as C_p or AIC, have often been used as criteria for screening different subsets and selecting the 'best' one, with implied false inference about thereby obtained predictors. To be able to get more adequate estimators one has to study the distribution of $\hat{Y}_W$ under that very exploratory procedure which has yielded this predictor.

As exact distributions are extremely difficult to study analytically, even for relatively simple selection procedures, prediction assessment requires data sets independent from the construction data V. To expedite this process, the independent sets could be replaced by pseudosamples which in some sense are close to the original set. There are different methods for construction of pseudosamples. One leads to different forms of splitting the data or cross–validation (e.g., Stone, 1974; Geisser, 1975). Cross–validation methods could be very helpful in realizing the behaviour of MSEP for some selection procedures as was demonstrated by Hjorth (1982), and Picard and Cook (1984) among others. Complications arise, though, for more complicated procedures, which examine different classes of models and sets of predictor variables that have not been planned in advance, but evolve in the process of selection itself. Proper cross–validation in this case requires excluding more and more observations for assessment of the ensuing results at each new iteration, which could become impractical.

Another method of generating pseudosamples, which does not have these limitations, is bootstrap. Recently bootstrap methods have been successfully used for solving various statistical problems including estimating the predictive ability for an *a priori* specified regression model (e.g., Bunke and Droge, 1984). In the discussion on applying bootstrap to estimating MSEP under subset selection (e.g., Miller, 1984), it is usually pointed out that this approach seems to suffer from the complication that the selected set of predictor variables will vary for different pseudosamples and could be quite different from that fitted for the given construction sample. The key to our present approach is just a small step from the usual bootstrap regression analysis but an important one nevertheless. We bring in the procedural approach and suggest that assessment of the efficiency of the predictor should rest on the bootstrap assessment of the exploratory procedure by which this predictor has been chosen, rather than the evaluation of any particular subset of variables. It is interesting to note that this approach has been implicitly applied in Efron (1983, 1986) and Gong (1986), where it was found better than cross-validation, and in Freedman et al. (1987), where it was declared a failure. We will briefly comment on this failure in Section 3 (see Remark 1).

3. Bootstrap Estimators.

The idea is to analyze performance characteristics of the procedure g on the data generated by a known random mechanism. The main requirement is that this mechanism, or as we will call it, pseudomodel, should simulate the unknown regression model (1). In other words, it should generate pseudosamples that are 'close' to the observed one with regard to their statistical structure. There are several ways to construct such a pseudomodel. One possible approach is to use

$$\tilde{Y} = \tilde{Y}^o + \tilde{\epsilon} = X\tilde{\beta} + \tilde{\epsilon}, \tag{8}$$

where $\tilde{\beta}$ and the distribution $\tilde{F}$ of $\tilde{\epsilon}$ are estimated from the data V. In a parametric bootstrap, when the form of the distribution F of ϵ is assumed known, $\tilde{F}$ is obtained by estimating the unknown parameters of F. In a nonparametric bootstrap, $\tilde{\epsilon}$ is usually a random sample (perhaps, times a weighted factor) from the empirical distribution $\tilde{F} = F_n(r_1, \ldots, r_n)$ of the residuals $r = Y_V - X_V\tilde{\beta}$. When the true model (1) is unknown, choosing $\tilde{\beta}$ seems to be one of the most sensitive elements of constructing a pseudomodel. If procedure g is rather complicated, e.g. involves subset selection, misspecification of the mean component $\tilde{Y}^o$ in (8) could lead to quite a different performance on pseudodata as compared to the real data.

In subset selection, it is usually assumed that X in (1) includes all relevant variables plus, perhaps, some extraneous variables. In this situation it seems reasonable to use the OLS estimate $\tilde{\beta} = (X_V' X_V)^{-1} X_V' Y_V$ based on the 'full' set of explanatory variables. By using the best unbiased estimator of β, one hopes to get a pseudomodel that is as close to the real one as possible (see also Remark 3 below). A different approach — resampling rows from the matrix V — does not seem to be appropriate here. First, it would lead to pseudodata with a matrix $\tilde{X}$ different from its counterpart X, which is assumed fixed. Second, if k is close to n, there is a high probability of getting fewer than k distinct rows and, thus, not a full rank matrix $\tilde{X}$.

Consider now two pseudosamples, a pseudoconstruction set $\tilde{V} = (X_V, \tilde{Y}_V)$ and a pseudotarget set $\tilde{W} = (X_W, \tilde{Y}_W)$, where $\tilde{Y}_V$ and $\tilde{Y}_W$ are $n \times 1$ and $n_W \times 1$ random vectors, respectively, independently generated from model (8). Applying the same selection procedure g that is used for the original construction set V to the pseudoset $\tilde{V}$, we get a pseudopredictor $\hat{\tilde{Y}}_W = g(\tilde{V}; X_W)$ and the pseudolosses $L_{\tilde{W}}(g; \tilde{V}) = \frac{1}{n_W}(\tilde{Y}_W - \hat{\tilde{Y}}_W)'(\tilde{Y}_W - \hat{\tilde{Y}}_W)$. As the pseudomodel (8) is completely known, Monte Carlo replications could be used to analyze the distribution of $L_{\tilde{W}}(g; \tilde{V})$. Characteristics of this distribution serve as estimates of their counterparts for the distribution of the real losses $L_W(g; V)$. This approach leads to the so called direct bootstrap estimator

$$\hat{R}_W^D(g) = E_{\sim}[MSEP_{\tilde{W}}(g; \tilde{V})], \tag{9}$$

where $E_{\sim}$ denotes expectation with respect to the random mechanism (8).

The idea of direct bootstrap estimation has recently been applied in Freedman et al. (1987). Disappointingly, the bootstrap did not perform very well. That $\hat{R}^D$ is generally biased may be easily illustrated with the procedure g_p.

PROPOSITION 1: *Let g_p be defined by (5), where $X_{V_p} = X_V D_p$ is fixed in advance and $X_W = X_V$. Let the parameters $\tilde{\beta}$ and $\tilde{\sigma}^2$ in the pseudomodel (8) be unbiased estimators for β and σ^2, respectively, in the model (1). Then the direct bootstrap estimator $\hat{R}_W^D(g_p)$ for $MSRP(g_p)$ has a positive (unless $p = k$) bias $(k-p)\sigma^2/n$.*

PROOF: By analogy with (7)

$$\hat{R}_W^D(g_p) = \frac{\tilde{\sigma}^2}{n}(n+p) + \frac{1}{n}\tilde{Y}_V^{o\prime}(I_n - P_p)\tilde{Y}_V^o$$

so that

$$\begin{aligned} E[\hat{R}_W^D(g_p)] &= \frac{\sigma^2}{n}(n+p) + \frac{1}{n}E[\tilde{Y}_V^{o\prime}](I_n - P_p)E[\tilde{Y}_V^o] \\ &\quad + \frac{1}{n}tr((I_n - P_p)\mathrm{VAR}\,[\tilde{Y}_V^o]) \\ &= \frac{\sigma^2}{n}(n+k) + \frac{1}{n}Y_V^{o\prime}(I_n - P_p)Y_V^o. \end{aligned}$$

Recalling $MSEP(g_p)$ from (7) gives $E[\hat{R}_W^D(g_p)] - MSEP(g_p) = (k-p)\sigma^2/n$.

REMARK 1: The above result explains why the direct bootstrap estimator performs well in the traditional regression framework, when the procedure g_k is applied to the 'true' model (1). Then $p = k$, and $\hat{R}_W^D$ is unbiased as has been demonstrated in the literature. On the other hand, the fact that the direct bootstrap estimator becomes biased even for such a simple procedure as g_p, may explain its failure in subset selection as was observed in Freedman et al. (1987).

Another possible approach to deriving MSEP estimators is based on using the pseudomodel (8) for evaluating 'overoptimism' of the autolosses $AL(g,V)$ in order to make an appropriate adjustment. At least two choices present themselves for representing average overoptimism: the difference

$$Q_W^A(g) = MSEP_W(g) - E[AL(g;V)]$$

and the ratio

$$Q_W^M(g) = MSEP_W(g)/E[AL(g;V)].$$

We will estimate $Q_W^A(g)$ and $Q_W^M(g)$ by their pseudo-counterparts

$$\tilde{Q}_{\tilde{W}}^A(g) = E_{\sim}\left[MSEP_{\tilde{W}}(g;\tilde{V}) - AL(g;\tilde{V})\right]$$

and

$$\tilde{Q}_{\tilde{W}}^M(g) = E_{\sim}\left[MSEP_{\tilde{W}}(g;\tilde{V})\right]/E_{\sim}\left[AL(g;\tilde{V})\right],$$

As a result, we arrive at the following two bootstrap estimators of MSEP: the additive estimator

$$\hat{R}_W^A(g;V) = AL(g;V) + \tilde{Q}_{\tilde{W}}^A(g), \tag{10}$$

and the multiplicative one

$$\hat{R}_W^M(g;V) = AL(g;V)\tilde{Q}_{\tilde{W}}^M(g). \tag{11}$$

The reason behind these estimators is to get more pivotal statistics as compared to the direct estimator (9). It is interesting to note that the additive estimator (10) was used in Efron (1983, 1986) and Gong (1986) in the framework of logistic regression, but its properties were not compared with those of the direct estimator. In what follows we will omit subscripts $W, \tilde{W}$, etc. and auxiliary arguments if it does not cause confusion.

If the selection procedure g is simple enough, estimators (10), (11) could be calculated analytically, without recourse to Monte Carlo. Consider again procedure g_p (5). This case seems to be especially interesting since, as is mentioned above, the conventional MSEP estimators have been originally derived for this particular procedure.

PROPOSITION 2: *Let the parameters $\tilde{\beta}$ and $\tilde{\sigma}^2$ in the pseudomodel (8) be unbiased estimators for β and σ^2, respectively. Then for the procedure g_p (5), the additive bootstrap estimator $\hat{R}^A(g_p)$ is an unbiased estimator for $MSEP(g_p)$.*

PROOF: By analogy with (6)-(7), $\tilde{Q}^A(g_p) = 2p\tilde{\sigma}^2/n$, so that from (10)

$$\hat{R}^A(g_p) = \frac{1}{n}Y_V'(I_n - P_p)Y_V + \frac{2}{n}p\tilde{\sigma}^2. \tag{12}$$

Then it follows from (6) that

$$E[\hat{R}^A(g_p)] = \frac{\sigma^2}{n}(n+p) + \frac{1}{n}Y_V^{o\prime}(I_n - P_p)Y_V^o,$$

which according to (7) equals $MSEP(g_p)$.

REMARK 2: Since the residual sum of squares for the p-subset is $RSS_p = Y_V'(I_n - P_p)Y_V$, it follows from (12) that the additive estimator $\hat{R}^A(g_p)$ coincides with the conventional $MSEP$ estimator

$$\hat{R}_p^c = \frac{1}{n}(RSS_p + 2p\hat{\sigma}^2), \tag{13}$$

where $\hat{\sigma}^2 = RSS_k/(n-k)$, described, e.g., in Seber (1977) . It is interesting to note, that $\hat{R}_p^c$ (and $\hat{R}^A(g_p)$ for this matter) is actually an unscaled extension of Mallows' C_p (e.g., Mallows, 1973) to the case when we predict Y_W instead of EY_W, since $\hat{R}_p^c = \frac{1}{n}\hat{\sigma}^2(C_p + n)$.

REMARK 3: The fact that the additive estimator $\hat{R}^A$ is unbiased, at least for the simple procedure g_p, if only $E[\tilde{\beta}] = \beta$ and $E(\tilde{\sigma}^2) = \sigma^2$, may be viewed as another rationale to use the full size set of predictor variables for constructing the pseudomodel (8).

PROPOSITION 3: *If, in addition to the conditions of Proposition 2, the fitted subset $X_{V\,p}$ represents the true model, that is, all components of β not included in β_p equal zero, and if $\tilde{\beta} = D_p(X'_{V\,p}X_{V\,p})^{-1}X'_{V\,p}Y_V$ in (8), the multiplicative bootstrap estimator $\hat{R}^M(g_p)$ is an unbiased estimator for $MSEP(g_p)$.*

Proof immediately follows from the fact that in the considered case $\tilde{Q}^M(g_p) = (n+p)/(n-p)$, $Y_V^{o\prime}(I_n - P_p)Y_V^o = 0$, so that according to (11), $E[\hat{R}^M(g_p)] = \frac{\sigma^2}{n}(n+p)$, which equals $MSEP(g_p)$.

REMARK 4: Under the conditions of Proposition 3, the multiplicative estimator $\hat{R}^M(g_p)$ coincides with another conventional estimator

$$J_p = \frac{RSS_p}{n}\left(\frac{n+p}{n-p}\right), \tag{14}$$

suggested by Rothman (1968) and Amemiya (1980).

4. Experimental Comparison of the Conventional and the Bootstrap Estimators.

To illustrate the effect of subset selection on conventional $MSEP$ estimators and to compare these estimators with the bootstrap estimators the following simulation study was conducted. In all the experiments the simulated data satisfied model (1), with $\epsilon \sim N(0, \sigma^2 I_n)$ where matrices X_V and X_W were orthonormal with the same number $n = n_W$ of observations. As was pointed out in Miller (1984), the orthogonal case is far from being the simplest one with respect to MSEP estimation and actually gives an example of intermediate deterioration of the conventional estimators under subset selection. An advantage of considering the orthogonal case (besides from obvious calculation simplifications) lies in the fact that all major subset selection procedures, such as the best subset regression, forward selection, backward elimination, and stepwise regression lead here to the same 'best' p-subset, for each $p = 0, 1, \ldots, k$, a property that does not hold for nonorthogonal predictor variables. Consider the best subset regression procedure g, which consists of screening all 2^k subsets and selecting the best one with regard to some criterion (e.g., Hocking, 1976). Usually such a criterion is based on one of the conventional adjusted estimators, mentioned above. Then for any fixed p, the 'best' p-subset is the one with minimum RSS_p, and g could be conceived of as a two-step procedure. At the first step, for each $p, p = 0, 1\ldots, k$, a p-subset corresponding to the minimum RSS_p is found. The second step consists in comparing these subsets and choosing the overall best according to the adopted criterion. In the experimental study the three bootstrap estimators (9)-(11) were compared with the two conventional estimators (13) and (14) with regard to the evaluation of the $MSEP$ for each of the best p-subset at the first step of the procedure. Then each estimator served as the criterion for the overall choice at the second step. It made it possible to compare these statistics not only as $MSEP$ estimators under the search process, but also as stopping rules in subset selection.

REMARK 4: In general, exploratory procedures often represent a multi-step model building strategy. Although the importance of having good $MSEP$ estimators for the finally chosen model is quite obvious, it is very essential to be able to assess the results of the intermediate steps. The latter would allow one to choose the best among the available alternatives at each step of the procedure before moving to the next step, thus improving the exploratory process itself.

The results of applying subset selection depend on the ratio k/n and on the value of the true parameter vector β scaled by σ^2. A small theoretical calculation might clarify matters. For any fixed p, let $S = S_p$ be the best p-subset selected by g. Then in the orthonormal case it follows from (3) that the unconditional risk equals

$$MSEP(g) = \sigma^2 + \frac{1}{n} E\Big(\sum_{i \in S} \epsilon_i^2 + \sum_{j \notin S} \beta_j^2 \Big). \tag{15}$$

Note, that S in (15) is a random set of indices. If the true model (1) were known, the best predictor with minimum $MSEP$ should be based on the subset $S^* = \{i_1, \ldots, i_p\}$ such that $\beta_{i_j}^2/\sigma^2 \geq 1$. This fact easily follows from (3) since for any two *fixed* subsets S_{p_1}, and $S_{p_2}, p_1 < p_2$,

$$MSEP(g_{p_2}) - MSEP(g_{p_1}) = \frac{1}{n}\Big[\sigma^2(p_2 - p_1) - \sum_{j \in S_{p_2} \setminus S_{p_1}} \beta_j^2\Big].$$

The situation, though, might be quite different when S_p is selected according to minimum $RSS_p = RSS_k + \sum_{j \notin S_p} \hat{\beta}_j^2$ or, equivalently, maximum $\sum_{i \in S_p} \hat{\beta}_i^2$. Here S_p includes indices of the first p order statistics from k independent random variables $\hat{\beta}_1^2, \ldots, \hat{\beta}_k^2$, where $\hat{\beta}_i^2 \sim \sigma^2 \chi_1^2(\beta_i)$.

REMARK 5: This consideration explains why a chosen subset S_p does not necessarily minimize measure (15) and, in fact, may lead to a higher $MSEP$ than the optimal subset S^* defined above.

In the experiments we put $n = 50$ and $\sigma^2 = 1$. As is often the case in applications the number of potential predictor variables may be quite large as compared to the number of available observations, so that we considered $k = 15$, 25, and 35. Since MSEP does not depend on the choice of X_V and X_W, as far as they remain orthonormal, we put $X_V = X_W = [e_1 e_2 \ldots e_k]$, where the i-th component of vector e_i equals one and all other components equal zero. Three values of the true vector β were considered: (i) $\beta_I = (0,0,\ldots,0)'$, which represents the model with no relevant variables; (ii) $\beta_{II} = (\beta_1,\beta_2,\ldots,\beta_q,0,0,\ldots,0)'$, where $\beta_i = E[Z(i;q)], Z(i;q)$ is the i-th order statistic from q $N(1,.25)$ random variables, q is the closest to $k/3$ integer. Here the first q values of the elements of β_{II} are near the resolving power of the system with the 'signal-to-noise' ratio $(\beta_i)^2/\sigma^2 \approx 1$; (iii) $\beta_{III} = (7.0,5.0,0,0,\ldots,0)'$, which represents the case with two very significant predictor variables with the signal-to-noise ratio being 49 and 25, respectively.

For each model specification, as defined by $\sigma^2 = 1$, $n = 50$, $k = 15,25,35$ and $\beta = \beta_I,\beta_{II},\beta_{III}$, 1000 basic data sets $V_m = (X_V, Y_V(m))$ and $W_m = (X_W, Y_W(m))$, $m = 1,2,\ldots,$ 1000, were generated following (1). To each data set V_m the best subset procedure g was applied, and for each $p = 0,1,\ldots,k$, predictor $\hat{Y}_W(m,p)$ based on the p-subset $X_{V_p}(m) = X_V(m)D_p(m)$ with the minimum RSS_p was found. The 'true' conditional MSEP was calculated by (2). The two conventional estimators were calculated by formulas (13) and (14) as based on subset $X_{V_p}(m)$. The adopted nonparametric pseudomodel was based on (8) with $\tilde{\beta} = (X_V' X_V)^{-1} X_V' Y_V$ and $\tilde{\epsilon}_i \sim \tilde{F} = F_{n-k}(r_{k+1},\ldots,r_n), r = Y_V - X_V\tilde{\beta}$. For each simulated set m, the direct, the additive, and the multiplicative estimators were calculated by generating 200 pseudosamples $\tilde{V}_i(m) = (X_V, \tilde{Y}_V(m,i))$ and $\tilde{W}_i(m) = (X_W, \tilde{Y}_W(m,i))$ from the chosen pseudomodel.

REMARK 6: The decision to generate $\tilde{\epsilon}_i$ from the empirical distribution of the last $n-k$ residuals, as opposed to the traditionally used empirical distribution of all the residuals, reflects a very important feature of an orthogonal regression. Due to the choice of matrix X_V, $r_1 = \ldots = r_k = 0$, so that when k is comparable with n, there is a high probability of getting too many zero pseudo-disturbances $\tilde{\epsilon}_i$. This would make pseudodata quite different from the real observations. Another advantage of the chosen $\tilde{F}$ lies in the fact, that it is based on independent residuals as opposed to the full set of components of r. Besides, $\tilde{\sigma}^2$ becomes an unbiased estimator for σ^2.

To judge the performance of each estimator $\hat{R}$ in the simulations, two criteria were used: the mean squared error

$$MSE_C(\hat{R}) = E(\hat{R} - MSEP(g;V))^2$$

about the conditional MSEP, and the mean squared error

$$MSE_U(\hat{R}) = E(\hat{R} - MSEP(g))^2$$

about the unconditional MSEP, where expected values were approximated by the averages, as based on the corresponding 1000 replications.

REMARK 7: There is no full agreement in the literature on whether conditional or unconditional MSEP is more appropriate for prediction assessment. The conventional statistics have been originally proposed for estimating the

TABLE 1.1 $\beta=\beta_I$. The means and standard deviations (as based on 1,000 experiments) of different MSEP estimators

p	MSEP(g,V)		J_p		$\hat{R}_p^C$		$\hat{R}^D$		$\hat{R}^A$		$\hat{R}^M$	
	E	SD	E	SD	E	SD	E	SD	E	SD	E	SD
0	1.00	0	.993	.192	.993	.192	1.49	.303	.995	.197	.995	.194
1	1.10	.044	.926	.182	.929	.183	1.48	.317	1.09	.227	1.03	.209
2	1.17	.063	.887	.178	.898	.183	1.48	.331	1.16	.256	1.06	.228
3	1.23	.077	.861	.178	.883	.186	1.48	.343	1.22	.280	1.10	.247
4	1.27	.087	.844	.179	.878	.192	1.48	.352	1.26	.300	1.13	.264
5	1.31	.095	.834	.182	.881	.200	1.48	.360	1.30	.316	1.17	.280
6	1.34	.102	.829	.186	.890	.208	1.48	.368	1.33	.330	1.20	.296
7	1.37	.108	.829	.190	.903	.217	1.48	.374	1.35	.342	1.23	.310
8	1.39	.112	.833	.196	.921	.226	1.48	.379	1.38	.353	1.26	.323
9	1.41	.117	.840	.202	.941	.236	1.49	.384	1.40	.362	1.29	.335
10	1.42	.121	.852	.209	.965	.246	1.49	.388	1.41	.370	1.32	.346
15	1.47	.133	.961	.254	1.11	.300	1.49	.402	1.47	.396	1.43	.387
20	1.49	.138	1.17	.318	1.29	.354	1.49	.407	1.49	.407	1.48	.406
25	1.50	.138	1.49	.408	1.49	.408	1.49	.408	1.49	.408	1.49	.410

TABLE 1.2 $\beta=\beta_{II}$. The means and standard deviations (as based on 1,000 experiments) of different MSEP estimators

p	MSEP(g,V)		J_p		$\hat{R}_p^C$		$\hat{R}^D$		$\hat{R}^A$		$\hat{R}^M$	
	E	SD	E	SD	E	SD	E	SD	E	SD	E	SD
0	1.19	0	1.18	.230	1.18	.230	1.68	.329	1.18	.242	1.18	.237
1	1.23	.071	1.07	.210	1.07	.210	1.63	.335	1.23	.256	1.17	.240
2	1.27	.093	1.01	.201	1.01	.202	1.60	.344	1.27	.277	1.18	.253
3	1.30	.103	.962	.196	.972	.201	1.57	.353	1.30	.296	1.19	.267
4	1.33	.112	.931	.194	.952	.202	1.56	.360	1.33	.311	1.21	.280
5	1.35	.115	.909	.194	.942	.207	1.55	.366	1.35	.325	1.23	.293
6	1.37	.118	.893	.195	.940	.213	1.54	.372	1.37	.337	1.25	.305
7	1.40	.122	.884	.198	.944	.221	1.53	.377	1.39	.347	1.27	.317
8	1.41	.123	.880	.201	.955	.229	1.52	.382	1.41	.356	1.30	.329
9	1.43	.125	.881	.206	.969	.238	1.51	.386	1.42	.365	1.32	.339
10	1.44	.128	.886	.211	.988	.247	1.51	.390	1.43	.372	1.34	.349
15	1.48	.134	.974	.253	1.12	.299	1.49	.402	1.47	.396	1.43	.387
20	1.49	.138	1.17	.318	1.29	.354	1.49	.407	1.49	.407	1.48	.407
25	1.50	.138	1.49	.408	1.49	.408	1.49	.408	1.49	.408	1.49	.410

TABLE 1.3 $\beta=\beta_{III}$. The means and standard deviations (as based on 1,000 experiments) of different MSEP estimators

p	MSEP(g,V)		J_p		$\hat{R}_p^C$		$\hat{R}^D$		$\hat{R}^A$		$\hat{R}^M$	
	E	SD	E	SD	E	SD	E	SD	E	SD	E	SD
0	2.48	0	2.48	.374	2.48	.374	2.98	.441	2.48	.383	2.48	.379
1	1.56	.142	1.52	.273	1.50	.268	2.00	.368	1.55	.305	1.53	.296
2	1.04	.065	1.03	.205	1.03	.205	1.52	.334	1.08	.241	1.04	.223
3	1.14	.062	.962	.194	.972	.198	1.50	.336	1.15	.255	1.06	.228
4	1.21	.077	.921	.190	.944	.200	1.49	.344	1.21	.277	1.09	.245
5	1.26	.085	.895	.190	.931	.204	1.49	.353	1.25	.297	1.13	.262
6	1.30	.094	.879	.192	.929	.211	1.49	.361	1.29	.315	1.16	.278
7	1.34	.101	.870	.195	.934	.219	1.49	.368	1.33	.330	1.20	.295
8	1.37	.107	.867	.199	.945	.228	1.49	.374	1.36	.343	1.23	.310
9	1.39	.112	.868	.205	.961	.237	1.49	.380	1.38	.354	1.26	.324
10	1.41	.117	.875	.211	.980	.247	1.49	.385	1.40	.363	1.29	.336
15	1.47	.132	.969	.254	1.12	.300	1.49	.400	1.46	.394	1.42	.383
20	1.49	.137	1.17	.318	1.29	.354	1.49	.407	1.48	.406	1.48	.406
25	1.50	.138	1.49	.408	1.49	.408	1.49	.408	1.49	.408	1.49	.410

unconditional MSEP (e.g., Hocking, 1976; Seber, 1977), but later most of them were looked upon as estimators of the conditional MSEP (e.g., Hjorth, 1982; Efron, 1983, 1986; Picard and Cook, 1984). Recently, though, the fashion seems to have been changing again. As is mentioned in Gong (1986) and Boudreau (1988), it is, perhaps, unfair to consider MSEP estimates as evaluating the conditional risk, so that they must be viewed as estimators of the unconditional MSEP. Without trying to resolve this dispute, it is worth mentioning that from the procedural point of view it is very important to estimate the unconditional risk. As was mentioned in Section 2, the unconditional MSEP is the only measure that evaluates the exploratory procedure, as opposed to any particular chosen model.

Since the results for the three considered k values were not qualitatively different, in Tables 1 - 4 we report three model specifications with $\beta = \beta_I, \beta_{II}$, and β_{III} for $k = 25$.

(i) **Bias.** From Tables 1.1 - 1.3 one can see that both conventional estimators J_p and $\hat{R}_p^c$ are considerably biased downward when the number of regressors in a subset exceeds the number of non–zero coefficients of β. For $p = 9$ the bias of J_p is about 40% and the bias of $\hat{R}_p^c$ is more than 30% of the actual MSEP values in all three considered examples. When p approaches the full model size, the bias gets smaller and becomes negligible when $p = 25$.

As was expected the direct bootstrap estimator is biased upward. When p is small this bias almost follows the same pattern as for the procedure g_p, i.e. $(k - p)\sigma^2/n$. When p exceeds 3 the bias gets smaller as compared to g_p; and it becomes almost negligible when $p > 10$. The 'indirect' bootstrap estimators have considerably smaller bias than the conventional estimators. The additive estimator $\hat{R}^A$ is a clear winner here, having the smallest bias throughout. For all three considered cases, its bias is less than 2% of the actual MSEP for p between 0 and 9, then it increases slightly and reaches about 3% of the MSEP when $p = 25$. The bias of the multiplicative estimator, $\hat{R}^M$, is higher, approaching 10% of the MSEP value when $p = 9$, but it still remains relatively small as compared to the bias of the conventional estimators.

(ii) **MSE.** As follows from Tables 1.1 - 1.3, both conventional estimators have smaller variances than the bootstrap estimators since they do not account for the prediction error due to selection. Still in terms of the MSE_U and/or MSE_C (Tables 2.1 -2.3), the indirect bootstrap estimators considerably outperform J_p and $\hat{R}_p^c$: their larger variances are more than compensated for by substantially lower biases. As compared between themselves, although $\hat{R}^A$ is less biased than $\hat{R}^M$, its lower bias trades off against its higher variability, so that the MSE values for both estimators are practically the same in the three considered examples. The direct bootstrap estimator occupies some middle ground except for small values of $p (p \leq 5)$, where $\hat{R}^D$ has the highest MSE due to a considerable bias.

(iii) **MSEP as a function of p.** What is even worse than biasedness, both J_p and $\hat{R}_p^c$ do not follow the actual MSEP and lead to wrong conclusions as to which subset is the overall best for prediction. From Tables 1.1 - 1.3 one can see that on average the best predictor (with regard to the actual minimum MSEP) includes no variables if $\beta = \beta_I$ and $\beta = \beta_{II}$ and has 2

TABLE 2.1 $\beta=\beta_I$. The conditional and unconditional mean squared errors (as based on 1,000 experiments) of different MSEP estimators

p	J_p		$\hat{R}_p^C$		$\hat{R}^D$		$\hat{R}^A$		$\hat{R}^M$	
	MSE_C	MSE_U	MSE_C	MSE_U	MSE_C	MSE_U	MSE_C	MSE_U	MSE_C	MSE_U
0	.037	.037	.037	.037	.331	.331	.039	.039	.038	.038
1	.062	.064	.061	.064	.243	.246	.050	.052	.047	.049
2	.112	.114	.108	.109	.204	.206	.066	.066	.063	.064
3	.166	.167	.154	.154	.182	.182	.081	.079	.079	.078
4	.216	.216	.194	.193	.170	.168	.094	.090	.093	.089
5	.262	.260	.227	.224	.164	.160	.106	.100	.105	.099
6	.300	.296	.252	.247	.161	.156	.116	.109	.115	.108
7	.331	.326	.269	.262	.161	.153	.126	.117	.123	.115
8	.355	.348	.280	.271	.162	.153	.135	.125	.131	.121
9	.372	.364	.284	.274	.164	.153	.143	.131	.138	.126
10	.382	.372	.284	.272	.166	.154	.150	.137	.144	.131
15	.344	.328	.238	.220	.179	.161	.175	.157	.170	.152
20	.228	.208	.185	.165	.186	.166	.186	.165	.186	.165
25	.188	.167	.188	.167	.188	.167	.188	.167	.189	.168

TABLE 2.2 $\beta = \beta_{II}$. The conditional and unconditional mean squared errors (as based on 1,000 experiments) of different MSEP estimators

p	J_p		$\hat{R}_p^C$		$\hat{R}^D$		$\hat{R}^A$		$\hat{R}^M$	
	MSE_C	MSE_U	MSE_C	MSE_U	MSE_C	MSE_U	MSE_C	MSE_U	MSE_C	MSE_U
0	.053	.053	.053	.053	.347	.347	.058	.058	.056	.056
1	.071	.070	.071	.071	.267	.266	.066	.066	.062	.061
2	.112	.111	.112	.111	.225	.223	.080	.077	.075	.073
3	.158	.155	.154	.150	.203	.198	.093	.087	.088	.083
4	.203	.198	.191	.185	.189	.181	.105	.097	.100	.093
5	.242	.236	.220	.213	.179	.171	.115	.105	.110	.101
6	.277	.270	.243	.234	.174	.164	.123	.113	.118	.108
7	.309	.301	.262	.252	.170	.159	.132	.121	.127	.116
8	.333	.324	.273	.262	.168	.157	.139	.127	.133	.122
9	.350	.340	.277	.265	.169	.157	.141	.133	.139	.127
10	.361	.350	.277	.265	.170	.157	.152	.138	.145	.131
15	.334	.318	.235	.217	.180	.162	.175	.157	.170	.152
20	.227	.207	.185	.165	.186	.166	.186	.165	.186	.165
25	.188	.167	.188	.167	.188	.167	.188	.167	.189	.168

TABLE 2.3 $\beta=\beta_{III}$. The conditional and unconditional mean squared errors (as based on 1,000 experiments) of different MSEP estimators

p	J_p		$\hat{R}_p^C$		$\hat{R}^D$		$\hat{R}^A$		$\hat{R}^M$	
	MSE_C	MSE_U	MSE_C	MSE_U	MSE_C	MSE_U	MSE_C	MSE_U	MSE_C	MSE_U
0	.140	.140	.140	.140	.445	.445	.147	.147	.144	.144
1	.084	.076	.084	.075	.340	.333	.096	.093	.092	.088
2	.045	.042	.045	.042	.347	.345	.061	.060	.052	.050
3	.068	.069	.067	.068	.240	.241	.064	.065	.058	.058
4	.118	.119	.110	.110	.199	.199	.078	.077	.074	.073
5	.169	.169	.152	.150	.178	.177	.092	.089	.089	.086
6	.217	.216	.187	.184	.168	.165	.104	.099	.102	.097
7	.259	.256	.215	.210	.163	.158	.116	.109	.113	.106
8	.293	.288	.235	.228	.162	.155	.126	.118	.122	.114
9	.320	.313	.248	.240	.163	.154	.135	.125	.130	.121
10	.338	.330	.255	.245	.164	.154	.143	.132	.138	.126
15	.331	.316	.231	.215	.178	.161	.172	.155	.167	.150
20	.227	.207	.185	.164	.186	.166	.186	.165	.185	.165
25	.188	.167	.188	.167	.188	.167	.188	.167	.189	.168

variables for $\beta = \beta_{III}$. At the same time $\hat{R}_p^c$ — the least corrupted of the two conventional estimators — has its minimum when $p = 4$ for β_I and $p = 6$ for β_{II} and β_{III}.

REMARK 8: The results for $\beta = \beta_{II}$ (Table 1.2) are especially interesting, because they illustrate the fact that subset selection is not the best way of deciding which variables should be included in the predictor equation, at least when some of the components of β are near the level of predictor significance. Due to selection, even when p equals the number of relevant regressors, the 'best' p-subset does not often include all the significant variables but contains some 'noisy' regressors with small or zero coefficients. As a result, on average we are still better off with the 'naive' prediction $\hat{Y}_W = 0$. Copas (1983) and Miller (1984), among others, make similar conclusions. Unfortunately, both conventional estimators, J_p and $\hat{R}_p^c$, do not reflect this very important feature of subset selection and mislead a researcher with an overoptimistic assessment.

The same remains true with regard to the direct bootstrap estimator. On the contrary, the indirect bootstrap estimators behave similarly to the true MSEP. $\hat{R}^A$ has the smallest average values for $p = 0$ when $\beta = \beta_I$ and $\beta = \beta_{II}$, and $p = 2$ for $\beta = \beta_{III}$. $\hat{R}^M$ differs only in the most difficult case of $\beta = \beta_{II}$, where it has the smallest average value when $p = 1$. The fact that the indirect bootstrap estimators closely follow the actual MSEP indicates that they could be used as criteria for choosing the overall best predictor. Tables 3.1 - 3.3 and 4 display characteristics of the final predictor selected from among $(k+1)$ best p-subsets, $p = 0, 1, \ldots, k$, according to a criterion based on each of the six statistics: actual $MSEP(g, V)$, $J_p \hat{R}_p^c$, $\hat{R}^D$, $\hat{R}^A$, and $\hat{R}^M$. Tables 3.1 - 3.3 contain the empirical distributions of the number of variables in these final predictors. It follows that the direct bootstrap estimator provides the worst criterion with respect to the distribution of the optimal p. The conventional estimators obviously lead to overfitting. The indirect bootstrap estimators provide distributions very similar to one based on using the actual MSEP, except for somewhat thicker right-hand tails. These long tails could be explained by the fact that all k components of $\tilde{\beta}$ in the pseudomodel (8) are always different from zero, as opposed to the three considered true vectors β in model (1). Table 4 contains the average MSEP values for the final predictors. It follows that although $\hat{R}^D$ still leads to some poor results, the two indirect bootstrap estimators, used as selection criteria, provide substantially better final predictors than those based on the conventional criteria.

5. Conclusion.

The theory behind conventional estimators for predictive efficiency is not valid when predictor selection and estimation are from the same data. The very selection process affects the distribution of those estimators and, in particular, leads to their substantial bias when the selection effect is not allowed for. It is suggested that each estimator should be developed for the selection procedure it is used with. As exact distributional results are extremely difficult to study analytically, even for relatively simple subset selection procedures, the indirect bootstrap assessment described above seems to be helpful in solving the problem. The bootstrap method appears general

TABLE 3.1 $\beta=\beta_I$. The empirical distributions (based on 1,000 experiments) of the number p of variables for the final subsets selected from among the (k+1) best p-subsets, p=0,1, ... k, by different criteria.

p	MSEP(g,V)	J_p	$\hat{R}_p^C$	$\hat{R}^D$	$\hat{R}^A$	$\hat{R}^M$
0	1000	6	25	372	868	710
1	0	14	94	62	49	99
2	0	38	119	50	18	55
3	0	54	179	27	11	36
4	0	115	161	26	5	19
5	0	127	164	25	13	21
6	0	156	86	16	4	15
7	0	153	74	17	2	7
8	0	101	47	12	3	4
9	0	87	27	8	1	7
10	0	56	12	10	2	2
11	0	43	5	13	1	4
12	0	25	5	8	1	3
13	0	13	1	4	0	4
14	0	9	0	12	2	2
15	0	2	1	15	0	1
16	0	0	0	6	1	1
17	0	0	0	10	0	1
18	0	1	0	14	2	0
19	0	0	0	17	0	2
20	0	0	0	17	2	0
21	0	0	0	24	2	1
22	0	0	0	29	3	3
23	0	0	0	34	1	0
24	0	0	0	42	7	2
25	0	0	0	130	2	1

TABLE 3.2 $\beta=\beta_{II}$. The empirical distributions (based on 1,000 experiments) of the number p of variables for the final subsets selected from among the (k+1) best p-subsets, p=0,1, ... k, by different criteria.

p	MSEP(g,V)	J_p	$\hat{R}_p^C$	$\hat{R}^D$	$\hat{R}^A$	$\hat{R}^M$
0	584	0	1	119	624	437
1	218	2	20	39	116	132
2	102	8	51	25	57	100
3	49	20	100	36	23	71
4	31	47	139	27	30	54
5	7	73	160	22	13	39
6	7	108	167	20	9	32
7	0	149	124	19	4	20
8	2	138	92	18	6	14
9	0	132	61	16	9	12
10	0	109	44	15	6	19
11	0	81	23	15	9	11
12	0	48	11	16	9	8
13	0	52	3	14	4	5
14	0	21	1	14	7	7
15	0	7	2	11	3	6
16	0	3	0	21	7	7
17	0	0	1	21	6	2
18	0	1	0	21	5	4
19	0	0	0	26	3	2
20	0	1	0	29	3	3
21	0	0	0	38	8	2
22	0	0	0	52	5	4
23	0	0	0	52	7	2
24	0	0	0	98	10	3
25	0	0	0	216	17	4

TABLE 3.3 $\beta=\beta_{III}$. The empirical distributions (based on 1,000 experiments) of the number p of variables for the final subsets selected from among the (k+1) best p-subsets, p=0,1, ... k, by different criteria.

p	MSEP(g,V)	J_p	$\hat{R}_p^C$	$\hat{R}^D$	$\hat{R}^A$	$\hat{R}^M$
0	0	0	0	0	0	0
1	0	0	0	3	26	13
2	993	8	31	252	786	640
3	5	16	115	104	78	143
4	0	45	136	58	33	64
5	2	78	177	51	14	45
6	0	130	185	27	11	21
7	0	148	148	25	7	21
8	0	151	70	25	3	14
9	0	148	69	15	5	7
10	0	99	37	13	2	4
11	0	70	17	17	3	5
12	0	52	8	14	0	4
13	0	30	4	13	2	3
14	0	13	1	10	1	3
15	0	8	1	9	1	0
16	0	3	0	12	0	2
17	0	0	1	11	0	2
18	0	0	0	14	3	0
19	0	0	0	19	2	1
20	0	1	0	20	2	1
21	0	0	0	23	4	2
22	0	0	0	32	2	1
23	0	0	0	44	1	2
24	0	0	0	48	8	1
25	0	0	0	141	6	1

TABLE 4 MSEP (averaged over 1,000 experiments) for the final subsets selected from among the (k+1) best p-subsets, p=0, 1, ... k, by different criteria

Criterion	MSEP		
	$\beta=\beta_{I}$	$\beta=\beta_{II}$	$\beta=\beta_{III}$
MSEP(g,V)	1.00	1.17	1.04
J_p	1.36	1.42	1.37
$\hat{R}_P^C$	1.29	1.38	1.30
$\hat{R}^D$	1.29	1.44	1.34
$\hat{R}^A$	1.05	1.27	1.11
$\hat{R}^M$	1.09	1.29	1.14

and flexible enough, and in principle could be used for any exploratory procedure. Indeed, after generating a necessary number of pseudosamples the same exploratory process that was used for the original data is applied to each one of them, and the corresponding empirical distribution of predictor errors provides all characteristics of interest.

One of the major problems in applying bootstrap to model building procedures consists in choosing an appropriate pseudomodel. Since the 'true' model is unknown, the usual bootstrap regression idea of using estimated parameters of the true model as the pseudoparameters in the pseudomodel does not work. This difficulty is aggravated by the fact that rather complicated exploratory procedures prove to be quite sensitive to the choice of the pseudomodel. In the framework of subset selection this problem seems to have been resolved by using the 'full' model estimates to construct a pseudomodel. But this approach is by no means mandatory, and other pseudomodels could and, perhaps, should be used in different situations.

It is also important to note that the direct bootstrap method does not work well enough in exploratory analysis, as was demonstrated by the performance of $\hat{R}^D$. The indirect approach, which tries to improve, or adjust, the existing estimators seems to be crucial in this case. As a result, the suggested method can be used to assess the efficiency of different regression procedures, to compare those procedures with each other, and to choose the most efficient one. It also may be helpful in correcting the existing procedures by providing, for example, a criterion for a stopping rule in multistep model building.

References

Akaike, H. (1973). Information Theory and an Extension of the Maximum Likelihood Principle, *2nd Int. Symp. on Information Theory* (B.N. Petrov and F. Csaki, eds.). Budapest: Akademiai Kiado, 267–281.

Amemiya, T. (1980). Selection of Regressors, *Int. Econ. Rev.*, **21**, 331-354.

Berk, K.N. (1978). Comparing Subset Regression Procedures, *Technometrics*, **20**, 1–6.

Boudreau, R. (1988). A Monte Carlo Assessment of Cross-validation and the C_p Criterion for Model Selection in Multiple Linear Regression, *Proceed. 20 th Symp. Interface*, 603-607.

Box, G.E.P. (1983). An Apology for Ecumenism in Statistics, *Scientific Inference, Data Analysis, and Robustness* (G.E.P. Box, T. Leonard, and C.-F. Wu, eds.,). New York: Academic Press, 51–84.

Bunke, O. and Droge, B. (1984). Bootstrap and Cross–Validation Estimates of the Prediction Error for Linear Regression Models, *Ann. of Statist.*, **12**, 1400–1424.

Copas, J.B. (1983). Regression, Prediction and Shrinkage (with Discussion), *J. R. Statist. Soc.*, **B**, **45**, 311-354.

Efron, B. (1982). *The Jackknife, the Bootstrap and Other Resampling Plans*; Philadephia: SIAM.

Efron, B. (1983). Estimating the Error Rate of a Prediction Rule: Improvement on Cross-Validation, *J. Amer. Statist. Ass.*, **78**, 316-331.

Efron, B. (1986). How Biased Is the Apparent Error Rate of a Prediction Rule? *J. Amer. Statist. Ass.*, **81**, 461-470.

Freedman, D. (1983). A Note on Screening Regression Equation, *Amer. Statist.*, **37**, 152-155.

Freedman D., Navidi W., and Peters, S. (1987). On the Impact of Variable Selection in Fitting Regression Equations, *Techn. Rep.* 87, Univ. California, Berkeley, CA.

Geisser, S. (1975). The Predictive Sample Reuse Method With Applications, *J. Amer. Statist. Ass.*, **70**, 320–328.

Gong, G. (1986). Cross-Validation, the Jackknife, and the Bootstrap: Excess Error Estimation in Forward Logistic Regression, *J. Amer. Statist. Ass.*, **81**, 108-113.

Hjorth, U. (1982). Model Selection and Forward Validation, *Scand. J. Statist.*, **9**, 95–105.

Hocking, R.R. (1976). The Analysis and Selection of Variables in Linear Regression, *Biometrics*, **32**, 1–49.

Lovell, M. (1983) Data Mining, *Rev. Econ. Statist.*, **LXV**, 1-11.

Mallows, C. (1973). Some Comments on C_p, *Technometrics*, **15**, 661-675.

Miller, A.J. (1984). Selection of Subsets of Regression Variables (with Discussion), *J. R. Statist. Soc.*, **A**, **147**, 389-425.

Picard, R.R., and Cook, R.D. (1984). Cross–Validation of Regression Models, *J. Amer. Statist. Ass.*, **79**, 575–583.

Pinsker, I.Sh., Kipnis, V. and Grechanovsky, E. (1985). The use of the F-Statistic in the Forward Selection Regression Algorithm, *ASA Proceed. Statist. Comput. Sect.*, 419-423.

Pinsker, I. Sh., Kipnis, V. and Grechanovsky, E. (1987). The Use of Conditional Cutoffs in a Forward Selection Procedure, *Commun. Statist. - Theor. Meth.*, **A**, **16**, 2227-2241.

Rothman, D. (1968). Letter to the Editor, *Technometrics*, **10**, 432.

Sawa, T. (1978). Information Criteria for Discriminating Among Alternative Regression Models, *Econometrica*, **46**, 1273-1291.

Seber, G. (1977). *Linear Regression Analysis*. New York: John Wiley.

Stone, M. (1974). Cross–Validatory Choice and Assessment of Statistical Predictions (with Discussion), *J. R. Statist. Soc.*, **B**, **36**, 111–147.

Thompson, M.L. (1978). Selection of Variables in Multiple Regression. Parts I and II, *Int. Statist. Rev.*, **46**, 1-19, 129-146.

BOOTSTRAPPING LIKELIHOOD RATIOS IN QUANTITATIVE GENETICS

Nicholas Schork†
University of Michigan

I. Introduction

Normal mixture distributions play a prominent role in the modeling of the genotypic determinants of quantitative traits [Morton, 1967; Schork and Schork, 1988; Elston, 1981]. Though the theory behind many models in genetics is quite complex, the following provides a simple introduction to the use of normal mixture distributions in quantitative genetic modeling.

Consider a genetic locus with two alleles A and a, whose allelic combinations form three distinct genotypes, AA, Aa (or aA), and aa occuring in the population with the Hardy-Weinberg equilibrium dictated frequencies $p_{AA} = q^2$, $p_{Aa} = 2q(1-q)$, and $p_{aa} = (1-q)^2$, where q is the frequency of A allele. If one assumes each genotype acts so as to produce a normally distributed phenotype with a mean, μ, and variance, σ^2, unique to that genotype, then one can model the overall distribution of the relevant trait, t, as a mixture of 3 normal distributions:

$$f(t|q, \mu_1, \mu_2, \mu_3, \sigma_1^2, \sigma_2^2, \sigma_3^2) = \sum_{i=1}^{3} p_i \cdot \phi(t|\mu_i, \sigma_i). \tag{1}$$

where ϕ is the normal density function and i indexes the genotype; i.e., AA (i=1), Aa or aA (i=2), and aa (i=3). Modeling the effects of dominance and recessivity at the assumed locus is straightforward; one simply assumes either $\mu_{Aa} = \mu_{aa}$ or $\mu_{AA} = \mu_{Aa}$. Often the assumption of equal variances can be made (i.e., $\sigma^2 = \sigma_{AA}^2 = \sigma_{Aa}^2 = \sigma_{aa}^2$). A graphical representation of the modeling of quantitative traits with normal mixture distributions is given in Figure 1 of the appendix, where dominance (of the a allele over the A allele) is assumed.

An oft-used statistical practice in quantitative genetics is the fitting of mixture distribution models (e.g., based on equation 1) to data with unknown

† Division of Hypertension, Department of Medicine and Department of Statistics, University of Michigan, R6592 Kresge I, Ann Arbor, Michigan 48109-0500

Exploring the Limits of Bootstrap. Edited by Raoul LePage and Lynne Billard.

genetic determinants in an effort to gauge the strength of single-locus, 2-allele models for those data. Traditionally, likelihood ratio tests involving 3 and 2 component normal mixture distribution models and single normal distribution models have been used as formal testing methods for relevant hypotheses. Many problems are associated with this modeling and hypothesis testing strategy. It is clear from Figure 1 that when the separation between the mean effects of the various genotypes is small, it is not obvious that a normal mixture distribution characterizes the data better than other models of "skewed" data (e.g., log normal distributions, Weibull distributions, etc.). In addition, Schork and Schork [1988] have shown that for data with given levels of skewness, one will almost always accept the normal mixture model as characterizing those data if a single *normal* distribution is taken as the alternative to which it is compared in a likelihood ratio test setting.

What is needed then is a test strategy which permits normal mixture distributions to be compared to skewed alternatives, such as the log normal distribution. In Section II we describe a strategy based on bootstrap principles for this purpose and examine its properties through some Monte Carlo studies. In Section III we examine the usefulness of this strategy in some complicated situations arising in quantitative genetics. In Section IV we offer some brief summary remarks.

II. Bootstrap Tests for Non-Nested Hypotheses

Section I introduced the problem of comparing the support given to normal mixture and lognormal distributional models by a given set of data. Such a problem involves hypotheses (e.g., models) from separate families of hypotheses. As such, many traditional asymptotic tests (such as the χ^2-based likelihood ratio tests) cannot be used since the underlying distribution of the likelihood ratio implicating normal mixture and lognormal models is unknown and potentially intractable mathematically. One can overcome this shortcoming, however, by simulating the unknown distribution of a likelihood ratio test statistic implicating both models and estimating critical values from this simulated distribution. More formally, given data $X = x_1, \ldots, x_n$ following an unknown distribution $f(x)$, a test for the hypothesis $H_0 : f(x) = g(x;\theta)$ vs. $H_1 : f(x) = h(x, \delta)$, where θ and δ are unknown parameter vectors, can be constructed as follows. For X compute $T = 2\sum log[g(x_i;\hat{\theta})/h(x_i;\hat{\delta})]$, the likelihood ratio statistic under $g(x;\theta)$, where $\hat{\theta}$ and $\hat{\delta}$ are consistent estimators of θ and δ, respectively. Draw data sets $X_1^*, \ldots, X_r^*$ from $g(x^*;\hat{\theta})$ and compute T_i^* $(i = 1, \ldots, r)$. From the T^*'s estimate a relevant critical quantile C. Compare C to T to make inferences about H_0. This test construction has been referred to as the "parametric bootstrap" test in the literature, and several authors have investigated its usefulness in nested hypotheses situations (see Hall and Titterington [1989]; Jockel [1986]; and Hope [1968]).

We examined the usefulness of the parametric bootstrap test for settings involving mixed normal and lognormal distributional hypothesis. To investigate the level of the test assuming the lognormal model is the correct model, we generated 100 data sets following lognormal distributions with mean 0 and differing variances for various sample sizes and computed a parametric bootstrap test with $r=19$ and a nominal level of 0.05 for each of the 100 replications. The number of rejections resulting from these simulations were tallied and taken as an estimate of the level of the test. These results are outlined in Table 1 of the appendix. This table suggests the parametric bootstrap test is at or near the nominal level of the test. The small number of replications (i.e., 100) used to estimate the significance levels was used to keep the computational burden assumed in the estimation of the mixture distribution parameters for each replication to a minimum.

To investigate the power of the test in this situation, we again kept the lognormal distribution as the null model, but generated data following a 2 component normal mixture with mixing weights of 0.75 and 0.25, standard deviation 1, and means separated at a distance varying from 1 to 4 "within" component standard deviation units. The rationale for this was that as the separation between the component means gets larger, "bimodality" (or the conspicuous presence of a normal mixture distribution) would become more pronounced (this is not the case in Figure 1) and therefore the lognormal hypothesis should be more easily rejected. Power estimates based on this strategy were computed from 100 replications assuming four different sample sizes. A plot of the results is given in Figure 2 of the appendix. It greatly suggests that, as expected, the power to reject lognormality increases as the separation between the means of the normal mixture increases. It should be emphasized that it has been our experience that a homoscedastic 2 component normal mixture with mixing weights 0.75 and 0.25 setting provides one of the poorest environments for rejecting the lognormal hypothesis at low mean component separations. That is, other settings (e.g., mixing weights of 0.5 and 0.5) produce higher power to reject the lognormal hypothesis at small mean component separations (which is intuitive, given that certain normal mixture distribution settings don't look or behave like lognormal distributions). Our investigations of the level and power properties of the test and test setting as described are by no means exhaustive; further studies involving other lognormal and mixture parameter settings are called for, as are studies investigating the level and power properties of the test when the roles of the hypotheses are reversed (i.e., the normal mixture is the null and the lognormal is the alternative hypothesis).

III. Quantitative Segregation Analysis

The goal of segregation analysis in genetics is to identify models that can explain the "transmission" of a trait from generation to generation. The unit of observation for segregation analysis is not an individual, but rather a "pedigree" comprised of, for example, parents, offspring, grandparents, cousins, etc. Because of the very detailed assumptions and modeling devices that go into segregation analytic tools, we forego an elaborate description of them (the interested reader is invited to read Elston and Stewart's [1971] excellent introduction to the subject and Elston's [1981] review). Instead we simply want to emphasize the role normal mixture distributions have in modeling the segregation of genes controlling quantitative traits. Figure 3 depicts a hypothetical nuclear family (i.e., parents and offspring only) segregating for a dominant gene controlling a quantitative trait. The arrows under the x-axes represent hypothetical trait values for each member of the family. From the figure it appears persons 2 (the mother, say) and 5 have trait values well within the normal component characterizing affliction (i.e., they have the "disease"). The question becomes, "how likely is it that a single locus, 2 allele model can explain the variation in trait values possessed by the family as a whole?" For our purposes it will suffice to say the null model, the assumption of a single locus with 2 alleles, is based on mixture distributions, whereas an alternative model, implicating, say, only environmental or polygenic (i.e., many loci) forces is not. In this way segregation models for quantitative traits parallel the hypothesis testing situation outlined in Section II.

In order to test hypotheses about single locus segregation for a quantitative trait, one can use the parametric bootstrap test. To illustrate this, we applied the parametric bootstrap test of segregation hypotheses to some published data. Penno and Vesell [1983] studied the elimination of the model compound antipyrine (AP) and other drugs in an effort to elucidate their genetic determinants, if any. They concluded, in part, that a single locus with 2 alleles appeared to control the variation in the elimination of N-demethylantipyrine (NDM-AP). Penno and Vesell did not base their findings on a formal segregation analysis of their data, but rather merely observed the consistency of their 13 pedigrees to single locus hypotheses after classifying each individual (based on their NDM-AP value) into 1 of 3 "disease" groups.

We subjected Penno and Vesell's data on NDM-AP to a segregation analysis using the parametric bootstrap test methodology. We fit single locus, 2 allele (i.e., normal mixture based) and polygenic (i.e., based on a single normal distribution) models to their data and computed likelihoods for each. We then simulated the distribution of the likelihood ratio first assuming the single locus model was correct and then assuming the polygenic model was correct, using $r = 99$. The simulated test statistic distributions and results are depicted in Figures 4 and 5, and suggest that though the major locus model is not rejected, the polygenic model is easily rejected. We conclude that there is evidence that a single locus with 2 alleles is, in part no doubt, responsible for variation in the rate of NDM-AP elimination.

IV. Conclusion

Often times researchers in genetics, as in other disciplines, derive models for phenomena so disparate that easy comparison of them is not straightforward. The non-nested nature of such models can be largely overcome through the use of the computer-intensive, though highly intuitive, parametric bootstrap test.

References

Morton NE (1967). The detection of major genes under additive continuous variation. *American Journal of Human Genetics* 19:23-34.

Schork NJ and Schork MA (1988). Skewness and mixtures of normal distributions. *Communications in Statistics* 17:3951-3969.

Elston RC (1981). Segregation Analysis. In, *Advances in Human Genetics*, ed. H Harris and K Hirshhorn. New York: Plenum, pp. 63-120.

Hall P and Titterington DM (1989). The effect of simulation order on level accuracy and power of Monte Carlo tests. *J.R. Statist. Soc. B.* 51:459-467.

Jöckel KH (1986). Finite sample properties and asymptotic efficiency of Monte Carlo tests. *Annals of Statistics* 14:336-347.

Hope AC (1968). A simplified Monte Carlo significance test procedure. *J.R. Statist. Soc. B.* 30:582-598.

Elston RC and Stewart J (1971). A general model for the analysis of pedigree data. *Human Heredity* 21:523-542.

Penno MB and Vesell ES (1983). Monogenic control of variations in antipyrine metabolite formation. *J. Clin. Invest.* 71:1698-1709.

Appendix

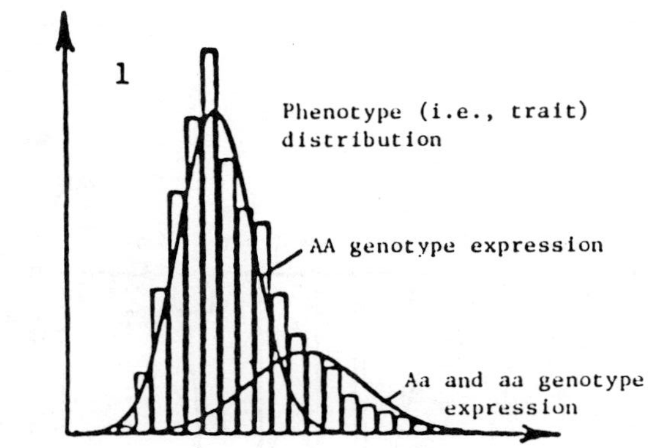

Table I

Estimated significance levels for Monte Carlo tests log-normal vs. homoscedastic 2-component normal mixture distributions, based on 100 replicates, using log-normal data with mean 0 and different variances.

Variance	Sample size / nominal level	25 .05	25 .10	50 .05	50 .10	100 .05	100 .10	250 .05	250 .10
.25		.04	.12	.04	.10	.04	.09	.02	.10
.50		.05	.08	.05	.06	.02	.07	.02	.04
1.00		.05	.16	.02	.07	.09	.21	.00	.01
2.00		.06	.22	.04	.20	.07	.12	.01	.06
4.00		.03	.06	.09	.12	.03	.10	.00	.00

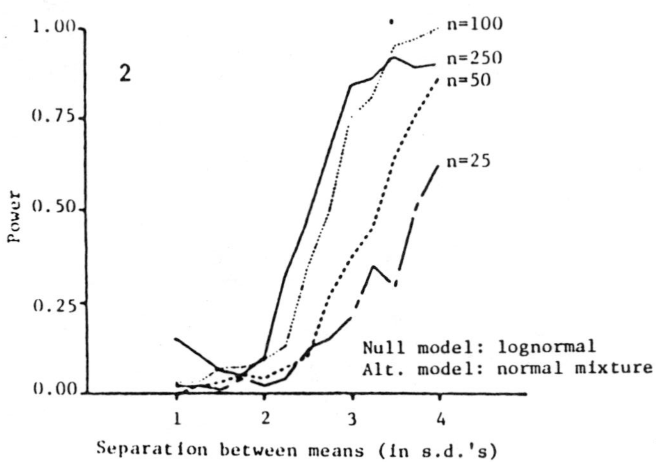

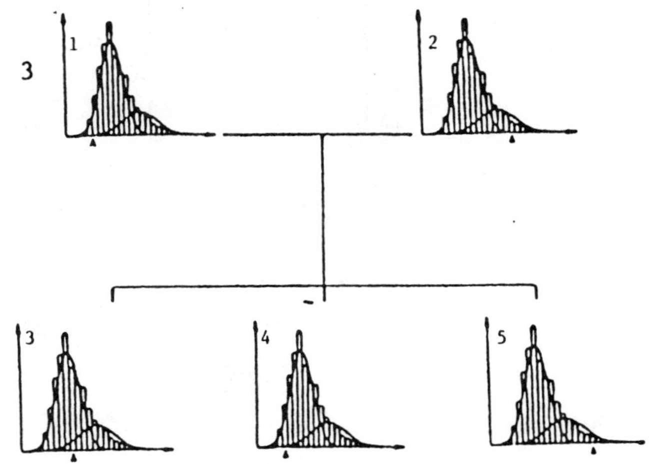

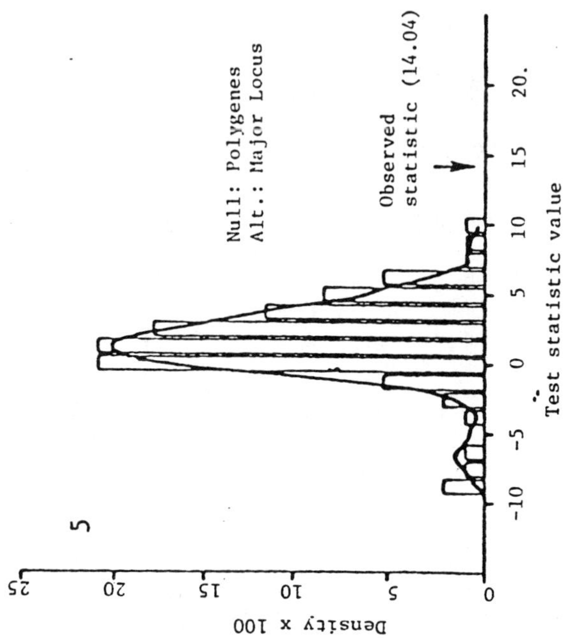
5
Null: Polygenes
Alt.: Major Locus
Observed
statistic (14.04)
Density x 100
Test statistic value
-10
-5
0
5
10
15
20.
0
5
10
15
20
25

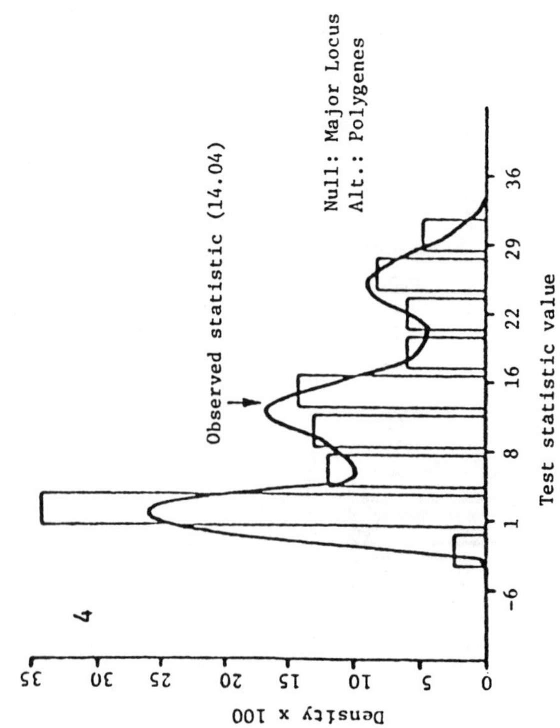
4
Null: Major Locus
Alt.: Polygenes
Observed statistic (14.04)
Density x 100
Test statistic value
-6
1
8
16
22
29
36
0
5
10
15
20
25
30
35

A Nonparametric Density Estimation Based Resampling Algorithm. [1]

Malcolm S. Taylor, US Army Ballistics Research Laboratory [2]
James R. Thompson, Department of Statistics, Rice University [3]

Abstract. The standard bootstrap algorithm may be viewed as the use of a Dirac-comb as a nonparametric probability density estimator from which a random number generator is created. In most cases, it will be more natural to use a continuous nonparametric probability density estimator to build the generator. Such an approach may be mathematically cumbersome and computationally inefficient. It is demonstrated that it is possible to create such a generator without explicitly obtaining the nonparametric probability density estimator.

Discussion. It is seldom the case that a graphical display of a nonparametric density estimator is the final end product needed. This is fortunate, since such a display becomes difficult for dimensions of three. Past a dimensionality of five, such a display is seldom practical.

There are cases, however, where the "curse of dimensionality" can be readily overcome. One of these is that of the creation of a pseudo-data set with properties very much like those of a real data set.

To motivate the SIMDAT algorithm of Taylor and Thompson, let us first consider the Dirac- comb density estimator associated with a one dimensional data set $\{x_i\}_{i=1}^n$. The Dirac-comb density estimator is given by

$$\hat{f}_\delta(x) = \frac{1}{n}\sum_{i=1}^{n}\delta(x - x_i). \tag{1}$$

We recall that $\delta(x)$ can be defined as

$$\delta(x) = \lim_{\tau \to 0} \frac{1}{\sqrt{2\pi}\tau} e^{-\frac{x^2}{2\tau^2}}. \tag{2}$$

Such a density is a "function" which is zero everywhere except at the data points. At each of these points, the density becomes a line stretching to infinity and with mass $1/n$. As a nonparametric density estimator, $\hat{f}_\delta(x)$ would appear

[1]This research was supported in part by the United States Army Research Office (Durham) under DAAL-03-88-K0131.

[2]Aberdeen Proving Ground, Forest Hill, Maryland 21005-5066
[3]Houston, Texas 77251-1892

Exploring the Limits of Bootstrap. Edited by Raoul LePage and Lynne Billard.
 ISBN: 0471-53631-8.

to be terrible. It says that all that can happen in any other experiment is a repeat of the data points already observed, each occuring with probability $1/n$.

Yet, for many purposes $\hat{f}_\delta(x)$ is quite satisfactory. For example, the mean of $\hat{f}_\delta(x)$ is simply

$$\mu_\delta = \int_{-\infty}^{\infty} x\hat{f}_\delta(x)dx = \frac{1}{n}\sum_{i=1}^{n} x_i = \bar{x}. \tag{3}$$

Similarly,

$$\sigma_\delta^2 = \int_{-\infty}^{\infty} (x-\bar{x})^2 \hat{f}_\delta(x)dx = \frac{1}{n}\sum_{i=1}^{n}(x-\bar{x})^2 = s^2. \tag{4}$$

The Dirac-comb density estimator may clearly be extended to higher dimensions. For example, if we have a sample of size n from a density of dimension p, $\hat{f}_\delta(x)$ becomes

$$\hat{f}_\delta(X) = \frac{1}{n}\sum_{i=1}^{n} \delta(X - X_i) \tag{5}$$

where

$$\delta(X) = \lim_{\tau\to 0}(\frac{1}{\sqrt{2\pi}\tau})^p exp(-\frac{\sum_{j=1}^{p} x_j^2}{2\tau^2}) \tag{6}$$

with x_j being the $j'th$ component of X.

We could, for example, for the two dimensional case, develop a 95% confidence interval for the correlation coefficient,

$$\rho = \frac{Cov(x,y)}{\sigma_x\sigma_y}. \tag{7}$$

Suppose that we have a sample of size n: $\{x_i, y_i\}_{i=1}^{n}$. Then we construct

$$\hat{f}_\delta(x,y) = \frac{1}{n}\sum_{i=1}^{n} \delta((x,y)-(x_i,y_i)). \tag{8}$$

Using this estimator, we construct 10,000 resamplings of size n, i.e., for each of the 10,000 resamplings we draw samples from the n data points (with replacement) of size n. For each of the resamplings, we compute the sample correlation

$$r_j = \frac{\sum_{i=1}^{n}(x_{ji}-\bar{x}_j)(y_{ji}-\bar{y}_j)}{\sqrt{\sum_{i=1}^{n}(x_{ji}-\bar{x}_j)^2\sum_{i=1}^{n}(y_{ji}-\bar{y}_j)^2}} \tag{9}$$

Then, we rank the sample correlations from smallest to largest. A 95 % confidence estimate is given by

$$r_{(250)} < \rho < r_{(9,750)}. \tag{10}$$

Such a resampling procedure has been proposed by Efron, who calls it the "bootstrap." Although it is clear that such a procedure may have some use for estimating the lower moments of some interesting parameters, we should never lose sight of the fact that it is, after all, based on the profoundly discontinuous Dirac-comb estimator $\hat{f}_\delta$.

It is quite easy to find an example where Dirac-comb nonparametric density estimator based techniques (i.e., the bootstrap) produce disastrous results. For example, suppose we have a sample of size 100 of firings at a bullseye of radius 5 centimeters. If the distribution of the shots is circular normal with mean the center of the bullseye and deviation one meter, then with a probability in excess of .88, none of the shots will hit inside the bullseye. Then any Dirac-comb resampling procedure will tell us the bullseye is a safe place, if we get a base sample (as we probably will) with no shots in the bullseye. We note that any realization of a bootstrap simulation will be different from the original sample. Some of the sample points will disappear. Others will be repeated multiple times. Indeed, the concatenation of a bootstrap followed by a bootstrap based on that bootstraped simulation, and so on, will lead ultimately to a simulated sample which consists of a single sample point. This is hardly desirable. It might be hoped that a single resampling would be of such a character that we would be almost indifferent as to whether we had this simulation or the original data set. But, of course, it would be dangerous to wander too far from the original sample. A resampling of a resampling of a resampling, etc., is not nearly as desirable as resamples which always point directly to the original sample.

Clearly, it would usually be preferred to use resampling schemes based on better nonparametric density estimators. One such would be

$$\hat{f}(X) = \frac{1}{n}\sum_{i=1}^{n} K(X - X_i, \Sigma_i), \tag{11}$$

where $K(X)$ is a normal distribution centered at zero with locally estimated covariance matrix Σ_i.

Although a simulation algorithm employing such an estimator has much to recommend it, it would appear to be extremely difficult of execution. It is at this point, that we should recall what it is we seek: not a good nonparametric density estimator, but a random sample from such an estimator. One is tempted to try a scheme which goes directly from the actual sample to the pseudo-sample. Of course, this is precisely what the bootstrap estimator does, with the very bad properties associated with a Dirac-comb. It is possible, however, to go from the sample directly to the pseudo-sample in such a way that the resulting estimator behaves very much like that of the normal kernel approach above. This is the SIMDAT algorithm of Taylor and Thompson.

We assume that we have a data set of size n from a p-dimensional variable X, $\{X_i\}_{i=1}^n$. First of all, we shall assume that we have already rescaled our data set so that the marginal sample variances in each vector component are the same. For a given integer m, we find, for each of the n data points, the $m-1$ nearest neighbors. These will be stored in an array of size $n \times (m-1)$.

Let us suppose we wish to generate a pseudo-sample of size N. Of course, there is no reason to suppose that n and N will, in general, be the same (as is the case generally with the bootstrap). To start the algorithm, we sample one of the n data points with probability $1/n$ (just as with the bootstrap). Starting with this point, we recall it and its $m-1$ nearest neighbors from memory, and compute the mean of the resulting set of points:

$$\bar{X} = \frac{1}{m}\sum_{i=1}^{m} X_i. \tag{12}$$

Next, we code each of the m data points about $\bar{X}$:

$$\{X_j'\} = \{X_j - \bar{X}\}_{j=1}^{m}. \tag{13}$$

Clearly, although we go through the computations of sample means and coding about them here as though they were a part of the simulation process, the operation will be done once only, just as with the determination of the $m-1$ nearest neighbors of each data point. The $\{X_j'\}$ values as well as the $\bar{X}$ values will be stored in an array of dimension $n \times (m+1)$.

Next, we shall generate a random sample of size m from the one-dimensional uniform distribution:

$$U\left(\frac{1}{m} - \sqrt{\frac{3(m-1)}{m^2}}, \frac{1}{m} + \sqrt{\frac{3(m-1)}{m^2}}\right). \tag{14}$$

We now generate our centered pseudo-data point X', via

$$X' = \sum_{l=1}^{m} u_l X_l'. \tag{15}$$

Finally, we add back on $\bar{X}$ to obtain our pseudo-data point X:

$$X = X' + \bar{X}. \tag{16}$$

The procedure above, as m and n get large, becomes very like that of the normal kernel approach mentioned earlier. To see why this is so, we consider the sampled vector X_l and its $m-1$ nearest neighbors:

$$\{X_l\}_{l=1}^{m} = \begin{bmatrix} x_{1l} \\ x_{2l} \\ \vdots \\ x_{pl} \end{bmatrix}_{l=1,\ldots,m} . \tag{17}$$

For a moment, let us treat this collection as m points from a truncated distribution with mean vector μ and covariance matrix Σ. Now, if $\{u_l\}_{l=1}^{m}$ is an independent sample from the uniform distribution in (14), then

$$E(u_l) = \frac{1}{m};\ Var(u_l) = \frac{m-1}{m^2};\ Cov(u_i, u_j) = 0, \text{ for } i \neq j. \tag{18}$$

Next, we form the linear combination:

$$Z = \sum_{l=1}^{m} u_l X_l. \tag{19}$$

For the $r'th$ component of the vector Z, $z_r = u_1 x_{r1} + u_2 x_{r2} + \ldots + u_m x_{rm}$, we observe the following relationships:

$$E(z_r) = \mu_r, \tag{20}$$

$$Var(z_r) = \sigma_r^2 + \frac{m-1}{m}\mu_r^2, \tag{21}$$

and

$$Cov(z_r, z_s) = \sigma_{rs} + \frac{m-1}{m}\mu_r \mu_s. \tag{22}$$

Note that if the mean vector of X were $(0, 0, \ldots, 0)$, then the mean vector and covariance matrix of Z would be the same as that of X, i.e., $E(z_r) = 0$, $Var(z_r) = \sigma_r^2$, and $Cov(z_r, z_s) = \sigma_{rs}$. Naturally, by translation to the local sample mean of the nearest neighbor cloud, we will not quite have achieved this result. But we will come very close to the generation of an observation from the truncated distribution which generated the points in the nearest neighbor cloud.

Clearly, for m moderately large, by the central limit theorem, SIMDAT comes close to sampling from n normal distributions with the mean and covariance matices corresponding to those of the n, m nearest neighbor clouds. If we were seeking rules for consistency of the nonparametric density estimator corresponding to SIMDAT, we could use the formula of Mack and Rosenblatt for nearest neighbor nonparametric density estimators:

$$m = Cn^{4/(p+4)}. \tag{23}$$

Of course, people who carry out nonparametric density estimation realize that such formulae have little practical relevance, since C is usually not available. Beyond this, we ought to remember that our goal is not to obtain a nonparametric density estimator, but rather to generate a data set which appears like that of the data set before us. Let us suppose we err on the side of making m far too small, namely, $m = 1$. That would yield simply the bootstrap. Suppose we err on the side of making m far too large, namely, $m = n$. That would yield an estimator which roughly sampled from a multivariate normal distribution with the mean vector and covariance matrix computed from the data.

Below, in Figure 1, we show a sample of size 85 from a mixture of three normal distributions with the weights indicated, and a pseudo data set of size 85 generated by SIMDAT with $m = 5$. We note that the emulation of the data is reasonably good. In Figure 2, we go through the same exercise, but with $m = 15$. There effects of a modest oversmoothing are noted. In general, if the data set is very large, say of size 1,000 or greater, good results are generally obtained with $m \approx .02n$. For smaller values of n, m values in the $.05n$ range appear to work well. A version of SIMDAT in the S language, written by E. Neely Atkinson, is available under the name "gendat" from the S Library available through email from Bell Labs. The new edition of the IMSL Library will contain a SIMDAT subroutine entitled "RNDAT" (beta versions may be obtained free by writing IMSL) .

In this paper, we have observed the usefulness of noting what we really seek rather than using graphically displayed nonparametric density estimators as our end product. The cases where we really must worry about obtaining the graphical representation of a density are, happily, rare. But, as with the case of SIMDAT, the context of nonparametric density estimation is very useful in many problems where the explicit graphical representation of a nonparametric density estimator is not required.

References

Efron, Bradley (1979). Bootstrap methods–another look at the jacknife, **Annals of Statistics,** 7, 1-26.

Mack, Y.P. and Rosenblatt, Murray (1979). Multivariate k-nearest neighbor density estimates, **Journal of Multivariate Analysis,** 9, 1-15.

Tapia, Richard A. and Thompson, James R.and (1978). **Nonparametric Probability Density Estimation,** Baltimore: Johns Hopkins.

Taylor, Malcolm S. and Thompson, James R. (1986). A data based algorithm for the generation of random vectors, **Computational Statistics and Data Analysis,** 4, 93-101.

NONPARAMETRIC RANK ESTIMATION USING BOOTSTRAP RESAMPLING AND CANONICAL CORRELATION ANALYSIS

Xin M. Tu, Harvard School of Public Health;
D.S. Burdick and B.C. Mitchell, Duke University *

Abstract

Canonical correlation analysis has proven to be remarkably successful as an alternative to the eigenvalue approach in rank estimation; a problem that has challenged the analytical chemists for more than a decade. A methodological advance of this new approach is that it focuses on the difference in structure rather than in magnitude in characterizing the difference between the signal and the noise. This structural difference is quantified through the analysis of canonical correlation, which is a well established data reduction technique in multivariate statistics. Unfortunately, there is a price to be paid for having this structural difference: at least two replicate data matrices are needed to carry out the analysis.

In this paper, we propose a bootstrap resampling method to extend the canonical correlation analysis to a single data matrix. Such a procedure not only removes the requisite for replicate data matrices but also leads to a robust estimator for the rank. With the percentile method, statistical inference about the rank can proceed without any distributional assumption about the noise. This "distribution-free" feature is especially desirable and useful in practice since it frees us from hinging results on some crutial assumptions about the random noise which are often difficult to justify and may even be erroneous. The procedure is illustrated with real as well as simulated mixture samples.

*X.M. Tu, Dept. of Biostat., Harvard School of Public of Health, 677 Huntington Ave., Boston, MA 02115; D.S. Burdick and B.C. Mitchell, Institute of Stat. and Decision Sciences, Duke University, Durham, NC 27706

Exploring the Limits of Bootstrap. Edited by Raoul LePage and Lynne Billard.
 ISBN: 0471-53631-8.

KEY WORDS: Rank estimation; Bootstrap resampling; Canonical Correlation; Excitation-emission matrix; Singular value decomposion.

INTRODUCTION

One of the problems which have challenged the chemists in analytical chemistry for more than a decade is to determine the number of components in a multicomponent mixture sample. Often the sample data is expressed in the form of a matrix whose rank, in the absence of noise, is equal to the number of components. However, the presence of noise in the data generally causes the rank to exceed the number of components in the mixture.

Most methods which have been proposed to estimate the rank in the presence of noise all, in essence, rely on the information summarized by the eigenvalues from the singular value decomposition of the underlying matrix (Rossi and Warner 1986; Malinowski 1990; Tway, Cline Love and H.B. Woodruff 1980; Wold 1978). Even though it is difficult to evaluate the amount of information which is lost due to this type of data summary (or reduction), it is not very hard to convince oneself at least heuristically that some information is lost since it ignores any information contained in the eigenvectors which may well provide much more information than the eigenvalues do. A different approach which also incorporates the information in the eigenvectors was therefore introduced recently as an alternative to the eigenvalue approach (Tu et al. 1989). A methodological turning point of this new approach is that it focuses on the difference in structure rather than in magnitude in characterizing the difference between the signal and the noise. This structural difference is then quantified through the analysis of canonical correlation, a well established data reduction methodology in multivariate statistics. Unfortunately, this new approach can only be applied to situations where replicate data matrices are available.

The objective of this paper is to continue to explore the potential and to extend the scope of this new approach. In particular, we propose a bootstrap resampling procedure to extend this new methodology to a single data matrix.

An important feature of the data matrix arising from such chemical experiments is the structure imposed on the data matrix by the signal. Such a signal structure plays a crutial role in rank estimation. As a consequence, a standard bootstrap resampling procedure, which relies on a random sample drawn with replacement from the elements of the data matrix, is not readily applicable since it will destroy this signal structure which depends both on the ordered rows and columns. We therefore discuss a variant of the bootstrap resampling to circumvent this problem when resampling the observed data matrix.

In the following discussion, we use the Excitation-Emission Matrix (Warner 1982; Warner, Neal and Rossi 1985) as a vehicle of examplifying the proposed methodology. However, the method can be applied to any matrix-formatted data where the rank of the matrix is determined by the number of components in the mixture.

RANK ESTIMATION BY CANONICAL CORRELATION

In this section, we briefly review the procedure RECCAMP (Rank Estimation by Canonical Correlation Analysis of Matrix Pairs) for analyzing replicate data matrices. A detailed discussion of the RECCAMP can be found in Tu et al. (1989).

Let the I by J matrix S be the EEM for an R-component mixture in the absence of noise. Then S can be expressed as:

$$S = XY^T \tag{1}$$

where $X = (x_1, \ldots, x_R)$ is an I by R matrix with the rth column being the excitation vector for the rth component and $Y = (y_1, \ldots, y_R)$ is an J by R matrix with the rth column being the emission vector for the rth component. The columns of X as well as those of Y will be assumed to be linearly independent. Under this assumption, the rank of S will be equal to R, which is the number of components in the mixture. In the absence of noise, the rank can be determined by the number of non-zero singular values from the singular value decomposition (SVD) of S.

In practice, however, since the measured EEM also contains a noise term N, i.e.,

$$M = S + N \tag{2}$$

the number of non-zero singular values will exceed the number of components because the random noise term N generally increases the rank. The eigenvalue approach for correcting this is to watch for a "large drop" between the singular values, which seems very reasonable since one would expect a relatively small singular value contributed from the low-magnitude noise. However, if these singular values corresponding to the noise are viewed as a summary statistic for the noise from a statistical point of view, which makes perfact sense since this is what is used in making inference about the rank when using the eigenvalue approach, it is not difficult to see that the random nature of the noise may not be well summarized by this statistic since it clearly ignores any information contained in the eigenvectors. An alternative to this eigenvalue approach was therefore introduced, hoping to provide a better statistic to summarize the information about the random noise. A methodological advance in this new approach is that it focuses on the random nature of the noise, which is manifested in an Euclidean subspace, and utilizes the analysis of canonical correlation, a well established data reduction method in multivariate statistics, to provide a statistic which fully reflects this random nature. Below, we briefly describe this approach.

Let M_1 and M_2 be two replicate EEM's. Then these matrices can be expressed as

$$\begin{aligned} M_1 &= S + N_1 \\ M_2 &= S + N_2. \end{aligned}$$

In the absence of noise, M_1 and M_2 will have a common column space which we denote by $Col(S)$ and a common row space which we denote by $Row(S)$, both of dimension R. In the presence of noise, let

$$\begin{aligned} M_1 &= U_1 D_1 V_1^T \\ M_2 &= U_2 D_2 V_2^T \end{aligned}$$

be the SVD's of M_1 and M_2, respectively. If the signal structure is retained at a given noise level, then since the entries of the noise terms N_1 and N_2 are independently distributed the subspaces $Col(U_1^{(R+1)})$ and $Col(U_2^{(R+1)})$ as well as $Col(V_1^{(R+1)})$ and $Col(V_2^{(R+1)})$ will be highly correlated only in an R-dimensional subspace, where the matrix $U_i^{(R+1)}$ (or $V_i^{(R+1)}$) is formed by the $(R+1)$ leading vectors of U_i (or V_i). Since canonical correlations provide a summary measure of closeness between two subspaces, we therefore expect R high correlations and one low correlation between $Col(U_1^{(R+1)})$ and $Col(U_2^{(R+1)})$ and also between $Col(V_1^{(R+1)})$ and $Col(V_2^{(R+1)})$. This pattern of the correlation structure provides a basis for estimating the rank or the number of components in the mixture.

Unfortunately, there is a price to be paid for having this correlation structure in the data. The trade-off is the requisite for replicate EEM's which provide realizations of a random noise that are at least not correlated. This might not be a stringent assumption in practice. However, it is at least a methodological deficiency compared to the eigenvalue approach. We therefore discuss a bootstrap resampling method in the next section to remove this restriction.

THE BOOTSTRAP RESAMPLING

With a single EEM, if we can somehow change the configuration of the noise in the observed data, we may still be able to obtain the structural difference as with replicate EEM's. This is possible since an observed EEM contains a total of $I \times J$ realizations of the random noise where I is the number of rows and J is the number of columns of the EEM. However, a standard bootstrap resampling procedure, which relies on a random sample drawn with replacement from the elements of the data matrix is not readily applicable here since it will clearly destroy the signal structure which depends on both the ordered rows (excitation wavelength) and the ordered columns (emission wavelength). We, therefore, describe a variant of the bootstrap resampling procedure below to avoid this problem, but still to allow us to change the configuration of the noise.

Let m_j denote the jth column of the EEM M for $1 \leq j \leq J$. It follows from (1) and (2) that

$$m_j = \sum_{r=1}^{R} x_r y_{j,r} + n_j \tag{3}$$

where x_r or y_r denotes the rth column of X or Y, $y_{j,r}$ denotes the jth element of the vector y_r and n_j denotes the jth column of N. In the absence of noise, if $A = [m_{k_1}, \ldots, m_{k_J}]$ is a matrix formed by any selected J columns of M, the subspace spanned by the columns of A and that spanned by the columns of M satisfy

$$Col(A) = Col(M) \tag{4}$$

provided that the matrix $A = [m_{k_1}, \ldots, m_{k_J}]$ is of full rank, or, equivalently none of the individual components is completely missing in the selected columns. Equations (3) and (4) provide a basis for our bootstrap resampling procedure. Suppose that we draw with replacement $2J$ vectors from the set of columns of M with each column in M being selected equally likely. Now arbitrarily group the selected columns into two sets of vectors each of size J and let A and B denote the matrices formed by these vectors from each set. It follows from (4) that in the absence of noise

$$Col(A) = Col(B) \tag{5}$$

provided that both of A and B are of full rank, or equivalently none of the components is completely missing in the columns of A and those of B. In the presence of noise, let

$$\begin{aligned} A^{(R+1)} &= U_A^{(R+1)} D_A^{(R+1)} [V_A^{(R+1)}]^T \\ B^{(R+1)} &= U_B^{(R+1)} D_B^{(R+1)} [V_B^{(R+1)}]^T \end{aligned} \tag{6}$$

be the rank $(R+1)$ fits of A and B from SVD's of A and B, respectively. It follows from (5) that $Col(U_A^{(R+1)})$ and $Col(U_B^{(R+1)})$ are highly correlated in an R-dimensional subspace whereas their orthogonal complements are not because of the random rearrangement in the columns n_j resulting from the resampling process. This gives rise to the same structural difference between $Col(U_A^{(R+1)})$ and $Col(U_B^{(R+1)})$ as between $Col(U_1^{(R+1)})$ and $Col(U_2^{(R+1)})$ as with replicate EEM's. We would therefore expect a large drop between the Rth and $(R+1)^{st}$ canonical correlation coefficients among the ordered coefficients calculated from $Col(U_A^{(R+1)})$ and $Col(U_B^{(R+1)})$. However, due to the randomness in selecting the columns for the two matrices, it is still possible to observe a relatively large $(R+1)^{st}$ coefficient which may make it difficult to observe this drop. It is therefore necessary to find the distribution for each of the coefficients if we wish to base our inference about the rank on these coefficients.

Here, it would be difficult to derive the distribution functions in closed form for these correlation coefficients. On the other hand, such an attempt would require some tenured assumptions about the noise distribution which may be difficult to verify and may result in limited applications. We therefore propose a criterion based on the empirical distribution of a bootstrap sample.

To obtain a bootstrap sample for the $(R+1)^{st}$ canonical correlation coefficient, we repeat the resampling process, say, k times and observe a sequence of correlation coefficients $\xi_i^{(R+1)}$ for $1 \leq i \leq k$. Because of the random resampling, we would expect each of these $\xi_i^{(R+1)}$ to appear equally likely between 0 and 1. We may, therefore, view this bootstrap sample as arising from a distribution similar to the uniform distribution between 0 and 1 with mean or median less than or equal to 0.5. As for statistical inference, we use the sample median, which is not only more robust but also more appropriate in our context since inference about the rank using this estimator can proceed without resort to any distributional assumption or Monte Carlo method.

The pth percentiles can be calculated using the ordered statistic

$$\xi_{(1)}^{(R+1)}, \ldots, \xi_{(k)}^{(R+1)}$$

In particular, the sample median is given by $m = \xi_t^{(R+1)}$, where t is the smallest integer such that $\sum_{i=1}^{t} \xi_i^{(R+1)} \geq 0.5$.

Following the above discussion, a null hypothesis at a given α-level can be constructed as

$$H_0: \quad m \leq 0.5 \tag{7}$$

Such a test in general requires some distributional assumption or the calculation of standard errors if the normal distribution is used. However, using the sample median, we can carry out such a test without making such parametric assumptions which are often difficult to justify and may even be erroneous in practice.

We note that the null hypothesis in (7) is equivalent to testing whether the lower limit of the $1-2\alpha$ confidence interval lies to the left of 0.5. With the bootstrap sample, we can calculate the conficence interval for the median using the percentile method (Efron 1982) without any assumption for the underlying distributon and without recourse to Monte Carlo simulation. Using the percentile method, the $1-2\alpha$ confidence interval is expressed in the form

$$[L, U] = [\xi_{(t_1)}^{(R+1)}, \xi_{(t_2)}^{(R+1)}]$$

Here the integers t_1 and t_2 are given by the smallest integers which satisfy

$$F(t_1) \geq \alpha$$

and

$$F(t_2) \geq 1 - \alpha.$$

where $F(t)$ denotes the cumulative density function of a binomial distribution $B(t|n,p)$ with the sample size parameter and the parameter for the probability of success fixed at $n = k$ and $p = 0.5$, respectively. We therefore reject the null hypothesis at the α-level if the interval lies to the right of 0.5.

Note that in the above discussion we have implicitely assumed that the matrices $A^{(R+1)}$ and $B^{(R+1)}$ constructed at each resampling step contain the signal for all the components. Even though it is possible for $A^{(R+1)}$ or $B^{(R+1)}$ or even both to contain only some of the components it is unlikely to occur in a random resampling process unless some components are only represented by a few columns of the EEM, or in other words, they have very narrow spectral range of excitation or emission. Moreover, since we use a large sample robust estimator, the sample median, in making statistical inference about the rank, a few such instances in the resampling process will not cause much difference in estimation.

Also, we have focused on the subspaces $Col(U_A^{(K+1)})$ and $Col(U_B^{(K+1)})$ which correspond to the process of resampling columns in the above discussion. Similar statements apply to the subspaces $Col(V_A^{(K+1)})$ and $Col(V_B^{(K+1)})$ as well when resampling the rows. In practice, it may be helpful to use both of the resampling processes, especially in dealing with samples which have a relatively low signal-to-noise ratio. For example, we may wish to reject the hypothesis that the rank is $R-1$ if the lower limit of the confidence interval for the median of the Rth canonical correlation coefficient from the process of resampling the columns barely falls below 0.5 while that from the process of resampling the rows stays above 0.5. Situations like that, on the other hand, may well depend on the data at hand. However, the more information we can extract from the data the more we will know about the sample, which could ultimately help reach a comprise if it is too difficult to ascertain the rank in such cases.

RESULTS AND DISCUSSION

The proposed bootstrap resampling procedure was applied to some real as well as simulated mixture samples. The real sample consists of two fluorescent emitters: 9, 10 diphenylanthracene (DPA) and anthracene (ANT) and the EEM is a 30 by 30 matrix. Plotted in Figure 1 is the EEM for this mixture at 5 MHz generated with the phase-modulation technique. The simulated mixture is a four component sample whose signal was extracted from some real sample data and has been analyzed by the RECCAMP (Tu et al. 1989). The signal strength of the simulated mixture sample is controlled by varying a pseudo-intensity parameter c according to

$$M = c * S + N \tag{8}$$

where N is generated by a normal deviate with fixed mean 0 and variance 0.01. With two replicate EEM's, the RECCAMP is able to determine the rank for c as low as 4 at which the noise level is very close to that of the signal (Tu et al. 1989). It is therefore interesting to see whether we would still be able to correctly estimate the rank at the same low signal level with only one EEM.

A consideration in implementing a bootstrap resampling procedure in general is to choose the right sample size so that minimum amount of computation can be performed without loss of much efficiency. As investigated by Efron (1982, 1983), a bootstrap sample of 500 or 1,000 is usually used as a guide. We have chosen a sample size of 500 for both mixtures after taking into consideration the amount of computation that is needed to compute the canonical correlation coefficients.

For each sample, canonical correlation coefficients were calculated for each pair of the rank K fits of the SVD's of the matrices corresponding to the column resampling and row resampling processes at each step of the resampling procedure for $K = 1, 2$, etc. So, for each value of K, we have a sample of 500 coefficients each for the two resampling processes, which we shall refer to as the bootstrap sample of canonical correlation coefficients. Shown in Tables I and II are the percentiles of the distributions of these bootstrap samples for each mixture sample for $K = 1, 2$, etc. Also listed in these tables are the confidence intervals (C. I.) for the respective medians, calculated by the percentile method.

For the real mixture, the drop in the percentiles between the second and third coefficients is obvious in all the resampling cases. This same idea is also graphically portrayed by the plots of the distributions from the column resampling process shown in Figure 2 (those from the row resampling process are similar and are therefore not shown). Following the discussion in the previous section, we can use confidence intervals to quantify our empirical assessment in the form of a statistical test at a given nominal level. In this example, the 90% and 95% confidence intervals shown in Table I clearly illustrate this point.

	Percentiles			Confidence Interval for the Median	
	25%	Median	75%	90% C.I.	95% C.I.
	Column Resampling				
K = 1	0.9986	0.9993	0.9996	(0.99921, 0.99932)	(0.99919, 0.999633)
K = 2	0.933	0.961	0.976	(0.957, 0.964)	(0.956, 0.965)
K = 3	0.207	0.418	0.649	(0.386, 0.458)	(0.379, 0.466)
K = 4	0.151	0.279	0.455	(0.264, 0.301)	(0.259, 0.302)
K = 5	0.111	0.239	0.402	(0.223, 0.255)	(0.217, 0.263)
K = 6	0.098	0.208	0.355	(0.188, 0.229)	(0.186, 0.235)
	Row Resampling				
K = 1	0.9982	0.9990	0.9994	(0.99891, 0.99907)	(0.99888, 0.99909)
K = 2	0.923	0.949	0.969	(0.945, 0.949)	(0.944, 0.950)
K = 3	0.248	0.476	0.661	(0.443, 0.508)	(0.439, 0.517)
K = 4	0.165	0.327	0.512	(0.298, 0.349)	(0.295, 0.359)
K = 5	0.119	0.252	0.429	(0.229, 0.279)	(0.226, 0.284)
K = 6	0.101	0.212	0.331	(0.197, 0.225)	(0.189, 0.228)

Table I: Percentiles and confidence intervals (C. I.) from the distributions of the bootstrap samples of canonical correlation coefficients for the mixture ANT and DPA.

	Percentiles			Confidence Interval for the Median	
	25%	Median	75%	90% C.I.	95% C.I.
			c = 4		
K = 1	0.9984	0.99904	0.9994	(0.99894, 0.99913)	(0.99892, 0.99915)
K = 2	0.9853	0.9888	0.9921	(0.9885, 0.9894)	(0.9884, 0.9895)
K = 3	0.8602	0.9048	0.9368	(0.9009, 0.9109)	(0.9008, 0.9122)
K = 4	0.7069	0.7936	0.8459	(0.7823, 0.8015)	(0.7796, 0.8022)
K = 5	0.1331	0.2881	0.4616	(0.2574, 0.3165)	(0.2541, 0.3195)
K = 6	0.0806	0.1625	0.2967	(0.1483, 0.1752)	(0.1452, 0.1775)
K = 7	0.0651	0.1367	0.2313	(0.1238, 0.1538)	(0.1214, 0.1577)
			c = 3		
K = 1	0.9981	0.9989	0.9993	(0.9988, 0.9989)	(0.9987, 0.9990)
K = 2	0.9776	0.9832	0.9870	(0.9824, 0.9836)	(0.9823, 0.9838)
K = 3	0.7168	0.8112	0.8704	(0.8056, 0.8197)	(0.8030, 0.8222)
K = 4	0.3257	0.5221	0.6609	(0.4993, 0.5445)	(0.4956, 0.5486)
K = 5	0.1157	0.2597	0.4225	(0.2426, 0.2821)	(0.2374, 0.2863)
K = 6	0.0765	0.1641	0.2934	(0.1486, 0.1778)	(0.1465, 0.1814)
K = 7	0.0503	0.1217	0.2106	(0.1077, 0.1353)	(0.1065, 0.1387)

Table II: Percentiles and confidence intervals (C. I.) from the distributions of the bootstrap samples of canonical correlation coefficients corresponding to the process of resampling the columns for the simulated mixture.

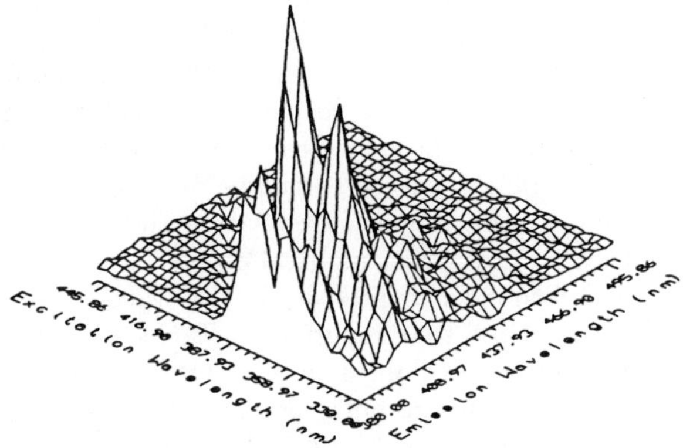

(a)

Figure 1: The EEM for the mixture of ANT and DPA.

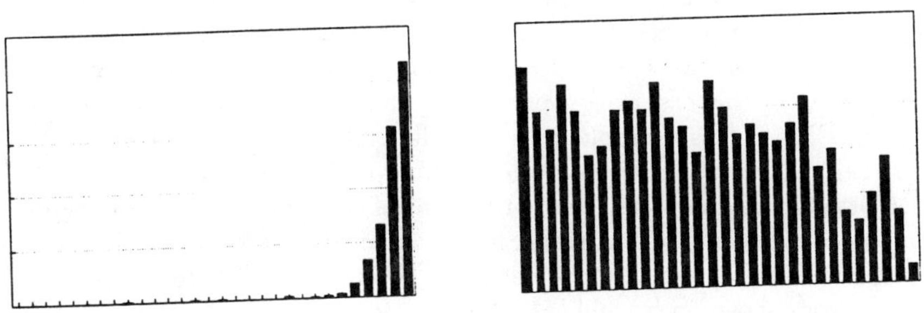

(a) (b)

Figure 2: The empirical distributions of the bootstrap samples of canonical correlation coefficients from the column resampling process for the mixture of ANT and DPA: (a) at $K = 2$; (b) at $K = 3$.

As for the simulated mixture sample, our major reason to study this sample is to see whether reduction in sample data would have any significant impact on the canonical correlation coefficients. In statistical inference, reduction in data reduces efficiency and may lead to bias in estimation. In our context, both could have influence on the canonical correlation coefficients on which our inference about the rank is based. It may be difficult, if not impossible, to theoretically characterize this impact. However, with this simulated mixture sample, we can at least make some empirical observation about this effect.

The percentiles and confidence intervals for the two critical intensity levels are shown in Table 2. Note that reported in Table 2 are the results from the process of resampling the columns since the results from resampling the rows are similar and therefore are not reported. At $c = 4$, the drop for the percentiles occurs between the forth and fifth coefficients. The lower limits of both confidence intervals for the median of the fourth canonical correlation coefficient are way above 0.5. On the other hand, even the upper limits of the confidence intervals for the median of the fifth canonical correlation coefficient fall below 0.5. Hence, had we not known the composition of this mixture, we would have correctly estimated the rank at this low signal level without difficulty. The percentiles and confidence intervals for $c = 3$, however, show no sign of the four-component nature of the mixture sample. As a matter of fact, the rank would be estimated to be three if based on the confidence interval, which coincides with the estimate from the analysis of the RECCAMP using two replicate EEM's in this case (Tu et al. 1989). So at least with this simulated data, it seems that there is no loss of efficiency in inference, which may not be surprising since the information of the noise distribution (normal distribution in this case) may very well be summarized by the 2500 elements in one EEM. In this example, it is quite likely that one of the components is overwhelmed by the noise when the signal level drops down to $c = 3$.

Plotted in Figure 3 are the distributions of some of the bootstrap samples. The drop between the forth and fifth canonical correlation coefficients for $c = 4$ can also be seen from the striking difference in skewness between the distributions shown in (a) and (b). However, for $c = 3$, it is difficult to tell from the shape of the histogram plotted in (c) whether the coefficients correspond to the signal even though the C.I. in this case does cover the point 0.5 as discussed earlier.

In summary, we have proposed a bootstrap resampling procedure to extend the canonical correlation technique to the analysis of a single data matrix. Also, since a robust estimator is introduced for inference, the procedure may be applied to a wide range of data without any restriction on the noise distribution. On the other hand, the procedure may be computationally expensive compared to the eigenvalue approach, especially in case of low signal-to-noise ratio. Our experience shows that if the signal-to-noise ratio is high as in the

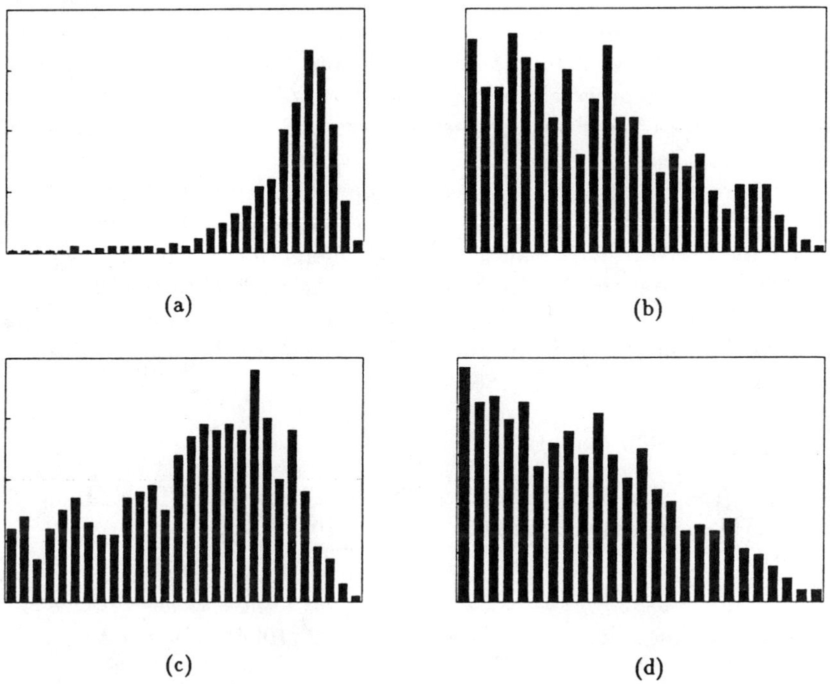

Figure 3: The empirical distributions of the bootstrap samples of canonical correlation coefficients from the column resampling process for the simulated mixture: (a) at $c = 4$, $K = 4$; (b) at $c = 4$, $K = 5$; (c) at $c = 3$, $K = 4$; (d) at $c = 3$, $K = 5$.

real mixture sample in our example, a sample size as small as 50 will suffice to make the correct inference. As a guide for general practice, we may start the procedure with a small bootstrap sample and then increase the sample size if necessary to improve efficiency for the boostrap estimate of the confidence interval. However, as discussed earlier, there is no point to increase the sample size beyond 1,000.

The procedure may also be used in conjunction with some other procedures such as those from the eigenvalue approach to reduce the amount of computation but also to make valid statistical inference. For example, an initial analysis using the eigenvalue approach may help expedite the search of rank to a small possible range of values. The bootstrap resampling may then be employed to quantify our uncertainty in terms of a valid statistical test.

References

[1] B. Efron, The Jackknife, the Bootstrap, and Other Resampling Plans, SIAM, monograph # 38, CBMS-NSF (1982).

[2] B. Efron and G. Gong, *American Statistician* **37**, 36 (1983).

[3] E.R. Malinowski, *J. Chemometrics* **4**, 102 (1990).

[4] T.M. Rossi and I.M. Warner, *Anal. Chem.* **54**, 810 (1986).

[5] X.M. Tu, D.S. Burdick, D.W. Millican and L.B. McGown, *Anal. Chem.* **61**, 2219 (1989).

[6] P.C. Tway, L.J. Cline Love and H.B. Woodruff, *Anal. Chim. Acta* **117**, 45 (1980).

[7] I.M. Warner, in Contemporary Topics in Analytical and Clinical Chemistry, Vol. 4, D.M. Hercules, G.M. Hieftje, L.R. Snyder and M.A. Evenson Eds., Plenum Press, New York (1982).

[8] I.M. Warner, S.L. Neal and T.M. Rossi, *J. of Research of the National Bureau of Standards*, **90**, 487 (1985).

[9] S. Wold, *Technometrics* **20**, 397 (1978).

Index

C

D

E

F

G

H

I

N

O

P

Q

R